Rodney Grapes
Pyrometamorphism

Rodney Grapes

Pyrometamorphism

With 192 Figures

 Springer

Author

Professor Dr. Rodney H. Grapes

Mineralogisch-Geochemisches Institut
Albert-Ludwigs-Universität Freiburg
Albertstr. 23 b
79104 Freiburg
Germany

Now at:
Department of Geosciences
Sun Yat-sen (Zhongshan) University
Guangzhou (Canton) 510275
China

Library of Congress Control Number: 2005934782

ISBN-10   3-540-29453-8   Springer Berlin Heidelberg New York
ISBN-13   978-3-540-29453-5   Springer Berlin Heidelberg New York

This work is subject to copyright. All rights are reserved, whether the whole or part of the material is concerned, specifically the rights of translation, reprinting, reuse of illustrations, recitations, broadcasting, reproduction on microfilm or in any other way, and storage in data banks. Duplication of this publication or parts thereof is permitted only under the provisions of the German Copyright Law of September 9, 1965, in its current version, and permission for use must always be obtained from Springer. Violations are liable to prosecution under the German Copyright Law.

Springer is a part of Springer Science+Business Media
springer.com
© Springer-Verlag Berlin Heidelberg 2006
Printed in Germany

The use of general descriptive names, registered names, trademarks, etc. in this publication does not imply, even in the absence of a specific statement, that such names are exempt from the relevant protective laws and regulations and therefore free for general use.

Cover design: Erich Kirchner, Heidelberg
Typesetting: Büro Stasch (stasch@stasch.com), Klaus Häringer, Bayreuth
Production: Christine Jacobi
Printing: Stürtz AG, Würzburg
Binding: Stürtz AG, Würzburg

Printed on acid-free paper   30/2133/CJ – 5 4 3 2 1 0

Preface

My interest in pyrometamorphosed rocks began in 1981–82 as a Humboldt Fellow at the Institute for Mineralogy, Ruhr University, Bochum, Germany, with Professor Werner Schreyer, where I began a study of basement xenoliths erupted from the Wehr Volcano, east Eifel. A comprehensive suite of xenoliths had already been obtained by Gerhard Wörner (now Professor of Geochemistry at Gottingen University) and I was fortunate to be able to select samples from his collection for further detailed study. The xenoliths, mainly mica schists, had all undergone various stages of reaction and fusion. Because the high temperature reaction products and textures were fine grained they were difficult to study petrographically and for me they presented an entirely new group of rocks that I had had almost no experience with. However, by extensive use of the backscattered electron image technique of the electron microprobe the "wonderful world" of mineral reaction/melting textures arrested in various stages of up-temperature transformation by quenching was revealed. Almost every image showed something new. I had entered the seemingly complex and at times contradictory realm of disequilibrium where metastable nucleation and crystal growth is the norm.

Pyrometamorphism is a type of thermal metamorphism involving very high temperatures often to the point of causing fusion in suitable lithologies at very low pressures. The high temperatures are provided by flow of mafic magma through conduits, by way of spontaneous combustion of coal, carbonaceous sediments, oil and gas, and through the action of lightning strikes. These conditions characterise the sanidinite facies of contact metamorphism. Although pyrometamorphic effects related to igneous activity are usually restricted to very narrow aureoles and xenoliths and to the point of impact in lightning strikes, pyrometamorphic rocks may be exposed over a surface area of hundreds to thousands of square kilometres in the case of combustion of gently dipping coal seams. In all these instances, temperature gradients are extreme, varying by several hundred degrees over a few metres or even centimetres. Relatively short periods of heating create an environment dominated by metastable melting and rapid mineral reaction rates driven by significant temperature overstepping of equilibrium conditions. This results in the formation of a large variety of high temperature minerals, many of which are metastable, are only found in pyrometamorphic rocks and are analogous to those crystallising from dry melts in laboratory quenching experiments at atmospheric pressure.

Compared with other types of metamorphic rocks, pyrometamorphic rocks are rare and volumetrically insignificant. This is probably the main reason why pyrometamorphism and sanidinite facies mineral assemblages have received scant attention in many modern petrology text books despite the fact that there is a considerable literature on the subject dating back to 1873 when the first buchite was described and named. In recent years, a number of papers have appeared in international Earth Science journals detailing field relations, microtextures, mineralogy and geochemistry of pyrometamorphic rocks and related phenomena and so it seemed to me that there was a timely need for a review/synthesis of the subject. This book is the result. It is not a textbook but essentially a compilation of available data relating to some 76 terrestrial occurrences of igneous, combustion and lightning strike pyrometamorphism of quartzofeldspathic, calc-silicate, evaporite and mafic rock/sediment compositions. Examples of anthropogenic pyrometamorphism such as brick manufacture, slag production, waste incineration, drilling and ritual burning are also given for comparison. The last chapter deals with aspects of high temperature disequilibrium reactions and melting of some common silicate minerals. My hope is that the book will stimulate further research into these fascinating rocks to help explain the many unanswered questions relating to processes and products of metamorphism under very high temperature/low pressure conditions.

Freiburg, September 2005   *Rodney Grapes*

Acknowledgements

I am indebted to Professor Kurt Bucher for discussion and comments on aspects of pyrometamorphism and who kindly put the resources, infrastructure and secretarial services of Christine Höher and Karin Eckmann of the Mineralogisch-Geochemisches Institut (Freiburg University) at my disposal and thus made the project possible. Emeritus Professor Wolfhard Wimmenauer kindly shared his extensive knowledge on fulgurites. Dr. Hiltrud Müller-Sigmund is thanked for her expertise in German to English translation and guidance in using the Institute's electron microprobe, and I'm grateful to Melanie Katt for producing superb polished sections. A special thanks to the Institute Librarian, Susanne Schuble, who promptly and uncomplainingly obtained the many references I requested, especially from the basement archive where everything old and forgotten is kept. Erica Lutz is thanked for her drafting assistance until I became familiar with the CANVAS drawing programme.

I am particularly appreciative for the information on pyrometamorphic rock occurrences in Russia and the Central Asia Republics, much of it, text and photographs, unpublished, that was kindly supplied by Dr. Ella Sokol, a leading researcher in combustion metamorphism at the Institute of Mineralogy and Petrography of the Russian Academy of Sciences, Novosibirsk.

Finally, I would like to thank my wife, Agnes, for support in all sorts of ways and above all, for keeping me at it.

Contents

1	**Introduction**	1
1.1	Terms	2
1.2	Sanidinite Facies	4
2	**Thermal Regimes and Effects**	11
2.1	Igneous Pyrometamorphism	11
	2.1.1 Aureoles	11
	2.1.2 Xenoliths	17
2.2	Combustion Pyrometamorphism	20
	2.2.1 The Burning Process	21
2.3	Lightning Strike Pyrometamorphism	28
2.4	Other Thermal Effects	31
	2.4.1 Columnar Jointing	31
	2.4.2 Microcracking	37
	2.4.3 Dilation	37
	2.4.4 Preservation of Glass	39
3	**Quartzofeldspathic Rocks**	42
3.1	Experimental Data and Petrogenetic Grid	42
3.2	Contact Aureoles and Xenoliths	49
	3.2.1 Metapsammitic-Pelitic Rocks and Schist-Gneiss Equivalents	49
	3.2.2 Sanidinite	81
	3.2.3 Granitoids	84
3.3	Combustion Metamorphism	92
	3.3.1 Sandstone, Siltstone, Shale, Diatomite	92
3.4	Lightning Strike Metamorphism	112
3.5	Vapour Phase Crystallisation	113
4	**Calc-Silicates and Evaporates**	115
4.1	Calc-Silicates	115
	4.1.1 CO_2-H_2O in Fluid Phase	119
	4.1.2 T-P-XCO_2 Relations	122
	4.1.3 Contact Aureoles and Xenoliths	129
	4.1.4 Combustion Pyrometamorphism	156
4.2	Evaporates	166

5	**Mafic Rocks**	167
5.1	Basaltic Rocks	168
	5.1.1 Contact Aureoles	168
	5.1.2 Amygdules and Mesostasis	172
	5.1.3 Weathered Mafic Rocks	178
5.2	Aluminous Ultramafic Rocks	183
5.3	Hydrothermally-Altered Andesite	185
5.4	Vapour Phase Crystallisation	188
5.5	Lightning Strike Fusion	189
6	**Anthropogenic Pyrometamorphism**	191
6.1	Bricks/Ceramics	191
6.2	Spoil Heaps	200
6.3	In-situ Gasification	205
6.4	Drilling	209
6.5	Slag	211
6.6	Ritual Burning, Vitrified Forts	215
6.7	Artificial Fulgurite	217
7	**Metastable Mineral Reactions**	219
7.1	Quartz	223
7.2	Plagioclase	223
7.3	Muscovite	227
7.4	Chlorite	232
7.5	Biotite	235
7.6	Calcic Amphibole	238
7.7	Clinopyroxene	240
7.8	Al-Silicates	242
7.9	Garnet	245
7.10	Staurolite	247
7.11	Cordierite	250
	References	253
	Index	269

Chapter 1

Introduction

Pyrometamorphism, from the Greek *pyr/pyro* = fire, *meta* = change; *morph* = shape or form, is a term first used by Brauns (1912a,b) to describe high temperature changes which take place at the immediate contacts of magma and country rock with or without interchanges of material. The term was applied to schist xenoliths in trachyte and phonolite magma of the Eifel area, Germany, that had undergone melting and elemental exchange (e.g. Na_2O) with the magma to form rocks consisting mainly of alkali feldspar (sanidine, anorthoclase), cordierite, spinel, corundum, biotite, sillimanite and (relic) almandine garnet ± andalusite. Brauns (1912a,b) considered the essential indicators of pyrometamorphism to be the presence of glass with the implication that temperatures were high enough to induce melting, pyrogenic minerals (i.e. crystallised from an anhydrous melt), replacement of hydrous minerals by anhydrous ones, the preservation of crystal habits of reacted minerals and of rock textures.

Tyrrell (1926) defined pyrometamorphism as pertaining to the "effects of the highest degree of heat possible *without actual fusion*" [authors italics] and considered that the term could be usefully extended to "all products of the action of very high magmatic temperatures, whether aided or not by the chemical action of magmatic substances". He regarded pyrometamorphic effects as "… conterminous with, and hardly distinguishable from, those due to assimilation and hybridization". Tyrrell's definition thus fails to include the presence of glass or fused rocks (buchites, see below) that are amongst the most typical products of pyrometamorphism.

A further term, *caustic (= corrosion) metamorphism* was introduced by Milch (1922) to describe the indurating, burning and fritting effects produced in country rocks by lavas and minor intrusions. It was replaced by Tyrrell (1926) with what he felt to be the more appropriate *optalic metamorphism* (Greek *optaleos* = baked [as bricks]) to describe such effects: "… optalic effects are produced by evanescent hot contacts at which heat is rapidly dissipated. The elimination of water and other volatile constituents, the bleaching of carbonaceous rocks by the burning off of carbon, the reddening of iron-bearing rocks by the oxidation of iron, induration, peripheral fusion of grains (fritting); in short, analogous kinds of alteration to those produced artificially in brick and coarse earthenware manufacture, are the most notable effects of this phase of metamorphism. Argillaceous rocks are often indurated with the production of an excessively hard material called *hornstone*, *lydian-stone*, or *porcellanite*. Some *hornstones* and *novaculite* are due to this action on siliceous clays and shales. The coking of coal seams by igneous intrusions, and the columnar structures induced both in coals and in some sandstones, are also to be regarded as the effects of optalic metamorphism". The term is now obsolete and the effects of combustion of coal and other organic matter, can be included under pyrometamorphism.

Pyrometamorphic rocks are thus extensively documented in relation to xenoliths in basic lavas and shallow intrusions, in narrow aureoles immediately adjacent basaltic necks and shallow intrusions, and as fragments in tuffs and volcanic breccias (Fyfe et al. 1959; Turner 1948, 1968). Most of the standard metamorphic petrology textbooks do not include the products of combustion of coal seams/organic-rich sediments, or lightning strikes, as pyrometamorphic products. The Subcommission on the Systematics of Metamorphic Rocks (SCMR) of the International Union of Geological Sciences (IUGS) (Smulikowski et al. 1997) refers to these variants of contact metamorphism as *burning metamorphism* (although the term *combustion metamorphism* is also in use) and *lightning metamorphism*.

In contrast to the localised (contact) occurrence of pyrometamorphism associated with igneous rocks and particularly so in the case of lightning strikes, it is important to note that pyrometamorphic products of burning coal seams and carbonaceous sediments can be of regional extent, e.g. outcropping over an area of some 200 000 square kilometers in the western United States. However, the process involves steep temperature gradients to produce burnt to completely melted rocks over a restricted interval of anything from a few centimeters to a few meters at or near the Earth's surface.

1.1
Terms

There are a number of rock terms commonly used in association with the phenomenon of pyrometamorphism, e.g. buchite, porcellanite, sanidinite, emery, paralava, clinker, fulgurite, together with the more general terms, vitrified or fused and burnt rocks. These can be characterised, together with related and mostly outdated terms, as follows:

- *Buchite* – A partly to almost completely vitrified rock resulting from intense contact metamorphism. Pyrometamorphic xenoliths and contact aureole rocks described as altered (*verändert*) or glassy (*verglast*) sandstone (buchites) associated with basaltic rocks were first described from a number of localities throughout Germany (e.g. Zirkel 1872, 1891; Mohl 1873, 1874; Lemberg 1883; Hussak 1883; Prohaska 1885; Rinne 1895). The term *buchite* appears to have been coined by Mohl (1873) presumably after the German geologist, Leopold von Buch (Tomkeieff 1940), to describe fused rock (*geglühte Sandstein*) in contact with basalt and replaced the term *basalt-jasper* (Fig. 1.1). Although initially applied to partially melted sandstone (Lemberg 1883; Morosevicz 1898; Bücking 1900; Harker 1904), *buchite* was later used to describe fused pelitic rocks (Flett 1911; Thomas 1922; Jugovics 1933). The rarely used term, *para-obsidian* describes a nearly holohyaline buchite containing a few microlites of phases such as mullite and tridymite.
- *Porcellanite* – A light coloured, very fine grained, completely recrystallised pyrometamorphosed clay, marl, shale or bauxitic lithomarge. Originally named from a naturally baked clay considered to be a variety of jasper (Werner 1789) which led to the adoption of the now obsolete term porcelain-jasper.

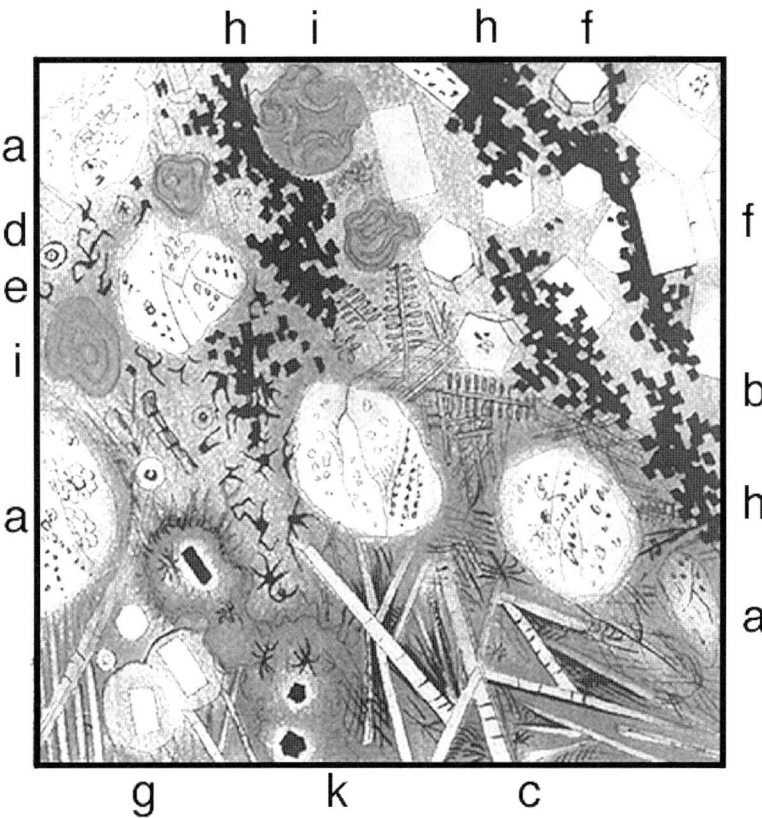

Fig. 1.1. Sketch of thin section of a tridymite-cordierite-clinopyroxene-spinel buchite from Otzberg, Germany (Fig. IV of Mohl 1873). This is the first rock to be termed *buchite*. The letters refer to: a = relic and cracked quartz overgrown by and with inclusions of tridymite; b = feathery-like/dendritic ?pyroxene (*centre of diagram*); c = needles of clinopyroxene; d = small vesicles with dark rims and centres; e = trichites of magnetite; f = rhombic and hexagonal crystals of cordierite (mistakenly identified as nepheline); g = glass surrounding cordierite; h = spinel; i = zoned pores filled with goethite; k = brown-coloured glass matrix

- *Sanidinite* – Sanidinite was proposed (Nose 1808) as a term for igneous rocks and segregations/inclusions therein consisting mainly of sanidine or anorthoclase (e.g. nosean sanidinite, hauyne sanidinite [*gleesite*], aegirine-augite sanidinite [*parafenite*], scapolite sanidinite [*hüttenbergite*]). The term is now mainly restricted to pyrometamorphic rocks (xenoliths) with a sanidine-syenite composition, but consisting of sanidine/anorthoclase, biotite, cordierite, orthopyroxene, sillimanite/mullite, spinel, corundum, ilmenite, Ti-magnetite (i.e. biotite sanidinite [*laachite*] of Brauns 1912a; Kalb 1935, 1936; Frechen 1947).
- *Emery* – A dark, hard, dense granular rock consisting mainly of corundum, spinel, magnetite and/or ilmenite-hematite (sometimes including pseudobrookite, mullite,

cordierite, sanidine) formed by high (near basaltic magma) temperature metamorphism of laterite (ferriginous bauxite), aluminous (pelitic) sediment, or as an aluminous residue (restite) resulting from removal of a "granitic" melt fraction from pelitic-psammitic rocks.

- *Paralava* – A name given to vesicular, aphanitic, fused shale and sandstone that resembles artificial slag or basalt in physical appearance, produced by the combustion of coal seams. Originally termed "para lavas" by Fermor (Hayden 1918) from observations in the Bokaro Coalfield, India, after previously regarding them as basaltic lava from their highly vesicular, ropy structure (Fermor 1914). The analogous term *parabasalt* has been adopted in Russian literature (e.g. Sharygin et al. 1999) to describe glassy and holocrystalline rocks that contain olivine, ortho/clinopyroxene, calcic plagioclase, Fe-Al spinels, ± leucite, resulting from melting by spontaneous combustion in waste heaps of the Chelyabinsk coal basin.
- *Clinker* – hard rock resembling burnt paving-brick (*klinkaerd*) used in Holland, that rings when struck with a hammer. Gresley (1883) defines clinker as coal altered by an igneous intrusion, but it is now applied to sedimentary rocks that have been baked and/or partially melted by the combustion of coal seams or bituminous sediments. In the United States, the term was first applied to hard red rocks overlying burnt coal seams in Eastern Montana by Prince Alexander Maximilian in 1833, because of their strong resemblance to the waste products of European brick kilns, and when struck, emitted a clear sound "like that of the best Dutch clinkers" (Sigsby 1966). Glassy clinkers (sometimes referred to as *scoria*) are effectively buchites.
- *Fulgurite* – (Latin *fulgur* = lightning). An irregular, glassy, often tabular or rod-like structure produced by the fusion of loose sediment, but also solid rock, by lightning.

1.2
Sanidinite Facies

As mentioned above, *sanidinite* is a term synonymous with pyrometamorphism and was first used by Brauns (1912a,b) to describe sanidine-rich xenoliths in volcanic rocks. The term was adopted by Eskola (1920; 1939, pp. 347–349) as his highest temperature, lowest pressure metamorphic facies (*sanidinite facies*) characterised by the occurrence of sanidine (typically with a high Na-content) and pigeonite (clinohypersthene). Compositions and notations of sanidinite facies minerals used throughout the text are listed in alphabetical order in Table 1.1.

In contact aureoles of shallow basaltic intrusions, in sediments overlying burnt coal seams and in combusted carbonaceous sediments, sanidinite facies rocks represent the end product of a continuous spectrum of contact metamorphism that occurs over very short distances ranging from centimeters to tens of meters. The effects of high temperature and chemical disequilibrium caused by incomplete reaction due to rapid heating and cooling that are indicated by pyrometamorphic mineral assemblages, their compositions, crystal habit, textures, and preservation of glass, are typically quite distinctive and do not fit well into other facies of contact metamorphism.

At low pressures (< 2 kb), differentiation of sanidinite facies from pyroxene hornfels facies rocks can be characterised by the absence of andalusite and pyralspite garnet in quartzo-feldspathic rocks (Chapter 3) and grossular in calc-silicate rocks. In silica-poor calcareous rocks, Turner and Verhoogen (1960) propose that the formation of monticellite from diopside and forsterite marks the transition from pyroxene hornfels to sanidinite facies (Chapter 4). In mafic and ultramafic rocks the distinction from pyroxene hornfels assemblages is difficult. Both may contain olivine, clinopyroxene, orthopyroxene, spinel, and exhibit a hornfelsic texture, so that the distinction, although not particularly relevant, might depend on recognition of diagnostic mineral assemblages in intercalated metasedimentary rocks (Chapter 5).

Although melting, disordered feldspars, and pigeonite may be stable under granulite facies conditions, they recrystallise, invert and/or unmix during slow cooling (Miyashiro 1973), and it is rapid cooling rates as evidenced by finer grain size and a low pressure geological setting that should also be used to distinguish sanidinite facies from granulite facies rocks. A case in point is the emery rocks of the Cortland Complex, USA, that contain a "sanidinite facies" assemblage of sillimanite, spinel, corundum and Ti-Fe oxides (emeries) and represent extreme metamorphism of quartz-poor rocks involving partial melting and melt extraction, as evidenced by networks of quartzo-feldspathic veins. Local silicification reactions in the proximity of these veins has led to the formation of sapphirine, pyrope-rich garnet, Al-orthopyroxene and cordierite with increasing silica activity. While thermobarometry indicates formation temperatures at ~1000 °C, a pressure of ~7.5 kb puts these rocks well within the granulite facies field (Tracy and McLellan 1985).

Occurrences of sanidinite facies rocks are typically insignificant compared with other facies of metamorphism. They clearly contain unusual and unique mineral compositions and exhibit features of paragenesis that indicate a merging with igneous rocks. Mineral assemblages developed from fusion of pelitic, psammitic and marl compositions in particular are closely analogous to the products of crystallisation of *dry* melts in laboratory quenching experiments at atmospheric pressure (Turner 1948).

As stated above, sanidine often with a considerable Na-component is a key mineral of Eskola's definition of sanidinite facies and reflects the continuous solid solution between K- and Na-feldspars at temperatures > 650 °C. The problem with the term is that many sanidinite facies rocks, e.g. calc-silicate, basic and ultrabasic lithologies, with K-absent, Al-Si-poor compositions, and even many buchites derived from fusion of quartzofeldsapthic rocks, do not contain sanidine. In buchites, the K-feldspar component commonly remains dissolved in the melt (glass). In the case of "non-igneous" sanidinite (as xenoliths), sanidine/anorthoclase form when the melt "modified" by Na-exchange with surrounding magma (especially trachyte or phonolite) crystallises during high temperature annealing to enclose typical sanidinite facies minerals such as Ti-rich biotite, mullite-sillimanite, spinel, corundum, ilmenite, etc.

Nevertheless, despite the above-mentioned problems, the "sanidinite facies" as a facies of contact metamorphism is still in widespread use and is retained as one of the ten recommended main metamorphic facies by Smulikowski et al. (1997). Based on *T-P* estimates of pyrometamorphosed rock occurrences described later, the sanidinite facies stability field is shown in Fig. 1.2. The upper pressure limit is somewhat arbitrary, although it can be significantly higher than the ~0.6 kb suggested by Turner and Verhoogen (1960; Fig. 79), and available data indicates lithostatic = vapour pressure conditions of < 3 kb. While maximum temperature conditions resulting from *igneous ac-*

Table 1.1. Pyrometamorphic (sanidinite facies) minerals, composition and notation

Mineral	Composition	Notation
Aegirine	$NaFe^{3+}Si_2O_6$	Aeg
Åkermanite	$Ca_2MgSi_2O_7$	Åk
Andradite	$Ca_3Fe_2^{3+}Si_3O_{12}$	Adr
Anorthite	$CaAl_2Si_2O_8$	An
Apatite	$Ca_5(PO_4)_3(OH)$	Ap
Biotite	$K(Fe,Mg)_3AlSi_3O_{10}(OH)_2$	Bt
Bredigite	$Ca_{14}Mg_{12}(SiO_4)_8$	Br
Brownmillerite	$Ca_2(Al,Fe^{3+})_2O_5$	Bm
Calcite	$CaCO_3$	Cc
Chromite	$(FeMg)Cr_2O_4$	Cr
Clinoenstatite	$Mg_2Si_2O_6$	Cen
Clinoferrosilite	$Fe_2Si_2O_6$	Cfs
Clinopyroxene (unspecified)	$(CaMgFeAl)_2(SiAl)_2O_6$	Cpx
Cordierite	$(Fe,Mg)_2Al_4Si_5O_{18}nH_2O$	Cd
Corundum	Al_2O_3	Co
Cristobalite	SiO_2	Cb
Diopside	$CaMgSi_2O_6$	Di
Dorrite	$Ca_2Mg_2Fe_4^{3+}Al_4Si_2O_{20}$	Dr
Enstatite	$MgSiO_3$	En
Esseneite	$CaFe^{3+}AlSiO_6$	Ess
Fassaite	$Ca(MgFe^{3+}Al)(SiAl_2)O_6$	Fas
Fayalite	Fe_2SiO_4	Fa
Ferrosilite	$FeSiO_3$	Fs
Forsterite	Mg_2SiO_4	Fo
Gehlenite	$Ca_2Al_2SiO_7$	Ge
Grossite	$CaAl_4O_7$	Gs
Grossular	$Ca_3Al_2Si_3O_{12}$	Gr
Hatrurite	Ca_3SiO_5	Ht
Hedenbergite	$CaFeSi_2O_6$	Hd
Hematite	$Fe_2^{3+}O_3$	Hm
Hercynite	$FeAl_2O_4$	Hc
Ilmenite	$FeTiO_3$	Ilm
Kalsilite	$KAlSiO_4$	Ks
K-feldspar (unspecified)	$KAlSi_3O_8$	Ksp
Kirschsteinite	$CaFeSiO_4$	Kr
Larnite[a]	Ca_2SiO_4	Ln
Leucite	$KAlSi_2O_6$	Lc
Lime	CaO	Lm

Table 1.1. Continued

Mineral	Composition	Notation
Magnetite	$FeFe_2^{3+}O_4$	Mt
Mayenite	$Ca_{12}Al_{14}O_{33}$	May
Melilite	(solid solution between Åk and Ge)	Mel
Merwinite	$Ca_3Mg(SiO_4)_2$	Mw
Monticellite	$CaMgSiO_4$	Mc
Mullite	$Al_6Si_2O_{13}$	Mu
Nagelschmidtite	$Ca_3(PO_4)_{2.2}$ (α-Ca_2SiO_4)	Ng
Native iron	Fe	Fe
Nepheline	$(NaK)AlSiO_4$	Ne
Oldhamite	CaS	Oh
Olivine	(solid solution between Fa and Fo)	Ol
Orthopyroxene	(solid solution between En and Fs)	Opx
Osumilite	$(KNa)(FeMg)_2(AlFe^{3+})_2(SiAl)_{12}O_{30}$	Os
Pentlandite	$(FeNi)_9S_8$	Pn
Periclase	MgO	Pe
Phlogopite	$KMg_3AlSi_3O_{10}(OH)_2$	Phl
Pigeonite	(low Ca clinopyroxene solid solution)	Pgt
Protoenstatite	$MgSiO_3$	Pen
Pseudobrookite	Fe_2TiO_5	Psb
Pseudowollastonite (β-form)[b]	$CaSiO_3$	Pwo
Pyrrhotite	$Fe_{1-x}S$	Po
Quartz (β)	SiO_2	Qz
Rankinite	$Ca_3Si_2O_7$	Rn
Rutile	TiO_2	Rt
Sanidine	$(KNa)AlSi_3O_8$	San
Sapphirine	$(FeMgAl)_4(AlSi)_3O_{10}$	Spr
Sillimanite	Al_2SiO_5	Sil
Spinel	$MgAl_2O_4$	Spl
Spurrite	$Ca_5Si_2O_8(CO_3)$	Sp
Tilleyite	$Ca_5Si_2O_8(CO_3)_2$	Ty
Tridymite	SiO_2	Td
Troilite	FeS	Tro
Ulvöspinel	Fe_2TiO_4	Usp
Wollastonite (α-form)	$CaSiO_3$	Wo
Wüstite	FeO	Wü

All Fe is Fe^{2+} unless otherwise indicated.
[a] β-form of α-Ca_2SiO_4.
[b] Also referred to as cyclowollastonite.

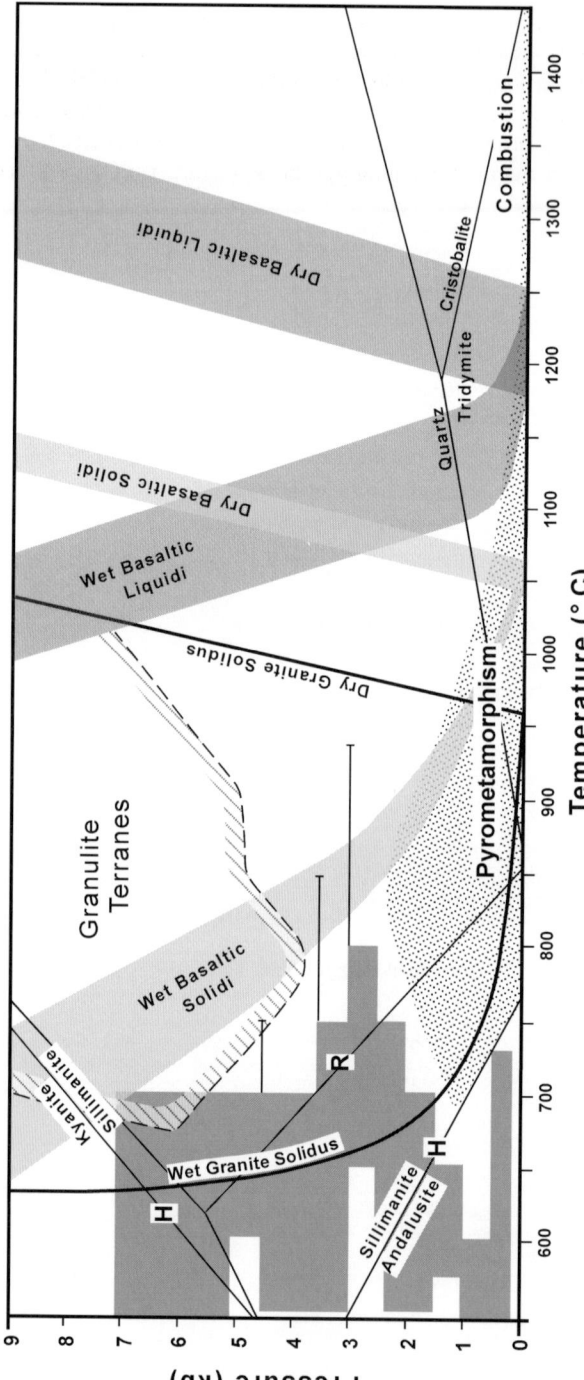

Fig. 1.2. Petrogenetic grid showing relationship of sanidinite facies rocks (*stippled area*) to contact metamorphic and granulite facies rocks. *Dark grey shaded area with horizontal line extensions* = P-T estimates of contact aureole rocks associated with shallow and deep intrusions listed by Barton et al. (1991). Field of granulite terranes delineated from Harley (1989). Wet/dry basalt solidi/liquidi of various basalt compositions after Yoder and Tilley (1962). Wet/dry granite solidi from Tuttle and Bowen (1958). Andalusite/sillimanite and sillimanite/kyanite transition curves from Holdaway (1971) (*H*) and Richardson et al. (1969) (*R*). Quartz/tridymite/cristobalite inversion curves from Kennedy et al. (1962) and Ostrovsky (1966)

tivity are defined by wet to dry basic magma liquidus temperatures (e.g. ~1190–1260 °C at 1 kb), at atmospheric conditions temperatures up to ~1500 °C can result from the combustion of coal and carbonaceous sediments and are up to 2000 K or greater for lightning strike pyrometamorphism. Details of mineral reactions, stabilities and fluid compositions that enable petrogenetic grids to be established for different bulk compositions are presented in chapters detailing pyrometamorphic rock occurrences.

Chapter 2

Thermal Regimes and Effects

Pyrometamorphism related to intrusion of mafic magma and the burning of organic material is expected to result in anisotropic thermal expansion and contraction during heating and cooling in contact rocks due to reaction and melting of the mineral phases. Columnar jointing, cracking and dilation are typical of the structures developed. Temperature variation can be extreme with thermal gradients of several hundred degrees developed over a few meters or even tens of centimeters, particularly in the case of combustion and lightning strike metamorphism.

2.1
Igneous Pyrometamorphism

2.1.1
Aureoles

Pyrometamorphism affects rock that is in contact with intrusion of typically mafic-intermediate magmas. Evidence of pyrometamorphosed metasediment, metabasite, and granitic host rocks intruded by shallow basaltic and andesitic plugs, sills and dykes has been reported from many localities commonly with the development of buchites in the case of psammitic-pelitic protoliths. Mineral reconstitution is largely restricted to within about 0.5 m of the igneous contacts, but is some cases the thermal effects and their resultant structures such as columnar jointing are locally developed up to 50 m from the contact.

The generation of very high temperatures, sometimes approaching 1200 °C, in contact aureoles results is steep thermal gradients and has been ascribed to flow of magma through conduits that represent feeders of lava flows, turbulent magma flow (e.g. in sills), convective circulation in larger magma bodies or relatively short filling times in the order of < 100 years, geologically 'instantaneous' (a few days) intrusion in the case of smaller bodies, and both convective and conductive heat transfer via water through the contact rocks. While various temperature profiles constructed normal to the igneous contacts have been attempted using appropriate mineral equilibria (see below), estimation of the timing of the pyrometamorphic event is more difficult, although the use of Ar and element diffusion profiles and modeled changes in thermal profiles with time have been made (e.g. Wartho et al. 2001).

The efficiency of heat transfer of magma through a conduit is critically dependent on the magma flow regime. Turbulent flow can dramatically increase temperatures in contact country rock to approach that of the magma (Huppert and Sparks 1985) resulting in fusion and the development of a wide aureole. Rocks containing substantial modal K-feldspar, Na-plagioclase, quartz and muscovite/biotite, e.g. pelite, arkose, granite-granodiorite, tend to fuse more readily than refractory lithologies such as quartzite, adjacent turbulent basaltic magma. Turbulence is commonly associated with high flow rate and reduced viscosity, particularly where the conduit acts as a feeder to lava flows. It may develop around areas of local irregularities in the wall rock, where there is a marked change in the attitude or a change in thickness of the intrusive body. In comparison, laminar flow typically results in chilling and a relatively low contact temperature of perhaps midway between that of the magma and the mean temperature of the country rock prior to intrusion (e.g. Delaney and Pollard 1982). Where heat conduction is the only mechanism of energy transport within the magma and contact rocks, the maximum temperature attained at a contact is equal to about two-thirds of the solidus temperature of the magma (Jaeger 1968; p. 520–23). On the other hand, convection in a narrow boundary layer along a vertical contact will maintain the contact temperature at or even slightly above that of the basaltic solidus during the early stages of crystallisation (Shaw 1974).

Thermal modeling of the contact rock considering a basaltic sill of half thickness h emplaced in country rock with the same thermal conductivity, specific heat capacity and density or thermal diffusivity κ as the magma, and ignoring the latent heat of magma crystallisation is detailed by Wartho et al. (2001) using three models:

1. Conductive cooling following instantaneous emplacement.
2. Convection driven by heat loss into the roof rocks of the sill while the floor remains insulated following instantaneous emplacement.
3. The magma/country rock contact is held at a constant temperature as in the case where prolonged turbulent flow of magma through the sill maintains the magma emplacement temperature along the contact.

The three models are illustrated in Fig. 2.1 in terms of time-temperature relationships at dimensionless temperatures (T), distances (y) and times (t) from the igneous contact. Temperature at any position in the country rock at a particular time can be found in terms of the following dimensionless variables:

- Dimensionless temperature:

$$\theta = \frac{T - T_{cr}}{T_{mag} - T_{cr}}$$

- Dimensionless distance:

$$Y = y/h$$

- Dimensionless time:

$$\tau = \kappa t/h^2$$

where T_{mag} = initial magma temperature and T_{cr} = initial contact rock temperature.

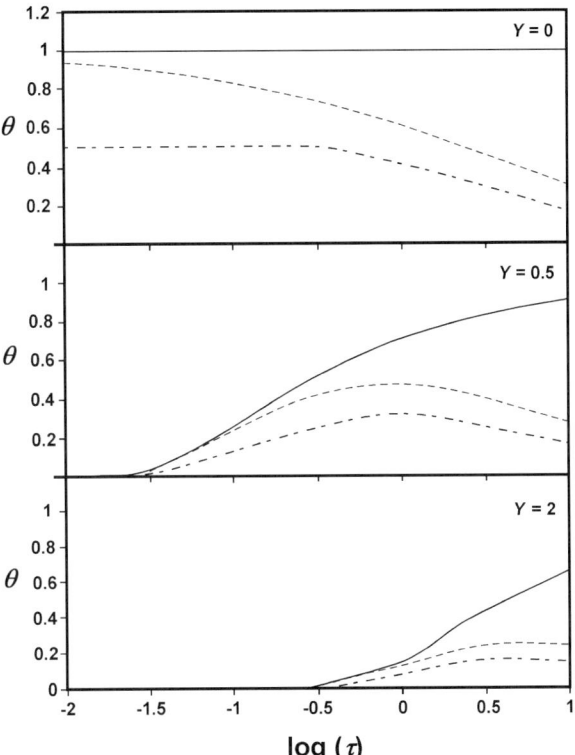

Fig. 2.1.
Temperature-time relationships for three thermal models at dimensionless distances ($Y = 0$, 0.5 and 2) from the contact of a sill. *Short-long dash line* = θ_{cond} (model 1); *dashed line* = θ_{conv} (model 2); *solid line* = θ_{cont} (model 3), (after Fig. 1 of Wartho et al. 2001). See text

Following Jaeger (1964), the temperature-position-time relations for the three models are:

- Conductive cooling (θ_{cond}) (model 1):

$$\theta_{cond} = 0.5\left\{\mathrm{erf}\left(\frac{Y+2}{2\tau^{1/2}}\right) - \mathrm{erf}\left(\frac{Y}{2\tau^{1/2}}\right)\right\}$$

- Convective cooling above upper contact (θ_{conv}) (model 2):

$$\theta_{conv} = \exp\left\{\frac{Y}{2} + \frac{\tau}{4}\right\}\left[\mathrm{erfc}\left(\frac{Y}{2\tau^{1/2}} + \frac{\tau^{1/2}}{2}\right)\right]$$

- Continuous turbulent intrusion (θ_{cont}) (model 3):

$$\theta_{cont} = \mathrm{erfc}\left(\frac{Y}{2\tau^{1/2}}\right)$$

The models predict, for example, that at the contact ($Y = 0$), the temperature remains at $\theta = 1$ for as long as the magma is flowing, whereas in the conductive model,

θ does not exceed 0.5. In the convective model, θ is initially 1 and then gradually decreases until reaching 0.5 at $\tau \sim 3$. At any given time the prolonged turbulent emplacement model gives the highest temperature and the conductive model the lowest temperature. Under conditions of $Y = 2$ the thermal histories are virtually indistinguishable after $\tau \sim 1$. It can also be noted that in all three models heating will be greater above than below the sill, whereas under conditions of conductive cooling or forced flow heating will be symmetrical. Wartho et al. (2001) point out that the symmetry of thermal metamorphism and the maximum temperatures reached at given distances, especially those near the contact, should be sufficient to constrain the dominant heat transfer process.

Koritnig (1955) has calculated a number of heating curves over a period of 180 days in sandstone adjacent the Blaue Kuppe dolerite, Germany, and also heating profiles developed within 100 cm of the contact over a period of three days (Fig. 2.2). In both cases a constant temperature of 1200 °C is maintained along the contact implying magma flow. The curves, for example, show that temperatures of 1000 °C in sandstone 0.5 m from the contact may be reached after ~20 days and after 180 days 2 m from the contact. After little more than one day, the temperature of sandstone 10 cm from the contact is 1000 °C. The temperature-distance relations are supported by the abundance of glass (buchite) present in the contact rocks. The calculated temperature curves ignore the role of the expulsion of intergranular fluid from the sandstone that is effective in transferring heat.

In Fig. 2.3 thermal profiles away from roughly circular gabbro (~50 m) and peridotite (200 m) plugs on Rhum, Scotland, are shown together with melt volume generated in their respective aureole arkose rocks (Holness 1999). The onset of melting occurs at a distance of ca.15 m from the gabbro but only 6 m away from the peridotite and in both cases the amount of melt increases abruptly with the beginning of melting, both reaching 75 vol.% within a few metres. Application of a simple, one-dimensional, two-stage model to both aureoles is made by Holness (1999), assuming:

1. The intrusions are vertical cylinders intruded into country rock with an initial ambient temperature of 30 °C.
2. 1st stage of thermal history; contact kept at constant temperature to simulate flow of magma through cylindrical conduits.
3. 2nd stage of thermal history; cooling only that occurs after end of magma flow.
4. Thermal diffusivity of both intrusions and country rock = 10^{-6} m^2 s^{-1} and heat capacity = 2.5 J cm^{-3} K^{-1}.
5. The latent heat of melting of a country rock arkose of 50 wt.% quartz, 25 wt.% albite, 25 wt.% orthoclase is modeled using expressions of Burnham and Nekvasil (1986) for latent heats of fusion of end member feldspar components. The latent heat of fusion of magma can be ignored in view of the fact that the latent heat of fusion of melting of country rock has no appreciable effect on T_{max} curves in Fig. 2.3.

The results of the thermal modeling are that for a peridotite magma temperature of 1000–1200 °C, intruded as a single stage crystal-rich mush, the heating lasted 3–10 years; for a gabbro magma temperature of 1200–1250 °C a constant contact temperature lasted

2.1 · Igneous Pyrometamorphism

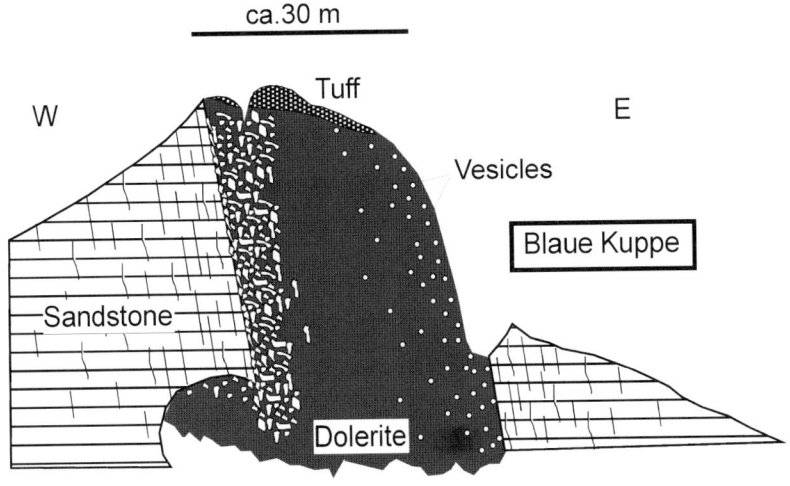

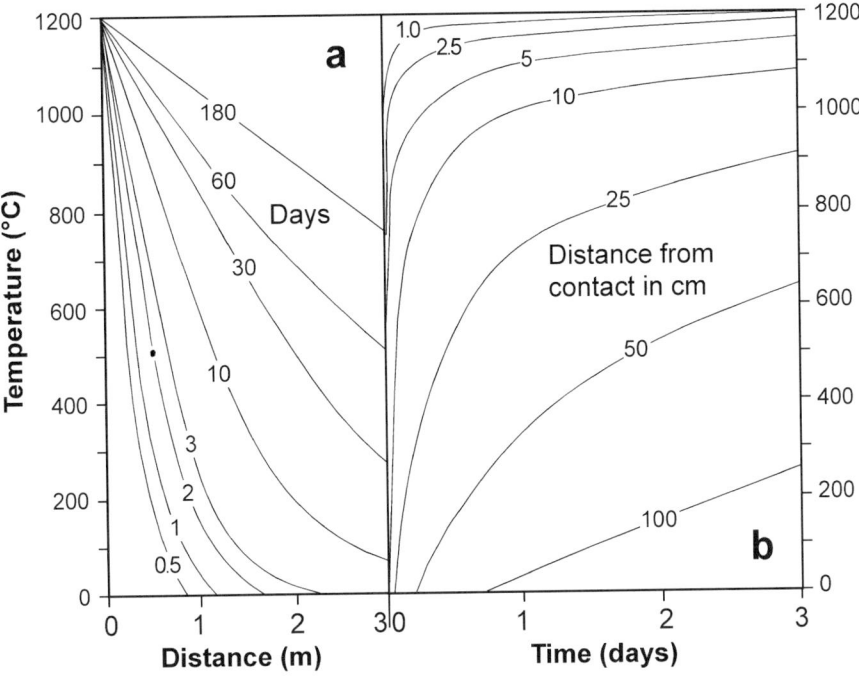

Fig. 2.2. *Above:* An E-W cross section of the Blaue Kuppe dolerite, near Eschwege, Germany (after Fig. 5b of Koritnig 1955). *Below:* **a** Heating curves for sandstone within 3 m of the dolerite contact assuming a constant contact temperature of 1200 °C over a period of 180 days; **b** heating curves for sandstone within 100 cm of the contact over 3 days assuming a constant contact temperature of 1200 °C (after Figs. 8 and 9 respectively of Koritnig 1955)

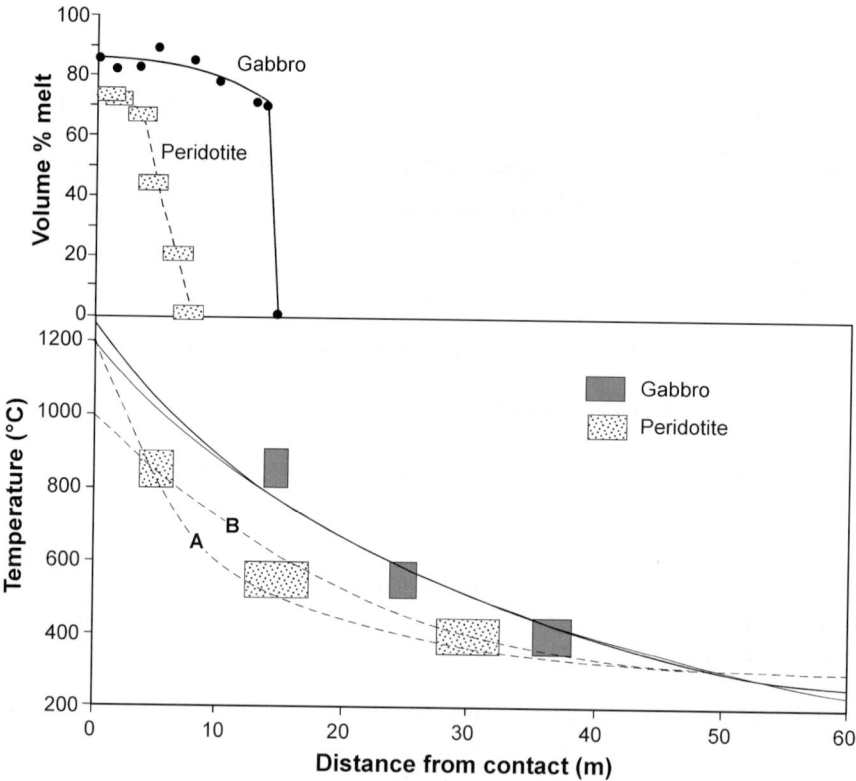

Fig. 2.3. Thermal profiles (T_{max}) (*below*) and melt contents (vol.%) (*above*) for arkosic country rock surrounding gabbro and peridotite plugs, Mull, Scotland (after Figs. 11 and 6 of Holness 1999, respectively). See text. Errors in temperature due to insufficient experimental data for melting (first appearance of melt), variable bulk Fe content relating to chlorite breakdown reaction, and uncertainly about kinetics of microcline-sanidine transition. Errors in distance due to exposure and finite sample spacing. *Curve A* = magma T 1200 °C, contact remains at constant T for 3 yr; *curve B* = magma T 1000 °C, contact remains at constant T for 10 yr. Unlabelled two curves fitting gabbro data = magma T 1250 °C, contact T remains constant for 35 yr, and for magma T 1200 °C, constant contact temperature for 40 yr

35–40 years due to magma flow through the conduit. The calculated contact temperatures also predict complete melting of the country rock within 10 m of the gabbro contact and within a few metres of the peridotite. This is not the case (Fig. 2.3) suggesting that either melting was inhibited by factors such as kinetics of the melting reactions or availability of H_2O, or the models are unrealistic close to the contacts. Within 10 m of the gabbro, the model predicts that all melt solidified within 10 years of the onset of the 2nd stage cooling. That cooling to temperatures below the solidus occurred within ca. 5 years is supported by evidence of destruction of millimetre-scale layers rich in detrital magnetite in the arkose within 10 m from the gabbro contact. From Stokes' law, the velocity at which spherical particles of radius r fall through a liquid of viscosity μ is given by

$$v = \frac{2gr^2\Delta\rho}{9\mu}$$

where g = acceleration due to gravity; $\Delta\rho$ = difference between density of solid (magnetite of 5.3 g cm^{-2}) and liquid (melt of 2.3 g cm^{-1}); melt viscosity (μ) = 10^6–10^7 poise. A magnetite crystal of 0.1 mm diameter will drop by 1 cm in 170–1700 days thus obliterating the fine-scale layering.

2.1.2
Xenoliths

The preservation of pyrometamorphosed suites of xenoliths characterised by different mineral assemblages within individual magmatic bodies implies mechanical disintegration of various source lithologies of a deeper contact zone. In areas where igneous contacts are irregular, mechanical erosion of wall rock of suitable composition may occur leading to concentrations of xenoliths such as observed in thin (up to 6 m wide) basic sheets of the Ross of Mull peninsula, Scotland (e.g. Kille et al. 1986; Fig. 2 of Holness and Watt 2002), or along the contact of a steeply dipping pipe-like body such as Blaue Kuppe, Germany (Fig. 2.2). Xenolith population sizes are typically small, a few centimetres to several tens of centimeters in diameter, although occasional rafts several meters across and length (rarely with outcrop dimensions of several hundred meters by several kilometers) also occur. In more homogeneous lithologies such as quartzite and sandstone, the resultant xenoliths tend to be blocky and are often larger compared with those of rocks that are layered on a scale of a few millimeters to centimeters, e.g. as in schist. Very large blocks, tens to a hundred meters or more in diameter, usually of coarsely crystalline rock, e.g. granite or homogeneous sandstone, are usually found close to their wall rock source.

Once rock temperatures attain values where melting (usually along quartz-feldspar contacts) occurs, the rigidity and mechanical integrity of the rock becomes significantly reduced and in wall rocks unstable portions tend to be dislodged and entrained within the magma (Fig. 2.4). Further disaggregation, for example of mica-rich and quartz/feldspar-rich parts (greater than a few mm thick) in schist blocks engulfed in magma, is generally the rule with each part undergoing different paths and rates of bulk melting and recrystallisation. Partially melted/reconstituted mica-rich layers tend to form rafts whereas interlayered quartz-feldspar rich layers may temporarily remain as scattered clots of partially melted (rounded and resorbed) grains of quartz and sometimes feldspar in a siliceous melt (Fig. 2.5).

Lovering (1938) has calculated that the temperature in the core of a 2 m thick xenolith of marble heated by conduction would approach that of the surrounding magma within a few weeks implying that the formation temperature of pyrometamorphic mineral assemblages in the xenolith can be equal to the magma temperature. An analysis of the heating and partial melting of a xenolith during cooling of a mafic magma is given in the schematic diagram shown in Fig. 2.6 after Wyllie (1961). The diagram shows a lag time between the heating of the xenolith (T_1) and formation of melt (t_2) at which time the temperature of the xenolith is increased to T_2 with an equilibrium melt for this

Fig. 2.4.
Schematic diagram showing formation and evolution of xenoliths in turbulent basaltic magma intruding a layered meta-sandstone–siltstone sequence (modified after Fig. 15 of Preston et al. 1999)

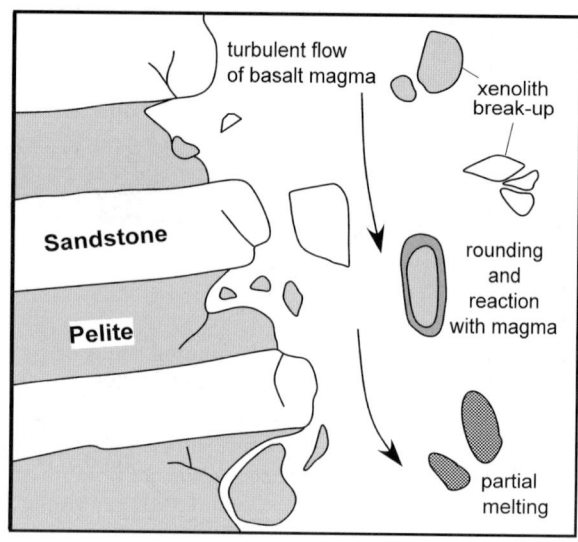

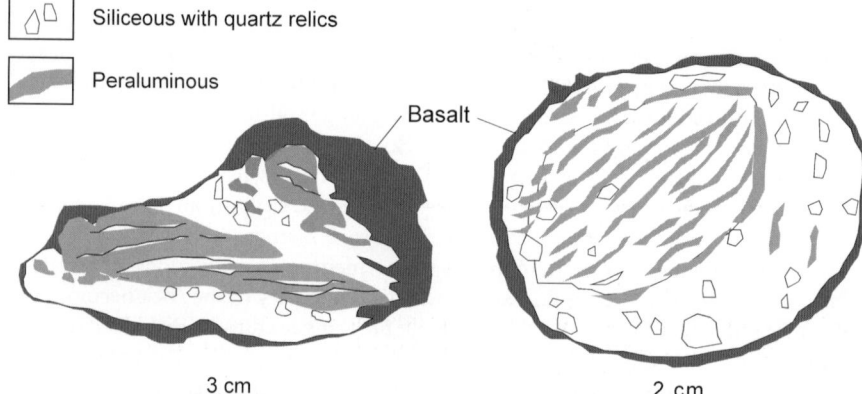

Fig. 2.5. Sketches of two partially melted xenoliths occurring in Recent hawaiites from Mt. Etna, Italy (after Fig. 11 of Michaud 1995). Siliceous parts contain cristobalite, tridymite in alkali-rich acidic glass. Quartz, zircon, apatite and titanite remain as relics. Peraluminous parts contain cordierite, spinel, Ca-plagioclase, magnetite-ilmenite and rutile in peraluminous (often K or K + Fe-rich) glass

temperature forming at t_4. At t_4 the temperature of the xenolith has increased to that of the cooling magma. By t_5 both magma and xenolith have cooled to T_3. While the magma cools from T_4 to T_3, the xenolith continues to melt and reaches its maximum state of fusion at T_3. After t_5 magma and melt within the xenolith cool together with crystallisation of minerals from the melt and with remaining melt quenched at t_q.

Because heating times during pyrometamorphism are fast and relatively short, the rate of crystal nucleation will also lag behind the change in temperature. Peak metamorphic temperatures are unlikely to be maintained for very long before cooling commences (T_3 at t_5) and as the rate of cooling is significantly less than the rate of

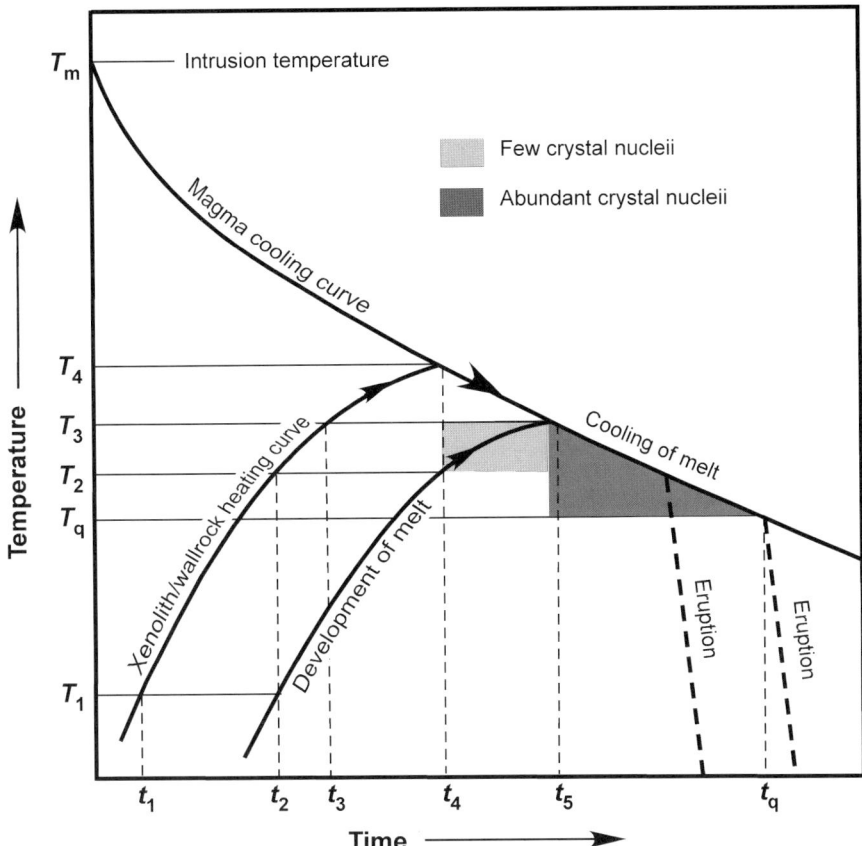

Fig. 2.6. Schematic diagram showing relationships between cooling curve of mafic magma, heating curve of a xenolith, and curve of equilibrium melt formation in xenolith. Horizontal lines between xenolith heating and melt formation curves represent the time-lag between the two (modified after Fig. 4 of Wyllie 1961). Additional T-t regions of formation of few and abundant crystal nucleii in the melt are also shown together with possible quenching on eruption curves. See text

heating, most crystallisation will probably occur during the period of cooling under nearly isothermal conditions as shown and cease with quenching on eruption.

Thermal modeling of a partially melted xenolith of granite in a cooling stock of trachyandesite and of granite surrounding the plug is analysed by Tommasini and Davies (1997). Dimensions and thermal conditions for the modeling are:

- *Plug*: diameter = 50 m, intrusion temperature = 1100°C, heat of crystallisation = 400 kJ kg^{-1} (effective thermal diffusivity κ of trachyandesite 9×10^{-7} – 2.6×10^{-7} m^2 s^{-1}).
- *Granite xenolith*: width = 4 m; beginning of melting = 850 °C at Pload < 500 bar; heat of fusion = 300 kJ kg^{-1} (effective thermal diffusivity κ of granite 1×10^{-6} – 4.4×10^{-7} m^2 s^{-1}).
- *Country rock granite*: maximum temperature = 50 °C at time of intrusion.

The resulting temperature-distance-time profiles shown in Fig. 2.7 indicate that the granite xenolith reaches solidus temperature within ~3 months and attains a maximum temperature of ~1000 °C after ca. 1.5 years. For melting to occur in the granite country rock, the initial intrusion temperature of the trachyandesite of 1100 °C would need to be maintained by a constant flow of magma. The time required to reach the granite solidus temperature obtained from

$$T_i = T_{magma}\left[1 - \mathrm{erf}\left(\frac{x_i}{\sqrt{i4\kappa t}}\right)\right]$$

where T_{magma} remains constant at the contact ($x_i = 0$ m); T_i = country rock temperature at distance x_i from the contact; κ = thermal diffusivity of the granite; t = time lapsed since intrusion, indicates that after 20 days of continuous flow the temperature is > 850 °C (~ temperature of biotite melting) 0.5 m from the contact.

2.2
Combustion Pyrometamorphism

Baked and fused rocks developed on various scales resulting from the combustion of organic and bituminous matter, coal, oil or gas in near-surface sediments are examples of this type of pyrometamorphism where thermal energy is provided through burning. Ground magnetic and aeromagnetic surveys across areas of such combustion

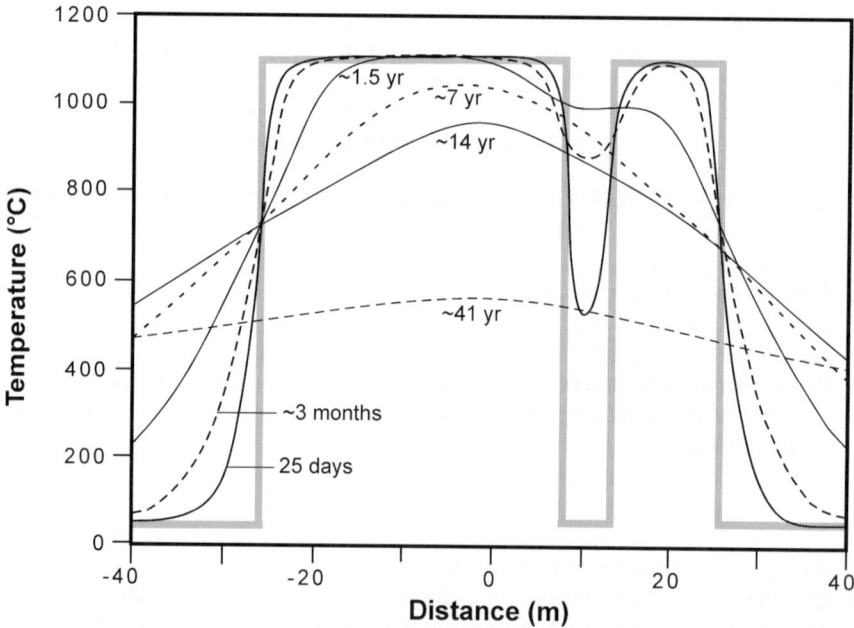

Fig. 2.7. Temperature-distance-time profiles of a cooling 50 m diameter trachyandesite plug intruding granite and containing a 4 m wide block of granite (after Fig. 4 of Tommasini and Davies 1997). See text

metamorphism indicate high anomalies in the range of several thousands of gammas (e.g. Cisowski and Fuller 1987). Temperatures attained during burning range from ~400 °C to as high as ~1600 °C resulting in a large variety of pyrometamorphic rock products, often on the scale of a single outcrop, from thermally altered but unmelted rocks, termed *burnt rocks* or *clinker*, to those that are fused, termed *paralava*, *scoria*, or *slag*.

2.2.1
The Burning Process

The agent of combustion pyrometamorphism is heat created by oxidation of organic matter and, in many cases associated sulphides (typically pyrite), through access of atmospheric oxygen into a rock sequence by way of joints, cracks and faults, or due to exposure by slumping. Heating occurs by way of the low temperature of oxidation combined with absorption of moisture. Because of its diverse composition and heterogeneous nature, oxidation of carbon, such as in coal, is a complex process, but a simplified exothermic reaction can be expressed as:

$$C + O_2 = CO_2 + heat\ (94\ kcal\ mol^{-1})$$

with the rate of reaction doubling for every increase of 10 °C (Speight 1983). Normally the generated heat is carried away by circulation of air, but where it cannot escape the temperature of the coal is raised to its "threshold temperature" of somewhere between 80–120 °C where a steady reaction resulting in the production of gases such as CO_2, CO and H_2O occurs. As the temperature continues to rise to somewhere between 230–280 °C, the reaction becomes rapid and strongly exothermic and spontaneous combustion occurs when the coal reaches its ignition point. While changes in moisture and oxidation can explain most spontaneously-generated heat in coal, there are several other important contributors to its pyophoricity such as:

- *Rank*. As coal rank decreases, the tendency for self-heating increases. Sub-bituminous coals tend to have a higher percentage of reactive macerals such as vitrinite and exinite that increase the tendency of coal to self heat.
- *Particle size*. There is an inverse relationship between particle size and spontaneous combustion; the smaller the particle size the greater surface area available on which oxidation can take place.
- *Pyrite content*. The presence of pyrite and marcasite (usually in concentrations > 2 vol.%) may accelerate spontaneous heating by reaction with oxygen according to the equation

$$4\,FeS_2 + 11\,O_2 = 2\,Fe_2O_3 + 8\,SO_2 + 6.9\ kJ\ g^{-1}\ pyrite$$

Both pyrite and marcasite swell upon heating. This causes the surrounding coal to disintegrate resulting in a reduction of the particle size involved in oxidation reactions.
- *Temperature*. The higher the temperature, the faster coal reacts with oxygen.
- *Air flow*. The flow of air provides the oxygen necessary for oxidation to occur and at the same time can also remove heat as it is generated.

- *Geological/environmental factors.* Sedimentary rocks which enclose coal seams or organic matter are poor conductors of heat. Faults and fractures in the rocks allow the influx of water and oxygen enhancing oxidation and heating. Areas of extensive combustion metamorphism are typically associated with anticlinal or rift structures where fracturing allows the access of oxygen. Drying and microbial decomposition of subsurface organic material will also result in heat production to the point of self ignition.

Coal Seams

The generally applicable nature of the burning process and the structures produced from the combustion of coal seams is discussed by Sigsby (1966) in relation to burning lignite seams in North Dakota. In the initial stages when fire spreads laterally along exposed lignite and back into the outcrop, air supply is unrestricted and provides

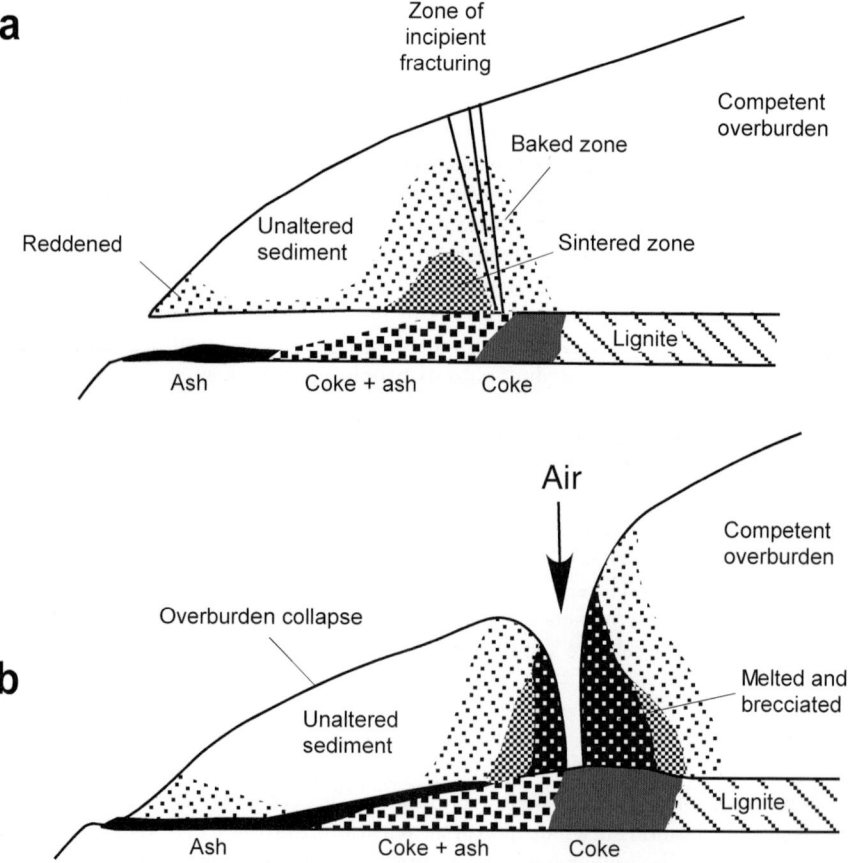

Fig. 2.8. Schematic diagrams showing the effects of combustion metamorphism of a lignite seam with respect to competent overburden (**a**) and collapsed overburden (**b**) (after Fig. 2 of Sigsby 1966)

strongly oxidizing conditions that causes reddening of the roof rocks for a few centimeters (Fig. 2.8a). With continued burning, a cavity may form where the lignite is burned out. If the overburden rocks are relatively thin and incompetent, they will tend to collapse exposing more lignite to the atmosphere (Fig. 2.8b). If the overburden is thicker and competent, the lignite continues burning inwards extending the cavity but at the same time the oxygen supply becomes progressively restricted creating more reducing conditions. Reducing conditions could also be enhanced if partial slumping of the overburden occurs and this could also extinguish the fire. Increasingly, reduced conditions coupled with conservation of heat by the competent overburden causes partial distillation of the lignite to form coke and resultant production of CO and H_2O which disassociates allowing the formation of more CO in addition to H_2. Along with SO_4 from the breakdown of gypsum and pyrite in the lignite, these gases move upward through cracks and fractures in the roof rocks, transmitting heat by contact and causing the wall rocks to melt. From the coal itself, combustion produces coal ash, typically a mixture of the non-combustable products, quartz, feldspar, clays (largely kaolinite), illite, pyrite and carbonates such as calcite and siderite, that melts from the heat of combustion to produce a slag or paralava containing new, high temperature minerals and glass.

As excavation continues by burning of the coal, structural resistance of the overburden to collapse by fracturing may be exceeded. When this happens, the fractures provide rapid access of air leading to a combination of newly supplied oxygen and accumulated combustion gases that provide enough heat to cause further and more extensive fusion (Fig. 2.8b). Some of the melt may flow into the fractures and in combination with collapsed rock fragments, clog the opening to produce a chimney assemblage and divert gases into other fractures. With erosion these chimneys remain as remnants of hard, fused rock masses projecting through partly baked rocks (Fig. 2.9; see also Plate 1A of Rogers 1917).

Fig. 2.9.
Anatomy of a chimney structure, an erosion remnant from burning of a coal seam, consisting of black and vesiculated paralava containing clasts of clinker. The coal-ash zone is composed of glassy, vesicular paralava (after Fig. 2 of Cosca et al. 1989)

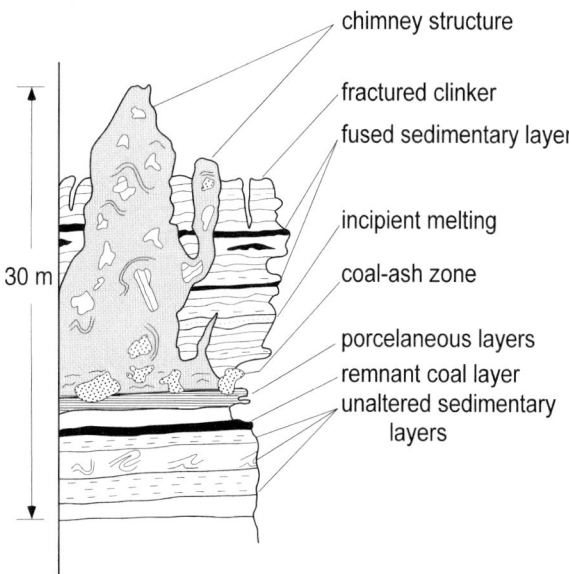

Combustion of lignite with a restricted supply of oxygen produces CO and H_2O as the main volatile products. Water quickly disassociates with the oxygen combining with C to produce more CO and the two combustion gases in combination with SO_4 from gypsum and pyrite in the lignite are strong reducing agents and move upward through fractures and interstices in the sediments transmitting heat by contact rather than conduction. Heat so transmitted may be high enough to melt to sediment close to the gas vents and rapid heat loss results in a gradation from melted, through fritted to baked rock (Fig. 2.8b). The gases also tend to cause reduction of hydrous iron oxides to ferrous compounds in the melted rock and to magnetite, as the gases tend to loose their reducing capacity in higher zones. With continued upward migration, H_2 and CO oxidize to H_2O and CO_2 and still retain sufficient heat to cause baking and resultant oxidation of iron compounds to produce the typical reddening effect commonly observed.

Whatever the cause (oxidation on exposure to the atmosphere, lightning, prairie fire, etc), combustion always starts at the surface, often where exposures are made by rapidly cutting streams, spreading first along the outcrop and then inwards. As the coal burns out, the overburden generally collapses to form a breccia or, where there are clay horizons a crumpling of slumped, coherent strata occurs. Into the outcrop heat tends to be more strongly conserved causing more intense pyrometamorphism until a point is reached where the combustion is smothered by a lack of oxygen. Rogers (1917) concludes that in general, where there is a cover of more than 15–30 m, and where the coal seam is horizontal, burning does not extend more than 60–90 m back from the outcrop. Field observations indicate that thin seams appear to be less commonly burnt than thick ones and seams of impure coal burn less commonly than those of cleaner coal.

An interesting example of pyrometamorphic cap rocks resulting from coal fires of Pliocene–Holocene age are recognised in NW China below river terraces (Zhang et al. 2004) (Fig. 2.10). The burnt rocks are between 100–150 m thick and characterised by reddish colour, millimeter-sized hexagonal columnar jointing, vesicular glassy rocks with microflow structure, porcellanitised kaolin, roof collapse and resultant brecciation, and with the coal layer reduced to an ash layer only a few centimeters thick typically rich in gypsum. Thermo-magnetic analysis of burnt rock indicates the presence of trace amounts of metallic iron implying minimum temperatures of 770 °C for the pyrometamorphism. Exposure of the coal and concomitant spontaneous combustion appears to be an interglacial phenomenon associated with uplift since the Pliocene when river downcutting and terrace formation occurred resulting in exposure of the coal to oxidation and resultant combustion.

Carbonaceous Sediments

For the heating and combustion of carbonaceous sediments, the situation is somewhat different and a good analysis is provided by Matthews and Bustin (1984) from variably weathered Cretaceous mudstones in the Smoking Hills area of Canadian Arctic coast (Fig. 2.11). Perhaps the earliest written report of pyrometamorphism of carbonaceous sediments in action was when smoke and sulphurous fumes were seen in 1826 issuing from what has since been termed the *Smoking Hills* bordering Franklin Bay, Arctic Inuvik Province of Canada, during the second overland expedition of John Franklin and John Richardson of 1825–1827. Richardson (1851) records at *"at Cape Bathurst (north-*

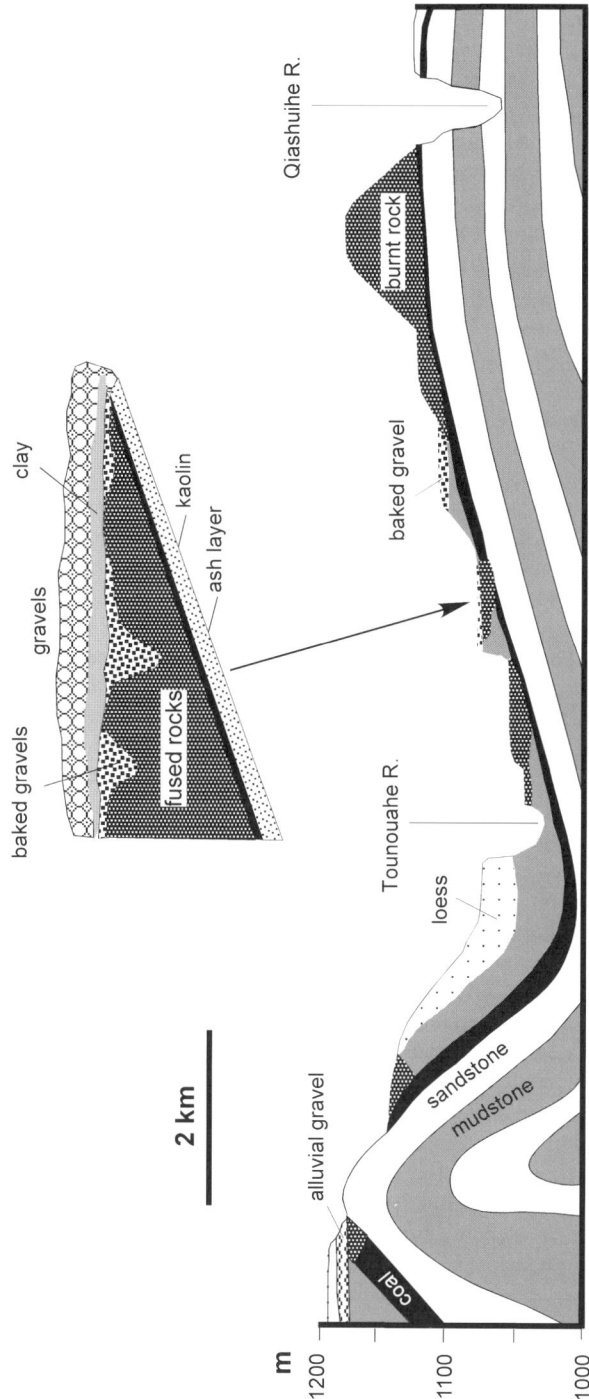

Fig. 2.10. Section centred on the Tounouahe River terrace sequence, NW China, showing occurrences of burnt rocks and a burnt rock profile (redrawn from Figs. 2 and 3 of Zhang et al. 2004). Several age groupings of combustion are identified: Pliocene-Early Quaternary at 200 m above present day flood plain; Middle Pleistocene at 90–70 m; Late Pleistocene at > 90 m; Holocene = current burning sites

Fig. 2.11.
Map of the Smoking Hills area, Canadian Arctic coast, showing the distribution of active and extinct *bocannes* (after Fig. 1 of Mathews and Bustin 1984)

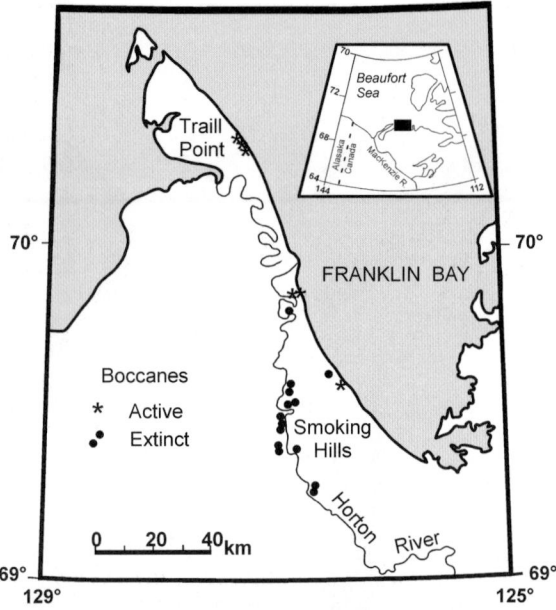

ern end of Franklin Bay) …, bituminous shale is exposed in may places, and in my visit there in 1826 was in a state of ignition; and the clays which had been thus exposed to the heat were baked and vitrified, so that the spot resembled an old brick-field". The Late Cretaceous bituminous shale with thin seams of jarosite is still burning although rarely are temperatures high enough to produce fused and vesicular clinker (Mathews and Bustin 1984). According to Yorath et al. (1969, 1975), an exposure may be quiescent one day, and the next it can be extremely hot producing noxious SO_4 fumes. Where the formation is burning or has been burnt, the shales are coloured bright yellow, orange, maroon and red. The red material consists of earthy hematite and at a number of localities large crystals of gypsum are scattered over burnt shale outcrops. Early geological explorations of British Columbia and the western Northwestern Territories also record localities of superficial burning (smoking) at sites of hot sulphurous gas emission, termed *bocannes*, in certain shales resulting in baking and reddening or bleaching (Crickmay 1967). As at Smoking Hills, extinct bocannes are marked by similarly coloured rocks but without smoke.

Mathews and Bustin (1984) have determined that the boccanes at Smoking Hills are fuelled by oxidation of fine grained, framboidal pyrite and/or organic matter (vitrinite and alginite) and that they are restricted to areas of glacially (?) disturbed strata and landsliding, e.g. landslide-prone coastal cliffs where currently active boccanes are situated (Fig. 2.11), indicating that disruption, rapid exposure, and access to atmospheric oxygen are required to produce combustion. Here, the mudstones contain between 1–8% carbon consisting of fine detrital vitrinite (humic matter) with a reflectance of 0.25% corresponding to the coal rank of lignite, probable alginite (sapropelic matter) and pyrite (between ~4–30%), in part as minute flamboidal aggregates. Calculated calorific values (kJ kg^{-1}) range from 360 to 2430 for carbon and

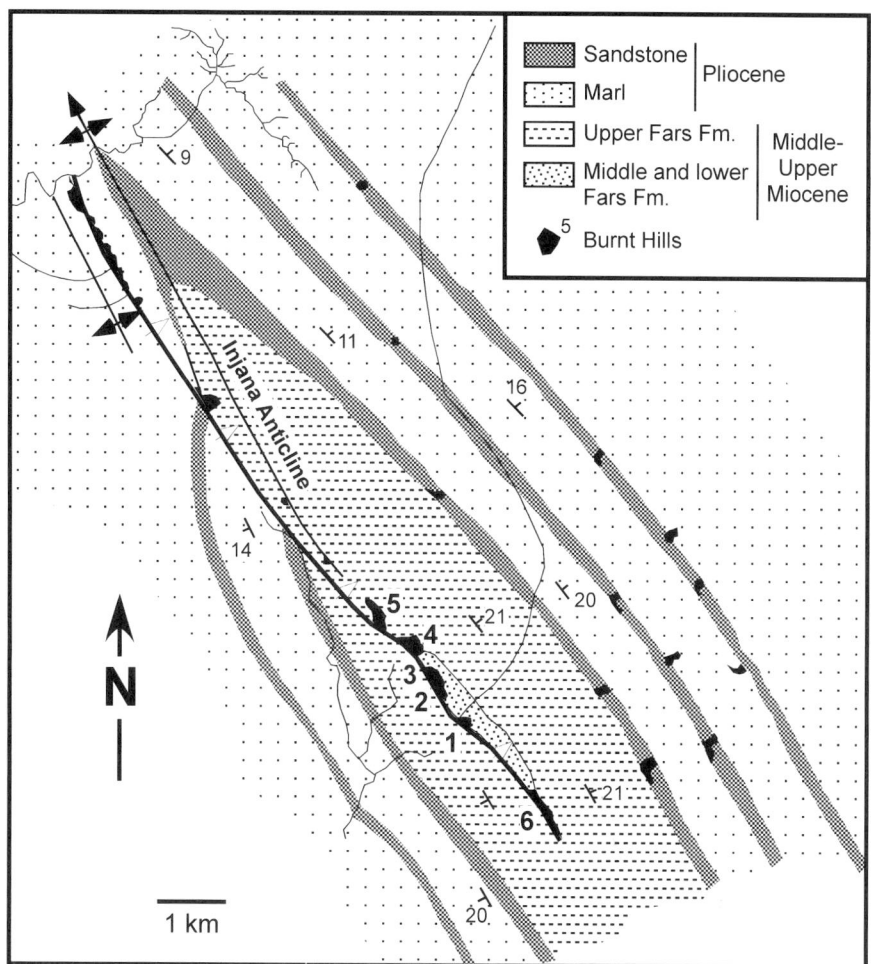

Fig. 2.12. Geological map of the Injana area, Iraq, showing structural setting of "burnt hills" (*labeled 1–6*) that indicate sites of combustion metamorphism (after Fig. 1 of Basi and Jassim 1974). See text

from 740 to 2100 for pyrite, and measured total unweathered mudstone calorific values range from 1200 to nearly 4000 kJ kg^{-1}. Compared to the heat capacity of a typical shale of 0.7–1.025 kJ kg^{-1} °C^{-1} (Handbook of Chemistry and Physics 1979), this is enough energy, given complete combustion at 100% efficiency, to raise the temperature of the mudstone by several thousand degrees. Exothermic peaks for the devolatilisation of combustible gases occur between 280–480 °C and if these gases were ignited at their point of emergence into the atmosphere they would be able to heat a small volume of the surrounding rock to melting point.

Pyrometamorphism of sediments by burning gas jets is also recorded from Iraq and Iran. In the Injana area of Iraq, burnt and melted rocks make up a line of remnant hills situated along a major thrust associated with an anticlinal structure (Fig. 2.12).

The hills rise 6 to 14 m from the surrounding plain and are capped by vesicular and completely crystalline rock that is commonly brecciated. The fused cap rocks are in abrupt contact with underlying baked and partly crystalline grey to yellowish to red marls that grade into unaltered rocks at the base of the hills. The sequence can be interpreted as forming by the action of burning gas seepages along the thrust fault and the hard, fused nature of the rocks means that they remain as erosion remnants.

Similar baked and fused rocks that occur as conical hills are developed along an anticlinal axis near the head of the Persian Gulf, Iran (McLintock 1932). The distribution of fused and brecciated rocks implies a vent-like structure originating from the explosive escape of gas and oil that became ignited. In this respect, McLintock (1932) records an instance of the formation of a mud voclano in Trinidad that was accompanied by violent eruption of enormous quantities of inflammable gas. This became ignited by sparks from abrasion of pyrite fragments to produce a ~90 m high flame that burned for 15 hours.

2.3
Lightning Strike Pyrometamorphism

Fulgurites, caused by lightning-induced melting and chemical reduction, provide the thermal limit of pyrometamorphism where temperature maxima are shorter and more extreme by several orders of magnitude than igneous or organic fire-related processes. Times of lightning-induced melting are on the order of a second with temperatures in excess of 2000 K. Voltages (and thus EMF's and chemical potential gradients) are extreme, with peak currents measuring 10 kA and up to 200 kA reached in microseconds (Unam 1969).

Fulgurite formation is common on the Earth's surface and there is a voluminous literature detailing occurrences. They typically consist of glass that often exhibits a fluidal texture and is seldom devitrified, and are usually tubular with hollow interiors with fragile, porous exteriors. Mostly, fulgurites are quite small with internal diameters in the mm to cm range. The longest fulgurites occur in unconsolidated material such as sand and soil and the longest recorded is a little over 4.9 m in length occurring in sandy soil, northern Florida (Fig. 2.13) where there is an average of 10–15 lightning strikes per square kilometer/year (Wright 1999). In such cases, the bottom limit of fulgurite formation is determined by the ground water table or a layer of wet sediment. As the tubes extend downward from the surface they decrease in diameter and become branched. Pebbles in the path of the discharge cause deflections. Outward projections of thread-like fused silica (lechatelierite) occur over short sections, especially in quartz-rich sand. Terminations of fulgurites vary from glassy, bulbous-like enlargements to an aggregate of loosely cemented partially fused sand grains.

Each lightning discharge follows the path of least resistance. This may be single or branching, and is controlled by changes in composition, moisture, compaction and bedding attitude. The tubular form of fulgurites has been attributed by Petty (1936) as the result of expansion of moisture present in the sand or soil, expansion of air along the path of the electrical discharge, or from the mechanical displacement of sand which is then fused around the resulting hole. Because many fulgarites form in

Fig. 2.13.
Sketch from a photograph (Fig. 4 of Wright 1999) of a ~2.9 m long fulgurite formed in cross-bedded sands, Florida, USA, where world record examples over 4.9 m in length have also been found

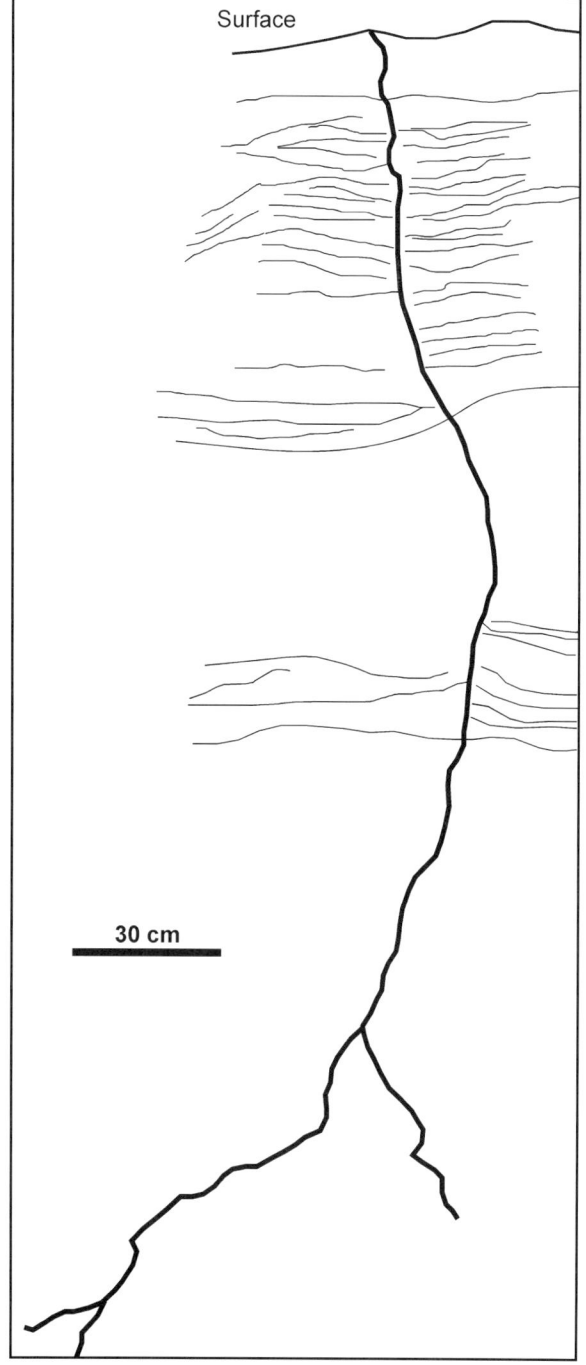

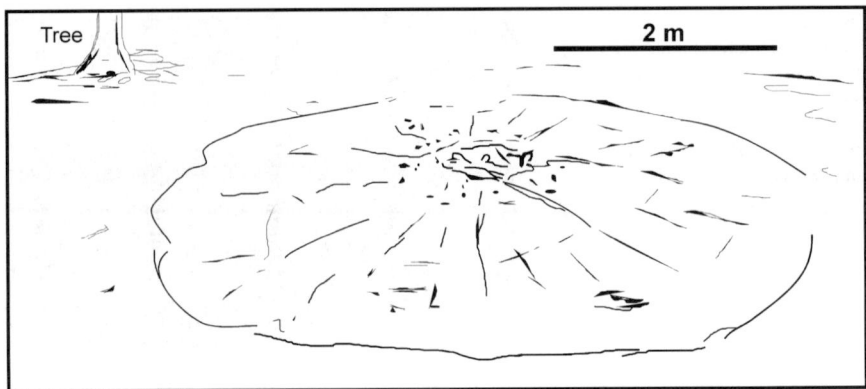

Fig. 2.14. Structure resulting from a lightning strike in clay soil at Avoca Downs, Eastern Goldfields of Western Australia (redrawn after Fig. 1. of Gifford 1999). The sub-surface soil has been expanded to a maximum of ~50 mm at the centre of the strike. See text

dry sand, expansion of air is probably the main factor of their formation. The amount of sand that is melted to form the tube will depend on the intensity of the discharge which, in turn, controls the energy expended in the form of heat. The melted sand acquires a cylindrical shape as a result of surface tension, the diameter of which is dependent on the amount of expansion of air and moisture along the discharge path. If the air/moisture expansion is large in proportion to the amount of sand melted, a large diameter thin-walled tube is formed.

While fulgurite formation in sand (*sand fulgurites*) is the most common type reported, *clay-soil fulgurites* are described from the Eastern Goldfields of Western Australia by Gifford (1999). These are different from lightning strikes in sand in that they cause notable disturbance of the soil such as that illustrated in Fig. 2.14. The point at which lightning entered the ground is indicated by a 50 cm diameter hole where the ground surface is considerably disturbed with clods of earth being thrown out onto the surface including a piece of vesicular fulgurite presumed to have been thrown into the air at the time of the strike and dropped back on top of the disturbed soil. The ground surface around the point of the lightning strike forms a low conical mound about 6 m in diameter that collapsed when walked on. It appears that the energy discharge of the strike was dispersed due to the electrical current spreading out radially from the point of impact. The fulgurite formed was small and consisted of a glassy vesicular core surrounded by baked clay which also had a vesicular structure. The nature and pattern of ground disturbance suggest that the path taken by the lightning discharge is controlled by moisture content of the clay soil. The most likely interpretation is that the lighting strike occurred after rain following an extended period of dry weather so that the soil at depth was dry with a shallow damp surface. This allowed the electrical discharge to radiate out from the point of impact within the shallow and more conductive damp surface layer of the soil. The discharge caused considerable heating resulting in the formation of steam and heated gases within the soil that expanded and created the low conical structure. Also, the radial dispersion of the electrical current reduced the development of glass and zoned structure in the fulgurite that formed.

Fig. 2.15.
Example of a fulgurite network pattern developed on an exposed quartz surface in granitic rock, Black Forest, Germany (Fig. 12 of Wimmenauer and Wilmanns 2004). Width of photo is 2.7 mm

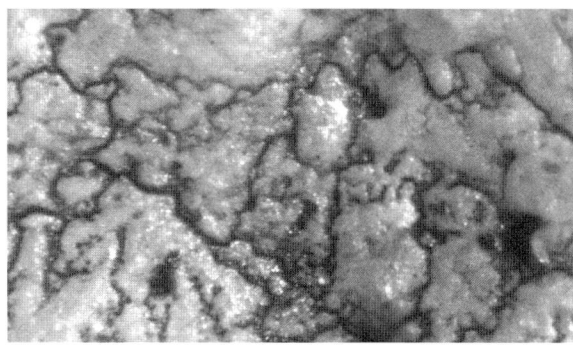

In rocks, particularly crystalline ones, the effects are different from sand or clay-soil fulgurites. In this case, fulgarites take the form of glass-lined holes drilled into and sometimes through corners of the exposed rock. Millimeter-sized gobules of melt may be ejected from these holes (e.g. Frenzel and Stahle 1984; Clocchiatti 1990). The tubes so formed often have a zonal structure on a mm scale, with the inner zone being iron-rich (Frenzel et al. 1989). In other cases, the path of the discharge is shown by a black filigrane network of melt threads or films along and over exposed grain boundaries, particularly quartz (e.g. Wimmenauer and Wilmanns 2004) (Fig. 2.15).

2.4
Other Thermal Effects

2.4.1
Columnar Jointing

Perhaps the first description of columnar jointing in sandstones was made by Glen (1873) on the Island of Butte, Scotland. He considered that the columnar structure was caused by steam, "or some other highly heated vapour" passing upward through a vertical fissure and affecting the sandstone for a few metres on either side; "The columns are all nearly vertical, none being more than 20 degrees from the perpendicular. Their size varies from six or seven inches [15.2–17.8 cm] in diameter down to half-an-inch [1.3 cm]. Some of the large columns break into smaller ones from exposure to the weather, and others branch into two, forming twin columns. The number of sides varies from four to eight or ten, six being the most common", e.g. such as that shown in Fig. 2.16.

Mohr (1873) provides a detailed description of columnar jointed sandstone in contact with basalt from localities in Germany:

> "The burned sandstones – long known from the Otzberg – are found in cubic meter sized blocks that are broken into 3 to 8 cm large and 4-, 5- or many fold squared prisms ... close to the surface some blocks are found which fall apart into quite sharp-edged small columns. Blocks of a white to whitish yellow sandstone which are very soft immediately after being brought above ground, harden in the air. These blocks are usually covered with a dark, chocolate-brown substance several centimetres thick. This material has a curved fractured surface, a nearly waxy lustre, is soft (H = 2) with a greasy touch, falls apart immediately with a cracking sound when immersed in water, becomes hard with transparent edges when heated, and melts on coal to a vesicular greenish-brown

Fig. 2.16.
Columnar jointed claystone occurring below a thick basalt flow, Egerli Dag near Iscehisar Province, Turkey (collected by Professor Joerg Keller, IMPG, Freiburg University, Germany)

yellow enamel. Microscopically, the grains that fall out of the substance when put into water as well as fragments of the fused material cannot but be interpreted as light bottle-green relics of augite full of tiny magnetite crystals. The rather imperceptible transition from this substance into a slaty friable, brittle basalt is in favour of interpreting the surrounding substance as a bole-like decomposition product of the basalt. Apart from the jointing no other changes of the sandstone are obvious. The sandstone consists of an argillaceous-carbonaceous cement and contains abundant quartz pebbles".

"Sandstone blocks in the basalt separate easily from it … the single small columns exhibit an extremely thin, whitish rim with a waxy lustre. This rim does not effervesce when treated with hydrochloric acid but detaches easily as thin films which consist of an aggregate of 0.008 mm diameter, six-sided, faintly polarising, tiny scales of tridymite within an opal-like substance. The sandstone is very tenacious and on cleavage planes partly still sandstone-like and spotted with small humps, partly pervaded by varnish-like shining dark veins, partly completely homogeneously aphanitic and black similar to the basalt itself, but always contains white to greyish, up to 1.5 cm diameter quartz pebbles".

"… One of the best examples of contact effects is displayed at Stoppelberg near Hünfeld. Here, there is a basalt quarry in the base of which ca. 1 m of red sandstone is exposed overlain by basalt. Vertical columns in the basalt, somewhat finer and porous/vesicular towards the base (at the contact), are intimately welded with the sandstone. In the sandstone, columnar jointing continues below the level of the quarry and from the overlying basalt downwards a gradual transition is observable from buchite to more or less sintered sandstone and horizontally-bedded sandstone."

Columnar jointing in coarse-grained arkosic sandstone is also reported by Poddar (1952) cropping out over an area of about 93 m^2 near Bhuj, northern India. In this case the columns are polygonal, have a thickness of ~8–20 cm and a length of 0.3–0.6 m. Although Poddar considered them to have resulted from contraction after dehydration due to being heated by basaltic dykes, they are in fact, related to combustion of coal seams. Prismatic columnar jointing caused by combustion of organic matter also occurs in bituminous marly rocks of the Mottled Zone, Hatrurim basin, Israel (see Chapter 4) (Avnimelech 1964). The columns are several centimeters in length, contain high temperature minerals such as spurrite, and are interpreted as a cooling-contraction phenomena with the long axes of the columns indicating the direction of the highest thermal gradient (Burg et al. 1992).

The mechanism of producing columnar jointing in partially melted sediment (buchite) adjacent an intrusion is described in detail by Solomon and Spry (1964) in relation to an inclined dolerite pipe at Apsley, Tasmania (Fig. 2.17a). Although the buchite columns plunge at shallow angles and tend to be radial to a small area around the centre of the plug, there is a general departure from the expected condition that they be normal to the igneous contact and there is also local irregularity (Fig. 2.17b). This is probably related to the form of the plug and associated pattern of heat-flow combined with variations in the physical properties (particularly permeability) of the sedimentary layers. The columnar buchites at Apsley contain > 40% glass (maximum of ~65%) and plunge mainly at low angles in various directions although the plunge is constant for each bed but differ from bed to bed. The columns are generally hexagonal with curved faces, but pentagonal, quadralateral forms also occur.

At a maximum melting temperature of the sediments of around 1000 °C at 50 bar PH_2O, the occurrence of buchites up to ~45 m from the southern contact of the dolerite and within ~5 m on its NE side, indicates a highly asymmetric heat flow regime suggestive of upward and outward transport of heat by steam (convection rather than conduction) in relation to a steep southward dip of the plug. Under convective conditions, heat-flow would be partly up the buchite zone but largely outward along the sedimentary layers giving rise to the mainly subhorizontal columns observed. This would also explain the irregular distribution of the columnar buchites with gas-phase H_2O penetrating further along those beds with higher permeability resulting in an irregular isotherm pattern within an estimated mass of sandstone converted to buchite of 3×10^{11} g.

The columns range in diameter from 2.5–50 cm and are related to lithological variation in the sediments (Fig. 2.18). Primary differences in the type of quartz packing, the composition of other components and their relative proportions and degree of heterogeneity are important factors controlling the shrinkage behaviour of the rocks. Furthermore, unlike the development of columnar structures in igneous rocks, in metamorphosed sedimentary rocks columnar jointing is related to heating as well as cooling. During heating contraction stresses may be established that can facilitate/accentuate the development of columnar jointing during cooling of the rocks below their softening point, i.e. numerically about half the melting point. Variable degrees of contraction during heating takes place by:

1. Mineral (e.g. clay, mica, chlorite) dehydration resulting in the development of allotropic forms. The proportion of phyllosilicates (mainly clays) is < 15% in the Apsley sediments and the effects of phyllosilicate contraction would be countered by the greater percentage of quartz that expands on heating.

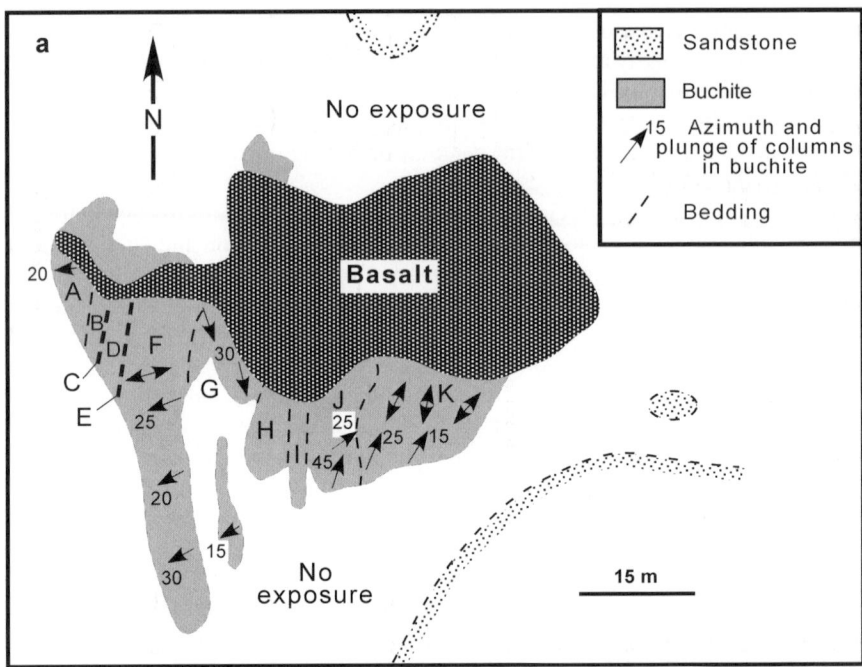

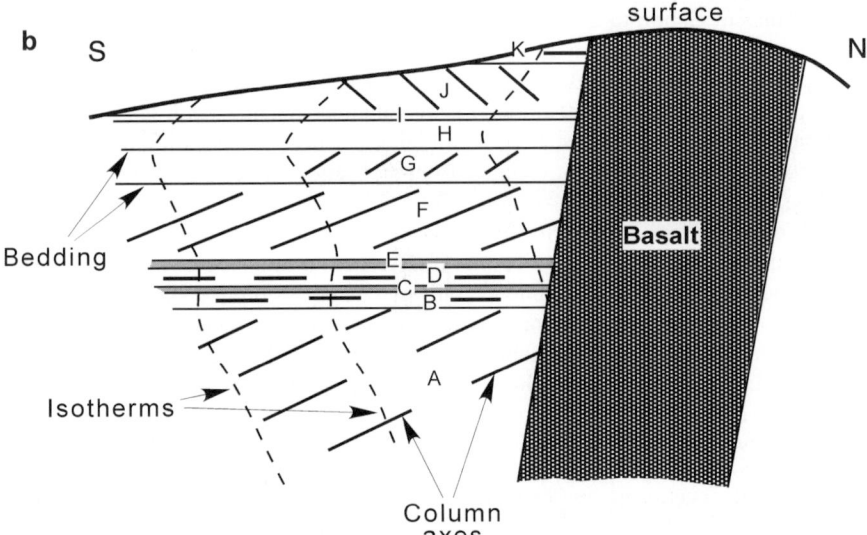

Fig. 2.17. a Geological map of the dolerite plug and distribution of columnar buchite at Apsley, Tasmania (after Fig. 2 of Spry and Solomon 1964). *Letters A–K* refer to rock-types in sandstone sequence shown in (**b**) and in Fig. 2.6. **b** Cross section of the dolerite-buchite association at Apsley, Tasmania, showing inferred buchite column orientation under convection conditions as discussed in text (after Fig. 8b of Spry and Solomon 1964).

2.4 · Other Thermal Effects

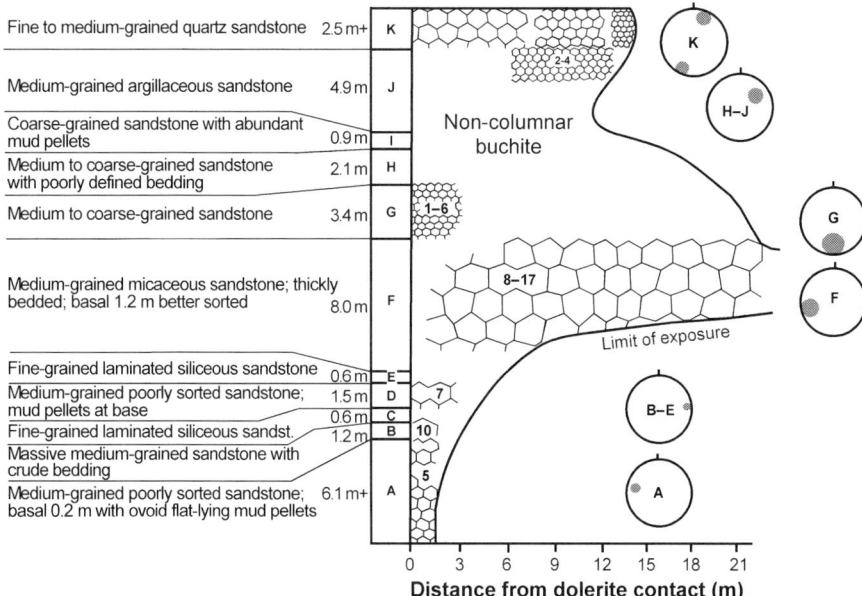

Fig. 2.18. Schematic diagram showing relations between sedimentary succession and distribution, size (diameter range in cm), and attitude (stereographic projections on a horizontal plane of column-axes) of buchite columns associated with the dolerite plug at Apsley, Tasmania (after Fig. 13a,b of Spry and Solomon 1964)

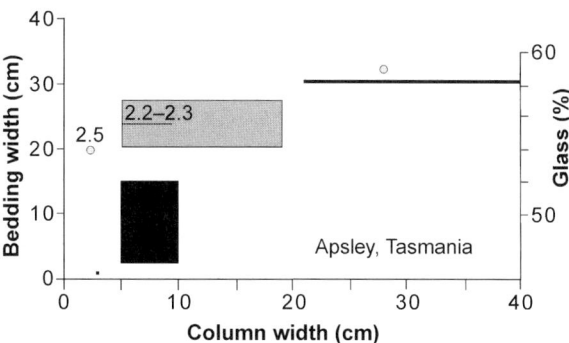

Fig. 2.19.
Plot of bedding width versus buchite column width and vol.% glass versus buchite column width, Apsley, Tasmania. Data from Spry and Solomon (1964). See text

2. Reaction to anhydrous minerals with greater density, e.g. muscovite, chlorite reacting to spinel, mullite, corundum etc.
3. Structural adjustment due to recrystallisation and consequent reduction in pore space, e.g. the change of sandstone to quartzite results in an increase in specific gravity from 2.2 (sandstone) to 2.6 (quartzite) and a shrinkage of 18%.
4. Formation of melt which allows repacking and fills pore-spaces. This is probably the main cause of shrinkage. The increase in specific gravity from ~2.1–2.2 in sandstone to 2.2–2.4 in buchites with increasing glass (Fig. 2.19) indicates a decrease in volume amounting to a shrinkage of 5–10%.

Larger columns occur in more massive, thickly bedded sandstones. They also contain more glass (Fig. 2.19) and hence have a greater tensile strength resulting in more widely spaced fractures. The more glass in a buchite, the less it will contract and fracture for a given temperature change. The smallest columns of ~2.5 cm occur in thinly-bedded sandstone (Fig. 2.18) and they do not form in beds that contain abundant mud pellets or in siltstones (porcellanites). It is probable that the mud pellets introduced sufficient irregularity in the tensile stress-pattern developed by shrinkage that columnar joints could not form. The original shaley nature of the porcellanites and

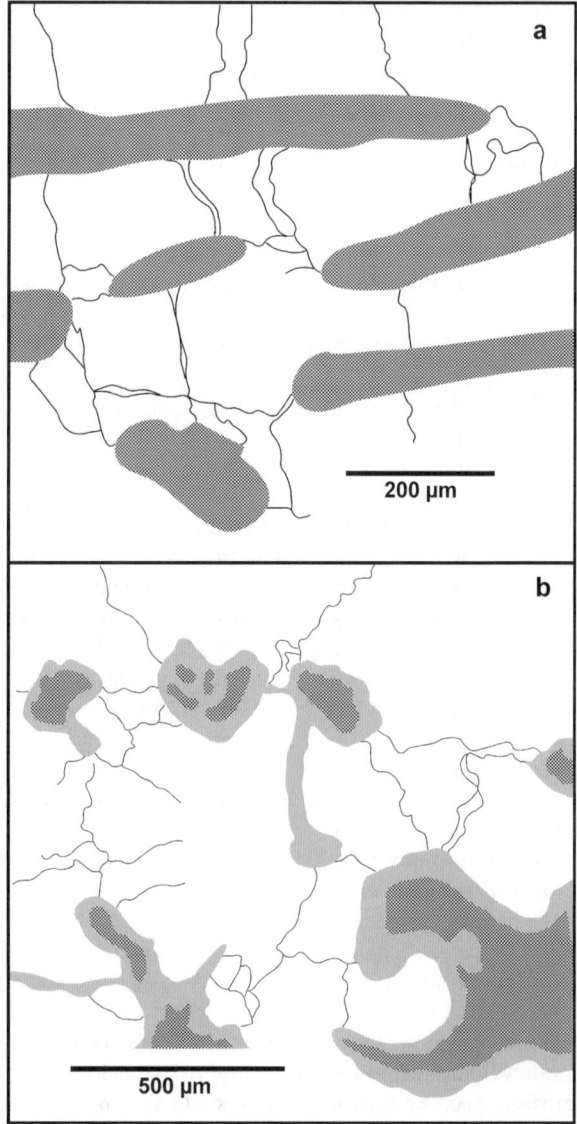

Fig. 2.20.
Diagrams showing microcracking during pyrometamorphism of sediments in contact with the Traigh Bhàn na Sgùrra sill, Isle of Mull, Scotland (see text).
a Psammite, 80 cm from the contact showing microcracks extending from muscovite grains (*shaded*), partly reacted around margins to cordierite, into quartz (*white*). Cracks have developed at a high angle to the foliation indicated by alignment of muscovite (drawn from a CL image Fig. 9 of Holness and Watt 2002).
b Psammite, 44 cm from the contact showing a largely interconnected pattern of cracks that are melt-filled and extend outwards along boundaries of quartz grains (*white*) from crystallised (granophyric) rims (*light grey shading*) that surround partly melted feldspar (*dark grey shading*) (after CL image Fig. 13d of Holness and Watt 2002)

possibly also closely spaced partings modified the stress-pattern and fracture-growth inhibiting the development of columns.

2.4.2
Microcracking

Because melting reactions generally involve a positive volume increase at the melting site, rapid melting causes the development of high melt pressures leading to hydraulic fracturing and migration of melt along these fractures. Several generations of such microfractures in gneissic psammitic and pelitic rocks within 3.5 m of a 6 m thick basic sill are described by Holness and Watt (2001). Up temperature (> 600 °C) crack development is caused by the breakdown of muscovite and melting along quartz-feldspar boundaries. The earliest generation of microcracks related to the pyrometamorphic event cut across quartz and feldspar crystals but rarely across their respective boundaries and are thought to be the result of anisotropic thermal expansion associated with a contribution from the α–β inversion in quartz which occurs at ~590 °C at 600 bar.

A second generation of cracking involving the formation of two sets of microcracks relates to the breakdown of muscovite according to the fluid-absent reaction

muscovite + quartz = biotite + K-feldspar + mullite ± cordierite + melt

This metastable reaction involves an increase in volume of perhaps ~5% to 7% that causes overpressuring and grain-scale fracturing (Brearley 1986; Connolly et al. 1997; Rushmer 2001). In the example described, the microcracks radiate away from muscovite grains and are melt-filled (Fig. 2.20). They also appear to re-use cracks that existed in the rocks prior to the pyrometamorphic event. No obvious microcracking is associated with the fluid-absent, metastable breakdown of biotite, possibly because this involves a much smaller positive volume change (Rushmer 2001).

The penultimate stage of microcracking occurs with melting at quartz-feldspar grain boundaries. Cracks form along the grain boundaries and extend into the grains and outwards across adjacent grains (Fig. 2.20). They are extremely common within 80 cm of the sill contact where temperatures were higher than 900 °C. Despite the high degree of melt-filled fracture connectivity, there is little evidence of melt segregation presumably because of the static nature of the pyrometamorphic event. The last stage of cracking occurred during cooling and produced cracks that cut across or reopen solidified melt-filled fractures. In addition to anisotropic thermal contraction of the rock, these cracks could have been produced by a reversal of the α–β quartz inversion at 575 °C.

2.4.3
Dilation

Heating at atmospheric pressure due to combustion of coal seams often results in dilation of sediments overlying the burning coal seam. Cavities may remain open and provide space for vapour phase crystallisation of high (sanidinite facies) and low tem-

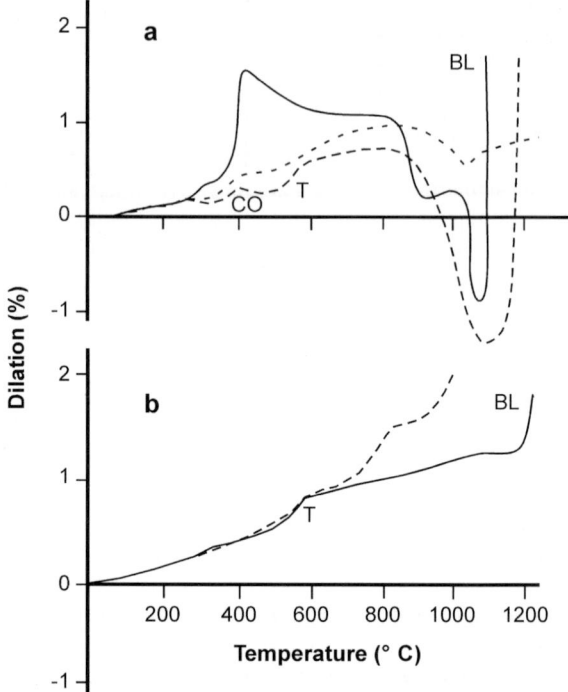

Fig. 2.21.
Dilation curves (warming-up rate of 0.5 °C min^{-1}) for:
a shale with dispersed organic matter (*solid line*); shale with dispersed organic matter and with silt and coal streaks parallel to the cleavage (*thin dashed line*); shale with dispersed organic matter and with silt and coal streaks normal to the cleavage (*thick dashed line*) (after Fig. 2 of Wolf et al. 1987);
b siltstone with carbonate matrix (*dashed line*); siltstone with clay matrix (*solid line*) (after Fig. 1 of Wolf et al. 1987). CO = coking; T = transition from α–β quartz; BL = bloating

perature (hydrous) minerals or become filled by melt. Heating experiments conducted by Wolf et al. (1987) on siltstone and shale that overlie coal seams in Belgium indicate the effects of devolatilisation, decarbonation and vitrification on rock dilation between 100–1200 °C related to combustion of the coal. The experiments were conducted at 25 bar with a heating rate of 0.5 °C min^{-1} and results are shown in Fig. 2.21.

In *clay-rich* siltstone (Fig. 2.21b), dilation is moderate but constant with increasing temperature except for an increase at the α–β quartz inversion at 587 °C. Vitrification at ~900 °C does not change the dilation pattern; surfaces of the heated samples become glassy and degasification channels and vesicles form. Bloating occurs at ~1200 °C resulting in a marked increase in dilation. In siltstone with a *carbonate-matrix*, decomposition of dolomite and siderite results in a marked dilation due to bloating between 600–900 °C.

In comparison with siltstones, carbonaceous shale exhibits much more dilatory variation (Fig. 2.21a). Between 100–200 °C, fracturing parallel to cleavage occurs in response to dehydration and dehydroxylation of clay minerals. From ~390 °C, coking of organic matter causes an increase in dilation, particularly where the organic material is dispersed rather than occurring as coal streaks. Formation of melt above ~850 °C results in shrinkage that continues up to ~1050 °C and is reversed above 1050–1100 °C when bloating associated with vesiculation occurs. Wyllie and Tuttle (1961) also report that half melted shales in their melting experiments are extremely vesicular and have a frothy, slaggy appearance.

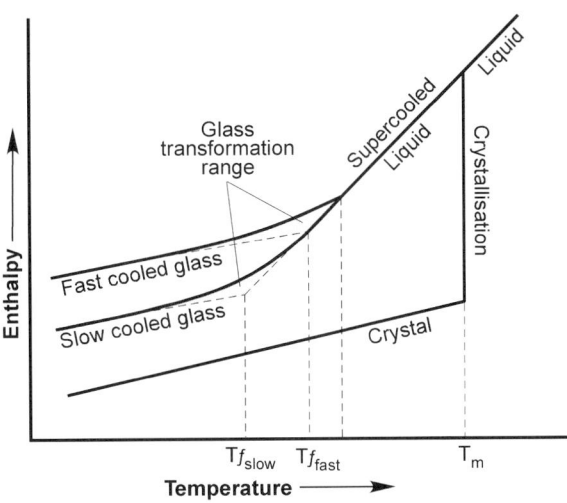

Fig. 2.22.
Diagram illustrating the effect of temperature on the enthalpy (or volume) of a glass-forming melt (redrawn from Fig. 1.1 of Shelby 1997). See text

2.4.4
Preservation of Glass

The terms *buchite* and *paralava* imply the presence of glass (quenched melt). While quenching would occur in partially fused xenoliths on eruption, this is not the case for slower cooled wall rocks at the contacts of sills, dykes and plugs and for xenoliths that they contain. Because rocks are poor conductors of heat, rapid cooling of melts does not seem possible in these circumstances and so the question of why glass remains largely undevitrified in many pyrometamorphic rocks needs explanation.

Figure 2.22 shows that above the 'glass transformation temperature' (Tg), a temperature dependent on cooling rate over which melt becomes solid (glass) and hence a temperature range, a supercooled liquid is stable. Below Tg, glass is stable. A melt cooled to the vicinity of the Tg region will rapidly devitrify to a crystalline aggregate. If cooling is very rapid (as in quenching) only glass forms. At progressively slower cooling rates more and more crystal nuclei are able to form as the melt passes through the Tg range. If sufficiently rapid, only a few crystal nuclei will form and the end product will be largely glass containing a few crystals. Glass will crystallise only if held for a certain time above Tg, i.e. the temperature at which viscosity is probably in the order of 10^{13} poises. At this viscosity glasses devitrify rapidly, i.e. in 10 to 10^3 seconds. At a higher viscosity, e.g. 10^{14} poises, glass could be expected to devitrify in 10^2 to 10^4 seconds. Liquids produced by melting of sediments are essentially alkaline aluminosilicates and would have high viscosity. If this were 10^{16} poises, then it would not crystallise if cooled from a high temperature over several months and would remain glass provided it was not subjected to the influx of water (e.g. Spry and Solomon 1964). Once formed, the preservation of glass in a contact aureole will also depend on whether fluid infiltration occurs during cooling.

An instructive case is the glass-rich quartzose metasediments that occur within 0.8 m of the Glenmore dolerite plug, Ardnamurchan, Scotland (Butler 1961; Holness et al. 2005). Given that metamorphism occurred at a depth of several hundred meters (~120 bar) it seems unlikely that the presence of glass was caused by rapid cooling but rather to the absence of pervasive fluid infiltration which would be expected to have caused extensive devitrification to a micrographic (granophyric) texture of quartz and feldspars. At Glenmore, late-stage hydrothermal fluid circulation was channeled through vein systems. Cooling rate was also important. In the relatively slowly cooled outer part of the aureole (see Fig. 2.1) the melt crystallised. In the transition zone to the inner aureole only the quartz component of the melt crystallised, while in the relatively rapidly cooled inner aureole, i.e. within 0.2 m of the contact, all the melt solidified to glass. On the basis of a thermal model developed by Holness et al. (2005) that ignores possible effects from latent heat of crystallisation (see Chapter 3), the critical cooling rate for glass formation is estimated to have been ca. 8 °C day^{-1}.

Chapter 3

Quartzofeldspathic Rocks

Pyrometamorphosed quartzofeldspathic rocks (sandstone, shale, claystone) and sediments (sand-silt, clay, glacial till, diatomaceous earth), their metamorphosed equivalents (phyllite, schist, gneiss), are characterised by the presence of tridymite, mullite/sillimanite, cordierite, orthopyroxene, clinopyroxene, sanidine, plagioclase (oligoclase–anorthite), corundum, hercynite-rich spinel, magnetite, ilmenite, hematite, pseudobrookite, sulphides and in carbonaceous protoliths, native metals. Ti-rich biotite and osumilite are less common; sapphirine is rare. These minerals are usually associated with acidic (granitic) to intermediate (granodioritic) glass that is frequently abundant enough for the rocks to be termed buchites and paralavas. Partly melted granite-granodiorite may contain tridymite, Ca-plagioclase, orthopyroxene and magnetite.

Eskola (1939) provides two examples of quartzofeldspathic sanidinite facies mineral assemblages in terms of, (*a*) an ACF diagram for rocks with excess silica (quartz/tridymite/cristobalite) and sanidine as possible additional phases (Fig. 3.1) and, (*b*) a MgO-Al_2O_3-SiO_2 plot (Fig. 3.2), that characterise typical mineral associations found in psammitic-pelitic rocks:

1. Anorthite-cordierite-mullite
3. Anorthite-clinopyroxene-orthopyroxene
4. Anorthite-cordierite-orthopyroxene
5. Cordierite-orthopyroxene
9. Corundum-mullite-spinel
10. Cordierite-mullite-spinel
11. Cordierite-mullite-tridymite

Silicate and oxide phases in pyrometamorphosed quartzofeldspathic rocks are plotted in terms of mol% $SiO_2 - (Al,Fe)_2O_3 - [TiO_2 + (Fe,Mg)O] - [CaO + (K,Na)_2O]$ in Fig. 3.3a. Bulk compositions are confined to the Qz-[Pl,San]-[Co,Hem]-[Rt,Wü]-Di volume within which detrital and authigenic reactants in quartzofeldspathic protoliths such as quartz, plagioclase, K-feldspar, kaolinite, illite, muscovite, chlorite, biotite, anatase, ilmenite, magnetite, hematite and goethite occur (Fig. 3.3b). Accessory detrital garnet, pyroxenes, amphiboles, epidote, tourmaline etc., may also be present in many clastic sedimentary protoliths and react during pyrometamorphism. Zircon, apatite, rutile, spinel and ilmenite are typically unaffected except in cases of lightning strike metamorphism and in rare cases phosphoritic sediments that originally contained

Fig. 3.1.
ACF diagram with excess silica after Eskola (1939) showing mineral compatibility fields 1, 3, 4, 5 in sanidinite facies quartzofeldspathic rocks

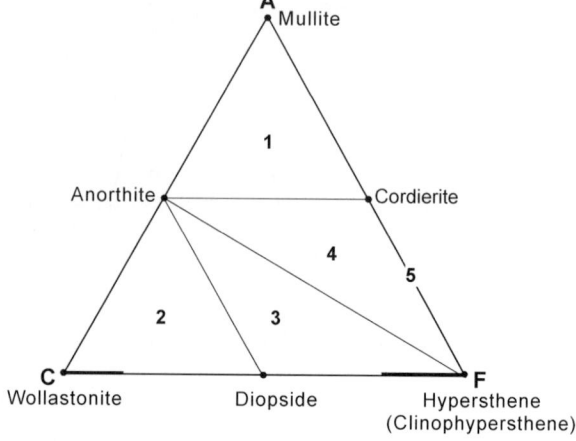

Fig. 3.2.
MgO-Al$_2$O$_3$-SiO$_2$ (mol%) plot of mineral compatibilities in sanidinite quartzofeldspathic rocks. *Shaded area* = field of typical quartzofeldspathic sediment compositions in terms of quartz, kaolinite/sericite and chlorite components

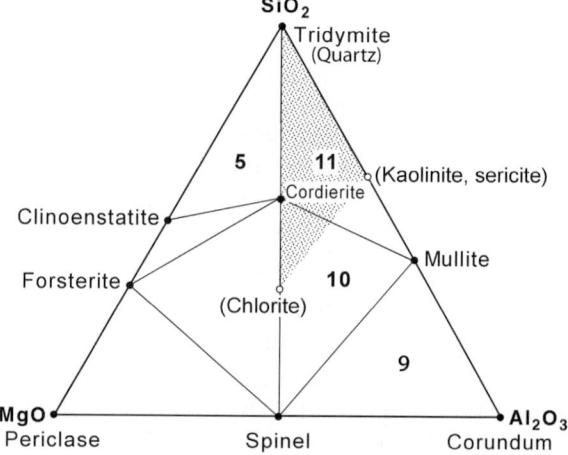

fluorapatite have undergone melting during combustion metamorphism. Addition of CO$_2$ allows for the presence of siderite and calcite, both of which occur in sedimentary rocks affected by combustion metamorphism and where melting of siderite has produced magnetite-hematite-sulphide-bearing paralavas.

3.1
Experimental Data and Petrogenetic Grid

The mineral assemblages in buchites and paralavas can be compared with those phases synthesised in atmospheric pressure experimental systems (Table 3.1) and in particular the systems SiO$_2$-Al$_2$O$_3$-FeO, SiO$_2$-Al$_2$O$_3$-MgO and SiO$_2$-FeO-MgO (Fig. 3.4). A pseudo-binary section within the system MgO-Al$_2$O$_3$-SiO$_2$ between metatalc and metakaolin compositions (line A–B in Fig. 3.4) illustrates the compositional limits of tridymite, mullite, cordierite and orthopyroxene formation (i.e. region of crystals + melt)

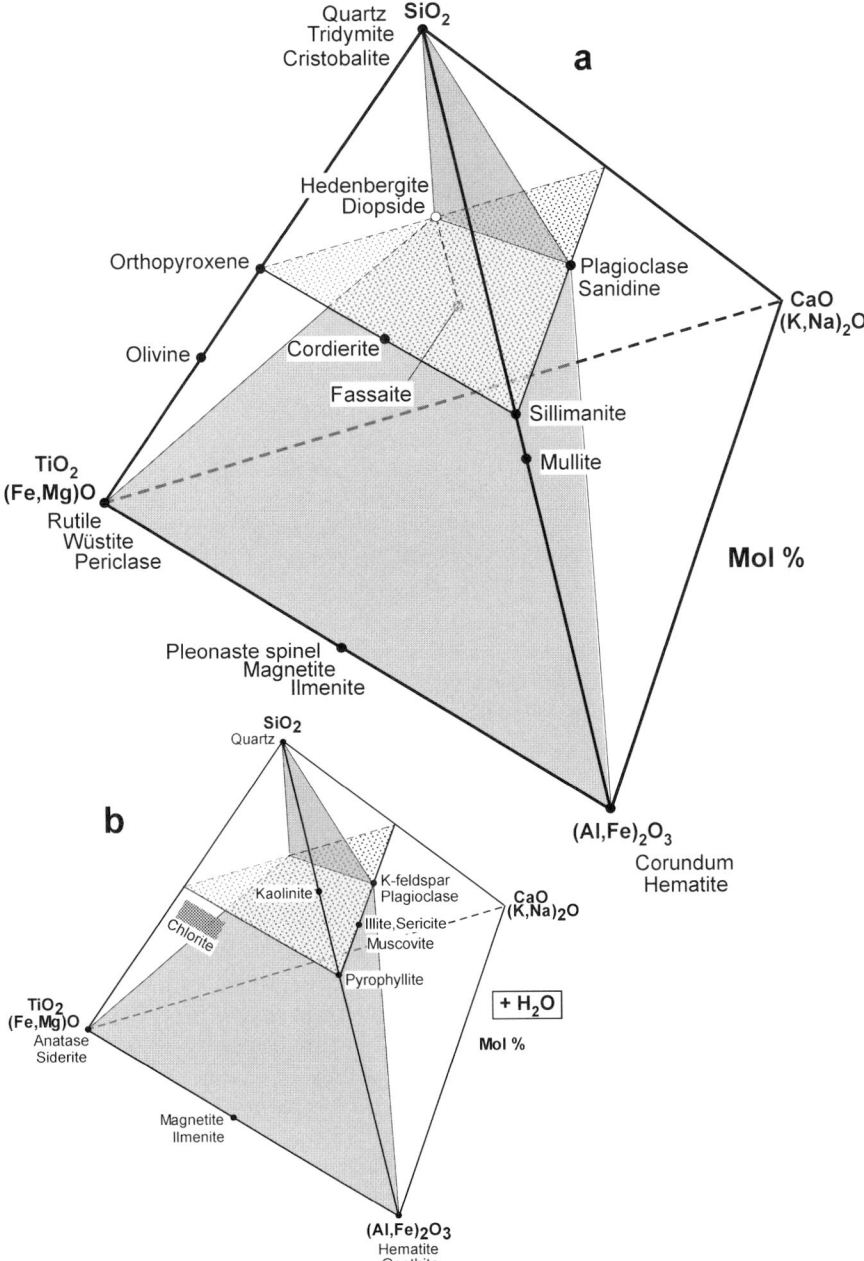

Fig. 3.3. Typical sanidinite facies silicate-oxide minerals plotted in terms of mol% $SiO_2 - (Al,Fe)_2O_3 - [TiO_2 + (Fe,Mg)O] - [CaO + (K,Na)_2O] + H_2O$ (**a**). Protolith minerals are shown in (**b**). The *dotted plane* at 50 mol% SiO_2 = ACF plane. The volume that contains bulk quartzofeldspathic compositions *within* the tetrahedron is delineated by the *shaded planes*

Table 3.1. Some invariant assemblages + liquid (L) in ternary oxide and mineral systems relevant to buchites derived by fusion of quartzofeldspathic compositions

Assemblage	Ternary system	T (°C)[a]	Reference
Quartzofeldspathic rocks (pelites, psammites)			
(i) Simple oxide systems			
En-Di-Tr-L	CMS	~1375	Osborn and Muan (1960)
En-Cd-Tr-L	MAS	1355	Schreyer and Schairer (1961)
Mul-An-Tr-L	CAS	1345	Schairer (1942)
Mul-Co-An-L	CAS	1512	Schairer (1942)
Mul-Co-An-L	MAS	1575	Rankin and Merwin (1918)
Mul-Cd-Tr-L	MAS	1440	Schreyer and Schairer (1961)
Mul-Co-Hc-L	FAS[b]	1380	Schairer (1942)
Mul-Tr-Hc-L	FAS[b]	1205	Schairer (1942)
Cb-Tr-Mul-L	FAS[b]	1470	Schairer (1942)
Fa-Tr-Hc-L	FAS[b]	1073	Schairer (1942)
Mul-Cd-Tr-L	FAS[c]	1210 ±10	Schairer and Yagi (1952)
Mul-Cd-Hc-L	FAS[c]	1205 ±10	Schairer and Yagi (1952)
Mul-Tr-L	FAS[c]	1470 ±10	Schairer and Yagi (1952)
Cd-Fa-Tr-L	FAS[c]	1083	Schairer and Yagi (1952)
Mul-Sp-Co-L	FAS[c]	1380 ±10	Schairer and Yagi (1952)
Mul-Sp-Co-L	MAS	1575	Keith and Schairer (1952)
Pen-Cd-Tr-L	MAS	1345	Keith and Schairer (1952)
Mul-Tr-Ksp-L	KAS	985 ±20	Schairer and Bowen (1947)
(ii) Mineral systems			
Cen[d]-An-Tr-L	SiO$_2$-Fo-An	1222	Anderson (1915)
Cen-Fo-An-L	SiO$_2$-Fo-An	1260	Anderson (1915)
An-Di-Fo-L	An-Di-Fo	1270	Osborn and Tait (1952)
Cd-Mul-Tr-L	Cd-Lc-Si	1435	Schairer (1954)
Cd-Fo-Spl-L	Cd-Lc-MgMS[e]	1370	Schairer (1954)
Mul-Spr-Spl-L	Cd-Lc-MgMS[e]	1490	Schairer (1954)
Mul-Spr-Cd-L	Cd-Lc-MgMS[e]	1455	Schairer (1954)
Cd-Spr-Spl-L	Cd-Lc-MgMS[e]	1450	Schairer (1954)
Mul-Cd-L	Cd-Lc-MgMS[e]	1470	Schairer (1954)
Mul-Spl-L	Cd-Lc-MgMS[e]	1490	Schairer (1954)
Cd-Pen-L	Cd-Lc-MgMS[e]	1364	Schairer (1954)
Mul-Co-Spl-L	Cd-Mul-Ksp	1478	Schairer (1954)
An-Tr-Fa-L	An-Si-Wu	1070	Schairer (1942)
An-Spl-Fa-L	An-Si-Wii	1108	Schairer (1942)
En-Tr-Ks-L	Lc-Fo-Si	985 ±20	Schairer (1954)

[a] ±5 °C or less unless otherwise stated. [b] Some Fe$_2$O$_3$ in all liquids. [c] Strongly reducing conditions. [d] Cen (clinoenstatite) should be Pen (protoenstatite). [e] Magnesium metasilicate.
Phase system: CAS = CaO-Al$_2$O$_3$-SiO$_2$; MAS = MgO-Al$_2$O$_3$-SiO$_2$; CMS = CaO-MgO-SiO$_2$; FAS = FeO-Al$_2$O$_3$-SiO$_2$; KAS = K$_2$O-Al$_2$O$_3$-SiO$_2$.

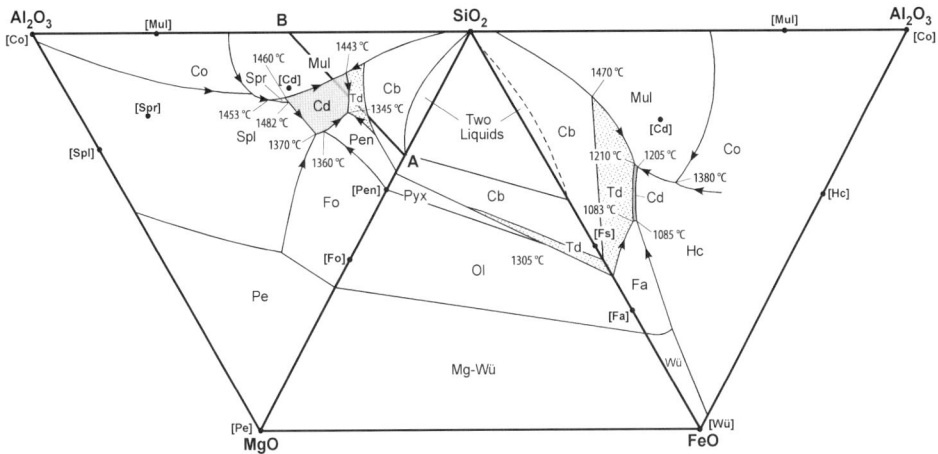

Fig. 3.4. Composite diagram of the atmospheric pressure systems MgO-Al$_2$O$_3$-SiO$_2$, MgO-FeO-SiO$_2$, FeO-Al$_2$O$_3$-SiO$_2$ (Keith and Schairer 1952; Bowen and Schairer 1935; Schairer and Yagi 1952, respectively). Tridymite and cordierite fields shaded for clarity

Fig. 3.5.
Crystalline phases + melt typical of buchites produced from compositions between metatalc (3 MgO · 4 SiO$_2$ + SiO$_2$) and metakaolin (Al$_2$O$_3$ · 2 SiO$_2$) represented by section line *A-B* in Fig. 3.4 (after Fig. 383 in Levin et al. 1956). *Stippled area* = liquid

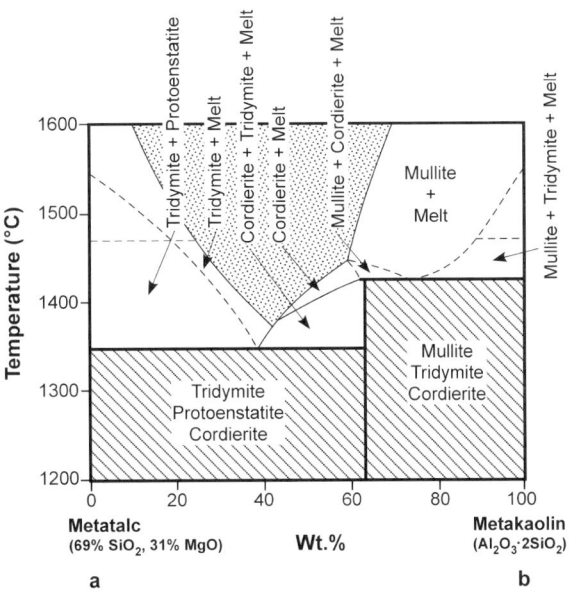

(Fig. 3.5). Similar sections constructed between metakaolin and compositions along the MgO-Al$_2$O$_3$ join would include spinel and corundum. Although some 200–300 °C above the temperatures at which many natural buchites form, the pseudo-binary section shows that mullite, mullite-tridymite, mullite-cordierite, cordierite, cordierite-tridymite, tridymite and tridymite-orthopyroxene buchites may be produced at almost the same temperature depending on composition of the sedimentary protolith. Other relevant phase diagrams are the systems FeO · Fe$_2$O$_3$-Al$_2$O$_3$-SiO$_2$

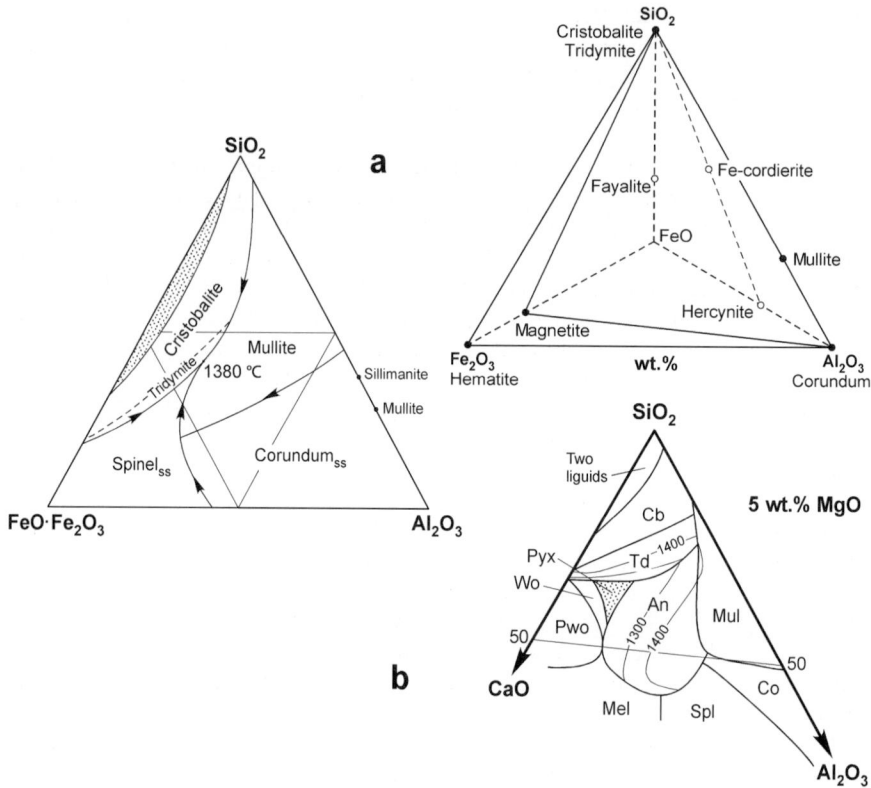

Fig. 3.6. a The system FeO · Fe$_2$O$_3$-Al$_2$O$_3$-SiO$_2$ after Muan (1956) and tetrahedron showing sanidinite facies mineral compositions in the system FeO-Fe$_2$O$_3$-Al$_2$O$_3$-SiO$_2$. **b** Section through the system CaO-Al$_2$O$_3$-SiO$_2$-MgO at 5 wt.% MgO

and particularly CaO-Al$_2$O$_3$-SiO$_2$ with addition of MgO (Fig. 3.6a and b, respectively) that allows for the assemblages, magnetite-corundum-mullite, magnetite-mullite-tridymite (a) and tridymite-clinopyroxene-anorthite, tridymite-mullite-anorthite, mullite-corundum-anorthite and spinel-corundum-anorthite (b). Increasing the MgO content, e.g. to 10 wt.%, in (b) causes an expansion of the pyroxene stability field relative to wollastonite, although quartzofeldspathic lithologies rarely contain > 5 wt.% MgO.

With respect to the system FeO-Al$_2$O$_3$-SiO$_2$, the narrow stability field of cordierite is of interest (Fig. 3.4). Experimental work on this system by Schairer (1942) failed to synthesis cordierite resulting in Hc-Mul-Td and Hc-Fa-Td ternary invariant points at 1205 °C and 1073 °C respectively. The later work by Schairer and Yagi (1952) required seed crystals of cordierite in appropriate quenching run compositions in order to delineate a stability field for cordierite. Also, no clinoferrosilite could be synthesised in this system, although there is a possibility that it might be a stable phase at subsolidus temperatures. The absence of cordierite and orthopyroxene in many buchites with appropriate bulk compositions could thus reflect nucleation difficulties in the melt due to the absence of "seeds" resulting from earlier reaction of, for example, biotite. With respect to direct crystallisation from buchite melts, the absence of cordierite and/or orthopyroxene could imply that the phases that are present are metastable.

Low pressure H_2O-saturated melting (synthesis) experiments using quartzo-feldspathic sedimentary rocks (usually in a powdered state to enhance mineral reactions) relevant to pyrometamorphic conditions and mineral formation (cordierite, anorthitic plagioclase, spinel) are reported in studies which define lower limits of melting, mineral-in and mineral-out reaction curves. An early study by Wyllie and Tuttle (1959) involved the melting of five shale compositions at pressures between ~600 bar and 2.8 kb at between 20–30 °C above the wet granite solidus. About 150 °C above the beginning of melting the shales contained ~50% melt (granodiorite composition) together with crystals of quartz, cordierite (frequently sector twinned), mullite, orthopyroxene, with anorthite forming from a calcite-bearing shale. In contrast to granitic liquids that contain many small bubbles when quenched, the half melted shales are highly vesicular and have a frothy, slaggy appearance not unlike those commonly produced during combustion metamorphism. Experimental studies of more feldspathic (greywacke sandstone–siltstone) compositions by Kifle (1992) using lightly crushed (rather than powdered) starting material delineate low pressure feldspar-, muscovite-, biotite-out and cordierite-, osumilite-, and orthopyroxene-in curves. Relevant data from these studies together with experimentally-determined mineral stabilities are used to construct a petrogenetic "grid" for high temperature metamorphism of quartzofeldspathic compositions that is given in Fig. 3.7. The β-quartz-tridymite transition is clearly important as it can be used to divide pyrometamorphosed (sanidinite facies) rocks into lower and higher temperature types. In lower temperature partly-fused examples, plagioclase, biotite, muscovite, chlorite etc., show evidence of melt-producing breakdown reactions; at higher temperatures within the tridymite stability field, extensive melting occurs, a few primary minerals, e.g. zircon, apatite, ilmenite, relics of unmelted quartz, remain and new minerals crystallise from the melt. Pyrometamorphosed partially melted granitoids may also contain tridymite so that the quartz-tridymite inversion in relation to the melting of quartz, K-feldspar, albite is of critical importance in evaluating T-P conditions of the melting reactions that usually also involve breakdown of biotite and, in granodiorite of hornblende (Figs. 3.37 and 3.38).

It can be noted that in synthesis experiments in silicate systems at atmospheric pressure, cristobalite often forms and persists metastably in the temperature field of tridymite, e.g. as observed in some paralavas. In the system FeO-Al_2O_3-SiO_2 (Schairer and Yagi 1952), tridymite only crystallised readily from appropriate compositions at temperatures just below liquidus temperatures. In this system, the 1470 °C isotherm separates the field of tridymite and cristobalite (Fig. 3.4). At lower temperatures, quartz may appear when the temperature is above the quartz-tridymite inversion temperature of 867 °C and once formed it can also persist metastably within the tridymite temperature field. Above 1050 °C, it invariably inverts to cristobalite rather than tridymite and at 1000 °C metastable quartz and cristobalite occur.

Chapter 3 · Quartzofeldspathic Rocks

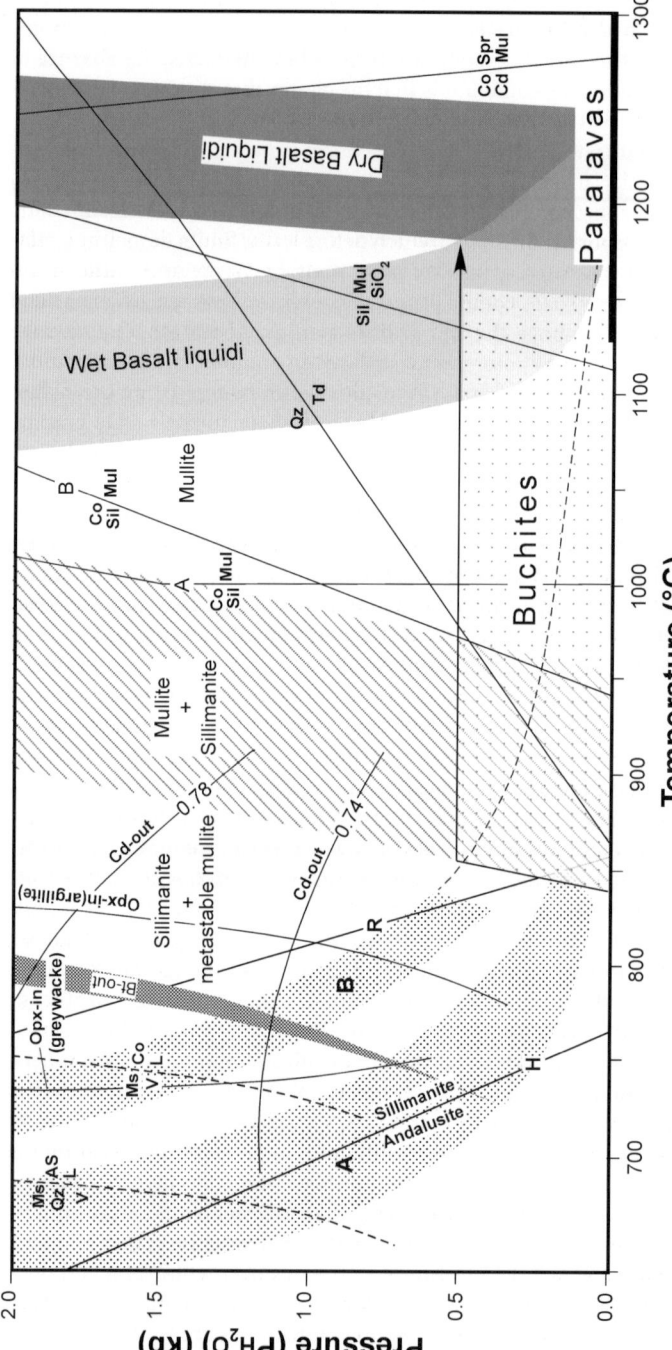

Fig. 3.7. Petrogenetic grid for pyrometamorphosed quartzofeldspathic compositions. *Stippled areas* labelled *A* and *B* are solidus curves for greywacke sandstone-siltstone compositions. *Stippled areas* labelled *A* and *B* respectively. Orthopyroxene-in, biotite-out and cordierite-out curves are from Kifle (1992). Muscovite disequilibrium breakdown curves are from Ruby and Brearley (1987). Andalusite-sillimanite curves are from Holdaway (1971) (*H*) and Richardson et al. (1969) (*R*). Fields of sillimanite + metastable mullite, mullite + sillimanite, mullite-only and buchites are from Cameron (1976a). Co Si = Mul reaction curves are from Holm and Kleppa (1966) (*A*) and Weill (1966) (*B*); Sil = Mul SiO$_2$ is from Holm and Kleppa (1966); Co Cd = Spr Mul is from Seifert (1974); Qz/Td transition from Kennedy et al. (1962), Ostrovsky (1966)

3.2
Contact Aureoles and Xenoliths

3.2.1
Metapsammitic-Pelitic Rocks and Schist-Gneiss Equivalents

South Africa

Fusion of flat lying arkosic sediments by the Karroo dolerite, South Africa, is a relatively rare phenomena but two examples have been carefully documented. Ackermann and Walker (1959) describe pyrometamorphism of a roof pendent associated with a 15 m thick sill exposed in an excavated pit at Baksteen, near Heilbron, Orange Free State (Fig. 3.8). The weakly metamorphosed sediments are coarse-grained arkose containing angular-subangular grains of quartz, plagioclase (An_{25-30}) and orthoclase with modal ratios of $quartz_{45}$: total $feldspar_{35}$ and $plagioclase_8$: $orthoclase_{27}$. Detrital accessories include almandine garnet, with lesser amounts of zircon and ilmenite, rare titanite, rutile, apatite, green spinel and orthopyroxene. The matrix of the arkose consists of microcrystalline quartz and goethite. Where the arkose contains flat discoid clay pellets these have been almost completely vitrified to form a black cordierite-spinel buchite. A hybrid or mixed zone is developed between dolerite and arkose with some parts being definitely igneous with clinopyroxene, Ca-plagioclase (possibly the results of cafemic transfusion) and dark glass, and other parts clearly sedimentary with pale glass containing relic quartz.

Towards the dolerite contact the arkose hardens due to an increase in the amount of glass which reaches 70% at the dolerite contact so that the rock resembles dark green pitchstone. Vitrification is accompanied by a series of progressive changes in the detrital mineral assemblage, glass content and specific gravity (Fig. 3.8) that are preceded by straining and cracking of quartz, feldspars, garnet and ilmenite. K-feldspar is the only detrital phase to be totally resorbed at about 0.6 m from the dolerite contact. With increasing temperature orthoclase becomes homogenised and inverts to sanidine which begins to break up along cleavage planes by the invasion of glass. Quartz and plagioclase decrease towards the dolerite but persist into the hybrid zone (Fig. 3.8). Microlites of cordierite develop at glass/feldspar contacts and later form within the feldspar itself. Melt in contact with orthoclase crystallizes (devitrifies) to a K-rich, sometimes fibrous, mantle and the margins of the grains recrystallise to feldspar microlites or form isotropic globules. Quartz persists as rounded relics after the resorption of sanidine. As the amount of glass increases, rock densities increase slightly (Fig. 3.8) indicating a decrease in volume (see Chapter 2) and needles and plates of tridymite fringes around quartz become more common. In the hybrid zone close to the dolerite contact, quartz is rimmed by acicular pyroxene and just prior to complete resorption, spherical grains develop an undefined isotropic mantle.

The other example of fusion of feldspathic sandstone that forms the roof of a ~17 km long inclined arcuate sheet of dolerite ranging in thickness from 30 cm up to 100 m is documented by Ackermann (1983). Buchites are developed within a maximum distance of 2 m, and more usually < 50 cm, of the dolerite contact and are localised in roof

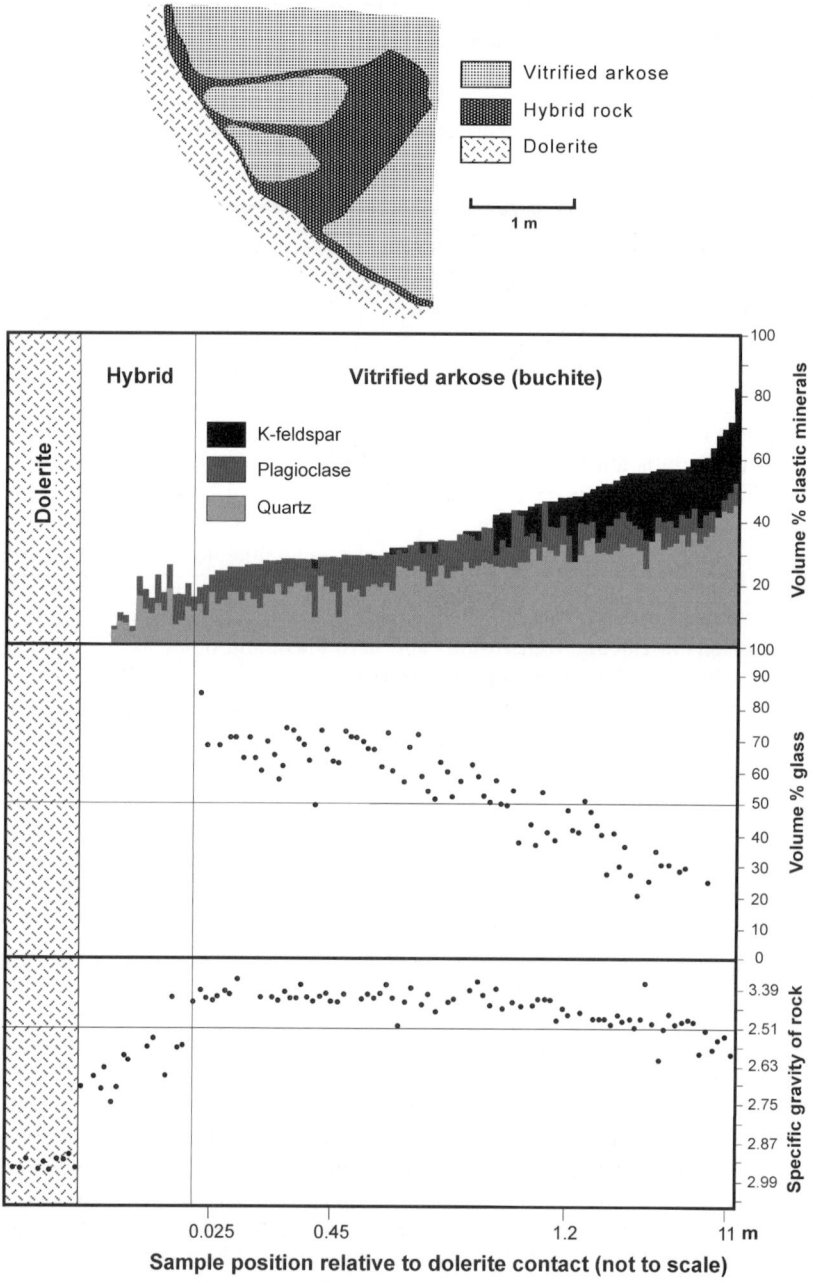

Fig. 3.8. Section of dolerite-arkose contact exposed in excavation pit at Baksteen, near Heilbron, South Africa (redrawn from Fig. 2 of Ackermann and Walker 1959), together with micrometric data for clastic minerals and glass, and specfic gravity changes with increasing vitrification of arkose towards the dolerite contact (redrawn from Fig. 4 of Ackermann and Walker 1959)

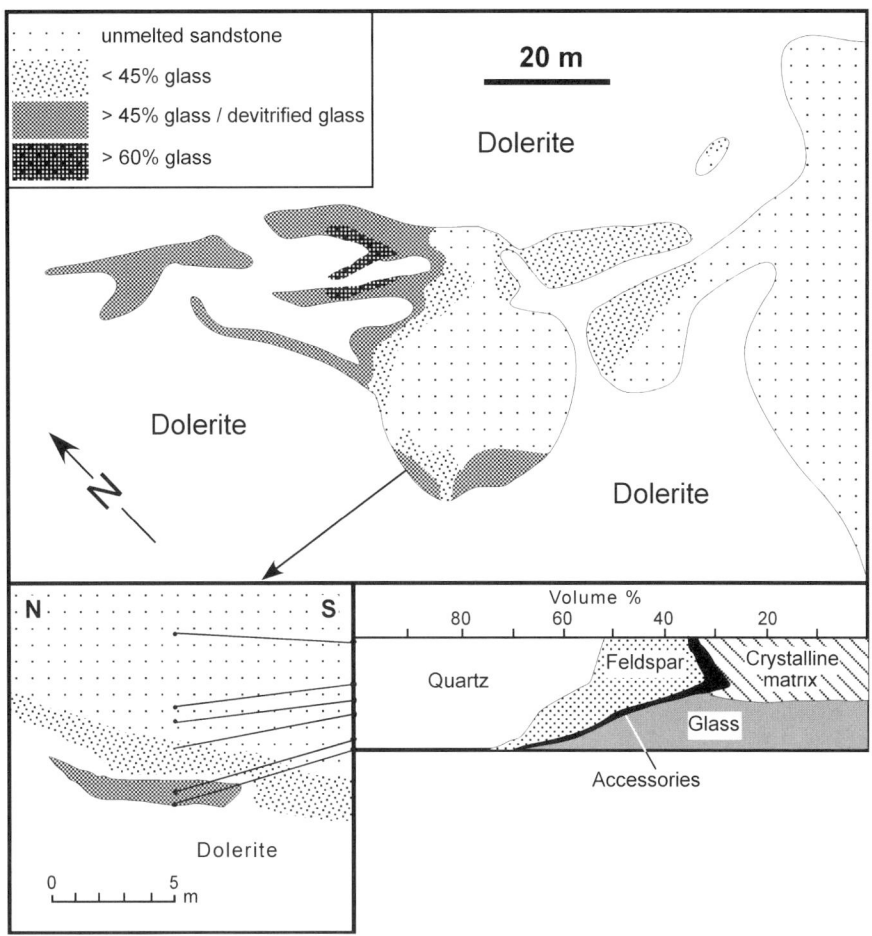

Fig. 3.9. Map, vertical section and modal data plot of an upper sandstone-dolerite contact, Sterkspruit Valley, South Africa (in part redrawn from Fig. 3 of Ackermann 1983)

pendants and slabs or lenses of sandstone ranging in size from 60 × 30 cm to 15 × 6 m within the dolerite (Fig. 3.9).

The fine grained feldspathic sandstone (0.02 to 0.16 mm detrital grain size) consists of up to 48% quartz, 28% feldspars and 44% microcrystalline matrix. Alkali feldspar is typically altered to sericite and kaolin, and also makes up a large part of the matrix. Accessories (< 5%) include magnetite and zircon with lesser amounts of green spinel, epidote and tourmaline. Fusion has produced buchites with melt contents ranging from < 45% to completely vitrified and can be characterised as follows:

- < 45% *glass.* Turbid, isotropic glass is confined to the microcrystalline groundmass and altered K-feldspar. In places devitrification has resulted in a fine intergrowth of quartz and K-feldspar. Quartz grains show various degrees of rounding and embayment in contact with the glass and some have partial fringes of needles of quartz after tridymite. K-feldspar is also rounded and frittered, whereas the plagioclase is largely unaffected although may show embayment when in contact with glass. In buchites with about 35% glass, the glass is clear and may contain spherical globules (~0.05 mm diameter) that have a lower R.I. (more K-rich) than the host glass. These may represent completely fused K-feldspar grains. Rectangular or hexagonal microlites of cordierite are present.

- **45–60% glass.** In these buchites most of the K-feldspar has largely disintegrated as evidenced by fritting or development of a fingerprint structure caused by melting (see Chapter 7). Cordierite microlites and quartz paramorphs after tridymite are abundant. With increasing amounts of glass its colour becomes brown and the amount of unfused K-feldspar decreases with respect to plagioclase and quartz which show clear evidence of melting along mutual grain boundaries.
- **60–80% glass.** Petrographic features in these rocks are more varied. Cordierite is associated with small amounts of remaining frittered K-feldspar. Glass is brown around Fe-oxides that have sometimes reacted to rutile? and hematite. With a decreasing amount of clastic grains the glass becomes darker brown and is colour streaked indicating flow.
- **80% glass.** Arcuate perlitic cracking is typical in buchites with > 80% brown or almost colourless glass. No cordierite or tridymite are present. The grain size of well rounded quartz and feldspar has decreased to ~0.05 mm and both mineral relics are sparsely distributed throughout the glass. Ellipsoidal nodules up to 50 mm across are composed of a microfelsitic intergrowth of quartz and feldspar that also occurs along cracks as a result of devitrification.

Assuming a thickness of 1.37 km of overlying Drakensburg basalts, pyrometamorphism of the sandstone at the two localities described above could have occurred at ~400 bar and with the crystallisation of tridymite in the buchites at temperatures above a minimum of ~950 °C and possibly between ~990–1050 °C in comparison with solidus temperatures of wet and dry tholeiitic magma (Fig. 3.7).

Apsley

Pyrometamorphism of flat-lying Triassic sandstones by a 60 × 20 m alkali basalt plug near Apsley, Tasmania, has resulted in their partial vitrification and the development of columnar jointing that is described in Chapter 2 (Spry and Solomon 1964). Sediments adjacent to the plug have melted to form buchites (after sandstones) and finely banded, flinty porcellanites (after fine grained clay-rich sediments). The unmetamorphosed sandstone consists of detrital quartz (~10%), feldspars (microcline, oligoclase-andesine; ~20%), muscovite, chloritised biotite, garnet, tourmaline, rutile, zircon, magnetite and graphite, in a clay matrix of kaolinite together with halloysite and probable illite and sericite. Clay pellets of chlorite and kaolinite also occur. In the least altered sandstones, the clay matrix and clay pellets are coarsened by recrystallisation with the formation of sericite.

At an advanced stage of thermal reconstitution (~900 °C), clay pellets have melted to a pale green glass although the clay matrix in the host sandstone remains unaltered.

Feldspars are unaffected but incipient melting has occurred in muscovite and biotite as evidenced by their blurred optical outlines. At higher temperature (~920–950 °C), the clay matrix is converted to a colourless glass that contains globules of more femic green glass with sharp contacts indicating immiscibility. Muscovite is replaced by colourless glass containing mullite needles, and biotite is replaced by yellowish glass that contains lines of Fe-oxide/hercynitic spinel. Feldspar melting is indicated by patches of clear glass containing mullite. Elsewhere, colourless glass contains variable amounts of newly-formed cordierite, spinel, mullite, corundum and tridymite. In rocks with up to 65% glass, quartz shows evidence of significant melting and is fringed with tridymite. In the system $FeO-Al_2O_3-SiO_2$ assemblages of corundum-spinel-mullite, spinel-mullite-cordierite and mullite-cordierite-tridymite form at 1380, 1205 and 1210 °C respectively (Fig. 3.4, Table 3.1). At 50 bar PH_2O (the estimated pressure of pyrometamorphism; Chapter 2), these temperatures are above the temperature range of the dry basalt liquidus. The occurrence of tridymite indicates a minimum temperature of ~870 °C at 50 bar and that of mullite coexisting with corundum and without sillimanite at ~945–1000 °C according to the Co Sil = Mul reaction curves shown in Fig. 3.7, i.e. 330–260 °C lower than the invariant point relevant assemblages in the oxide system (Fig. 3.4; Table 3.1).

Glenmore

Mineral changes and partial vitrification in schistose arkose and pelite within 1.2 m of the contact of a dolerite plug at Glenmore, Ardnamurchan, Scotland, are documented by Butler (1961). The contact dolerite is not chilled against arkose, and the absence of xenoliths together with a relatively wide thermal aureole for the small 44 × 26 m diameter intrusion suggests that the plug was a feeder to a lava flow(s). The pyrometamorphosed arkoses are grey, black or greenish-black, hard vitreous-looking rocks with a brittle fracture; an interbedded grey to black pelite within the examined sequence is tough and splintery.

The following mineral changes are noted towards the dolerite contact:

- *1.2 m.* Detrital micas in arkose and pelite are altered. Muscovite is yellow and turbid; biotite is largely opaque due to the presence of spinels. K-feldspar (sanidine) forms a reaction rim between both micas and quartz. Original detrital microcline in the arkose remains unaffected although some large feldspar grains are granulated.
- *0.9 m.* K-feldspar reaction rims surrounding mica grains in the pelite are more distinct and basal sections of muscovite contain hexagonally-arranged needles of mullite. In the arkose, quartz/feldspar grain boundaries are diffuse and in some cases are lined with glass. Original K-feldspar is converted to sanidine.
- *0.76 m.* Glass forms a distinct rim between quartz and feldspars in the arkose. In the pelite, sanidine rims around micas are ~0.025 mm wide although the original micas have disappeared. Although some relic garnet persists in the pelite, most has reacted to Fe-oxide and a fine-grained micaceous material that could represent pinitised cordierite.
- *0.6 m.* Sanidine reaction rims surrounding original micas in the pelite are widened up to 0.1 mm and exhibit a felsitic texture suggestive of glass devitrification from which the feldspar has crystallised. The schistose texture of the pelite remains. In

the arkose, brownish glass lines quartz-feldspar and relic mica grain boundaries. Edges and cleavage planes of plagioclase grains are altered to a turbid, brown mosaic producing a typical fingerprint structure indicative of melting (see Chapter 7).
- *0.4 m.* Schistose texture of the pelite is lost. Trains of spinel mark the position of former biotite grains. Sanidine is abundant as fine laths and skeletal crystals set within devitrified glass that contains rare amygdules of carbonate as in the contact dolerite.

Closer to the contact both arkose and pelite contain abundant glass and have the black, pitchstone-like appearance typical of buchites. The glass contains dark brown spinel, ortho- and clinopyroxene, cordierite, inverted tridymite and sanidine. Orthopyroxene is strongly pleochroic and occurs as prismatic crystals sometimes arranged in fan-shaped sprays of curved needles. Clinopyroxene forms fringes of short prismatic crystals around residual quartz. Cordierite exhibits characteristic rectangular and hexagonal cross sections and is often associated with recrystallised feldspar. Inverted tridymite forms fringes around quartz relics and is associated with orthopyroxene in glass.

At the contact itself, a 2 cm wide white rock occurs between the dolerite and vitrified arkose. It consists of embayed aggregates of quartz and recrystallised feldspar in a felsitic groundmass that contains crystals of igneous clinopyroxene (sometimes rimming quartz) and twinned plagioclase suggesting that it is a hybrid product between the dolerite and arkose. In places the white rock blends into patches of igneous appearance containing a larger proportion of clinopyroxene and plagioclase. From this contact rock rheomorphic veins up to 4 mm wide extend up to 20 cm out into the normal buchite. They consist of a felsitic groundmass containing skeletal K-feldspar, orthopyroxene needles and embayed quartz grains and aggregates. In one place the pelitic layer is intruded by a rheomorphic vein derived from enclosing partly fused arkoses. Here, the vein consists of brown glass containing green orthopyroxene microlites and relic quartz grains derived from the arkose.

Re-examination of melting in the Glenmore aureole by Holness et al. (2005) allows construction of a maximum temperature profile with distance from the contact at an estimated 120 bar PH_2O and 1190 °C basalt intrusion temperature (Fig. 3.10). Towards the contact there is a steady increase in the width of melt rims developed between quartz and feldspar as well as an increase in the total volume fraction of melt to 60 vol.% at ~0.28 m from the dolerite contact (Fig. 3.10). At the same time, there is a decrease in relic feldspar and an overall increase in the amount and size of relic quartz grains and aggregates. This has the effect of decreasing the melt-producing potential of the rock because melting occurs along the boundaries between quartz, feldspar and muscovite as seen by a flattening of the volume % melt curve near the contact and also the inverse relationship between the volumes of melt and quartz (Fig. 3.10). The thermal model considered by Holness et al. (2005) and described in Chapter 2 indicates that the magma conduit (i.e. the plug) may have been active for about one month.

Soay

Fusion of feldspathic sandstone/grit within 60 cm of a small picritic sill and as xenoliths within it and another sill was studied by Wyllie (1959, 1961) from the Island of

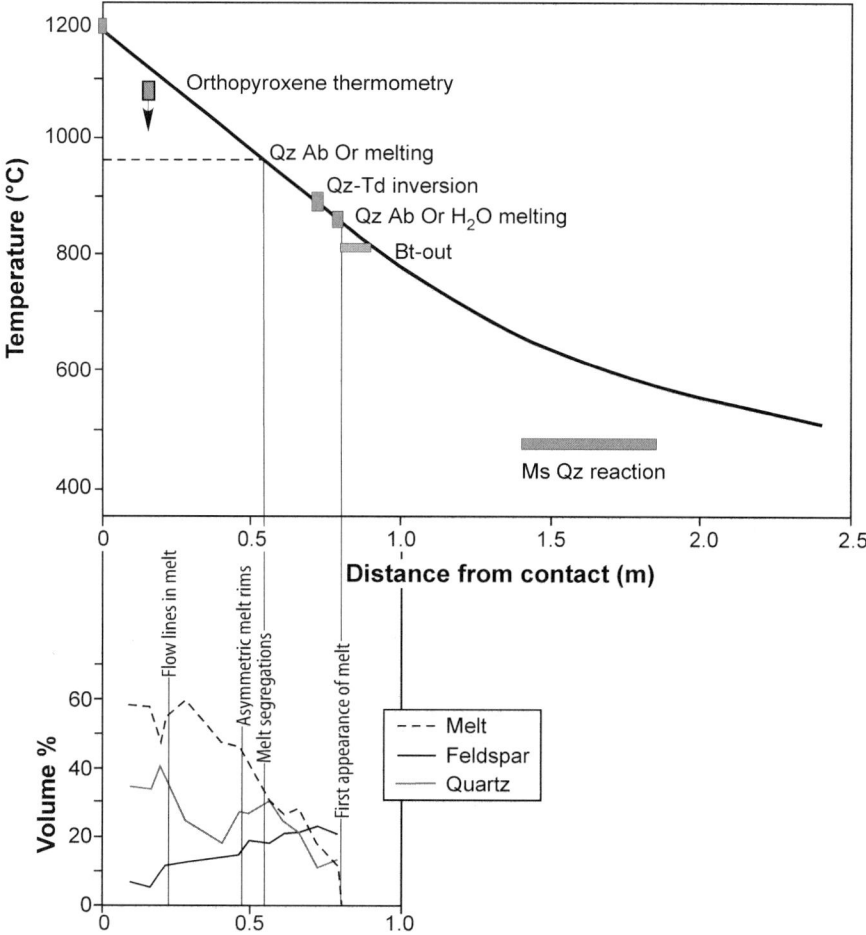

Fig. 3.10. Thermal profile for the Glenmore aureole, Ardnamurchan, Scotland (after Fig. 10 of Holness et al. 2005). *Below* is a plot showing volume % of melt, residual quartz and feldspar with respect to distance from igneous contact (after Fig. 7 of Holness et al. 2005)

Soay, Hebrides, Scotland (Fig. 3.11a). An unmetamorphosed red sandstone from 60 cm below the igneous contact consists of detrital strained quartz, turbid orthoclase and microcline containing micro hematite inclusions, a small amount of plagioclase, rare muscovite that contains magnetite, within a matrix of sericite and silica cement containing disseminated hematite. Progressive alteration/fusion of the sediment occurs towards the basal contact of the sill and involves the following changes:

- Reduction in modal quartz, alkali feldspar, and plagioclase due to melting.
- Loss of the sericite/silica matrix due to melting.
- A colour change from red to black partly resulting from increasing amount of glass and partly from reduction of hematite to magnetite.

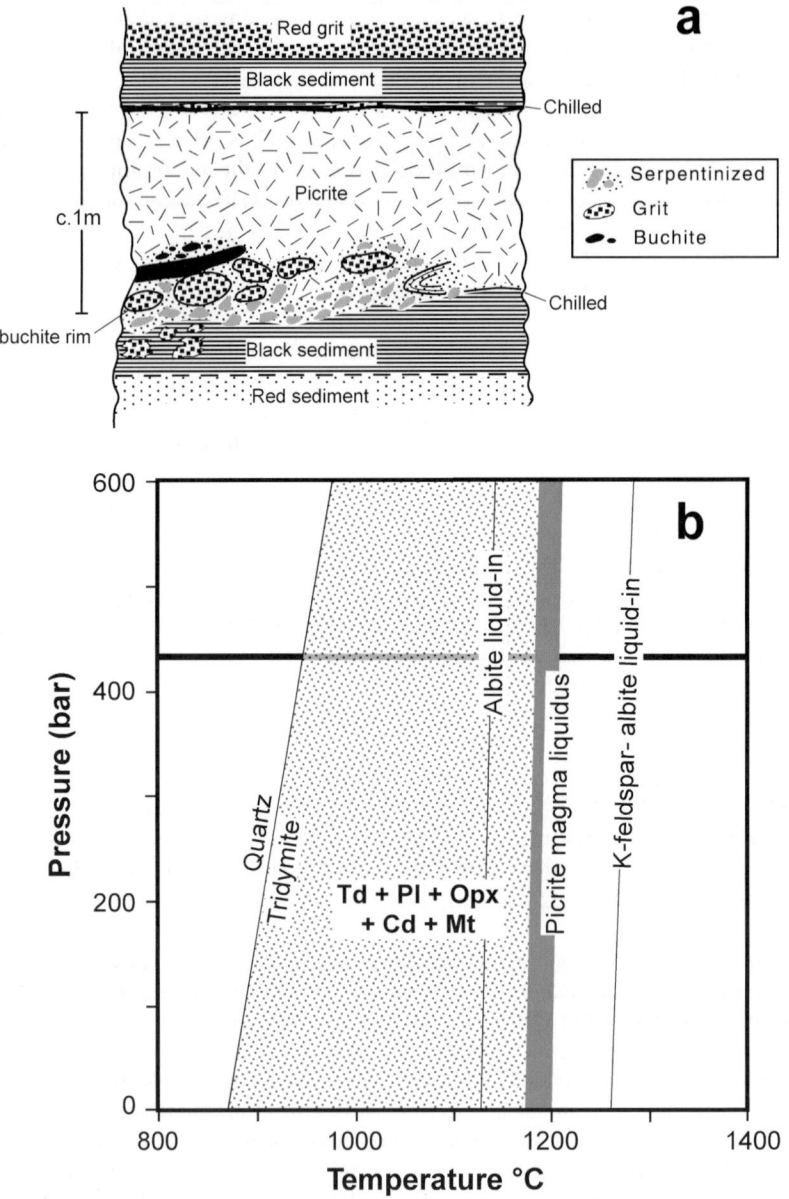

Fig. 3.11. a Diagrammatic section showing field relations of a ca. 1 m thick picrite sill with basal buchite and grit xenoliths, Soay, Hebrides, Scotland (redrawn from Fig. 1 of Wyllie 1961). **b** *P-T* diagram relevant to conditions of pyrometamorphism. Field of Td, Pl, Opx, Cd, Mt, albite liquid-in and albite-K-feldspar liquid-in curves calculated from composition of fused xenolith (sample 129, Table 2 of Wyllie 1961) using Theriak/Domino software (de Capitani and Brown 1987). The *thick horizontal black line* indicates estimated pressure of 420 bar for the pyrometamorphic event

- At 20 cm below the contact and closer, the glass contains abundant new magnetite, together with orthopyroxene and cordierite. Quartz has fringes of tridymite that are partially inverted to quartz.
- Within the sill, xenoliths of grit have rinds of black glass. Detrital feldspar, quartz and rounded quartzite of the xenoliths occur in a glass matrix that contains magnetite. An elongate buchite xenolith occurring above the grit fragments (Fig. 3.11a) has a pale green colour reflecting the presence of additional (secondary) carbonate, rare aegirine-augite and magnetite. A xenolith in another sill shows evidence of almost complete fusion with over 90 vol.% glass that contains cordierite, orthopyroxene, inverted tridymite and magnetite.

Wyllie (1961) infers an emplacement temperature of the picrite magma between 1175–1200 °C assuming a depth of intrusion of about 1.7 km (430 bar). The computed assemblage stability of a buchite composition (sample 129 in Table 1 of Wyllie 1961) using the Theriac/Domino software of de Capitani and Brown (1987), indicates that the temperature of the buchites could have as high as 1125 °C, i.e. position of appearance of an albite liquid about 50 °C below the minimum magma temperature, with cordierite, orthopyroxene, plagioclase, tridymite and magnetite crystallizing on cooling (Fig. 3.12b). The relatively high proportion of inverted tridymite in the buchites suggests that it continued to form during most of the cooling interval prior to quenching. The inferred maximum temperature of fusion at Soay is 220 °C lower than the invariant point assemblage cordierite-protoenstatite-tridymite of 1345 °C in the system $MgO-Al_2O_3-SiO_2$ (Fig. 3.4, Table 3.1).

Arran

Another example of the fusion at the contact of a minor intrusion is the buchite associated with a 1.3–1.6 m thick composite dyke of tholeiite intruding schistose-grit on the Isle of Arran described by Holgate (1978). The dyke consists of olivine tholeiite intruded by olivine-free tholeiite. Along part of the contact where the later olivine-free tholeiite abuts directly against grit, textural and mineralogical modification of the basalt occur over a width of 5 to 10 mm. This zone is abundantly variolitic and contains grains of quartz in a matrix of platy labradorite with prisms of orthopyroxene in place of clinopyroxene. Cordierite euhedra also occur and become more abundant towards the buchite contact. Coarser grains of quartz may be fringed by fibrous plagioclase and quartz associated with orthopyroxene and an unidentified opaque mineral. Interstitial areas are occupied by colourless glass, irregular feldspar-quartz spherulites, magnetite and pale brown biotite. Plagioclase and orthopyroxene disappear at the buchite contact.

In the buchite itself, the usual light-grey colour of the country rock is darkened by the presence of blackish glass that increases in abundance towards the dyke over a distance in excess of 20 cm. The rock is largely composed of a pale brownish glass ($D = 2.36$ g cm^{-3}) that contains patches and trains of corroded quartz grains representing original quartz-rich laminae of the schistose-grit. Granules of magnetite and occasional abraded zircon remain as relics. Small euhedra of cordierite and hercynite, felted patches of mullite (probably after muscovite), and tridymite fringes on quartz are the newly-formed high temperature phases. Away from the igneous contact, relic quartz is more abundant, cordierite is smaller, and hercynite occurs as fine grained wispy patches apparently replacing chlorite.

Investigation of the temperature of fusion of powdered buchite glass with a hot-stage microscope at atmospheric pressure under reducing conditions indicated incipient fusion (sintering) at 1180 °C and complete fusion at 1250 °C. In this case, fusion temperatures are consistent with the cordierite-mullite-tridymite and hercynite-cordierite-mullite invariant point temperatures of 1210 and 1205 °C respectively in the system $FeO-Al_2O_3-SiO_2$ (Fig. 3.4; Table 3.1). At the temperature of complete fusion, the melt has an irregular surface and together with the presence of gas bubbles indicates a high viscosity. This supports the viscous nature of the melt in the natural buchite inferred from the presence of localised steakiness and the formation from, and preservation in the glass of lines of cordierite. In comparison to the Soay example described above, the development of the Arran buchite implies comparatively long term heating indicating that the inner tholeiite was probably a feeder for lava extrusion.

Tieveragh

The pyrometamorphic effects of a plug of olivine dolerite intruding Old Red Sandstone at Tieveragh in Co. Antrim, Northern Ireland, were first recognised by Tomkeieff (1940). Although only small parts of the contact rocks are exposed because of scree cover, they have been re-examined by Kitchen (1984). The intrusive contact is a complex association of contaminated dolerite and brecciated buchite that has been mobilised and mixed, as acidic melt (glass), with the basaltic magma. Acid melt has also penetrated outwards from the contact zone into the buchite, causing brecciation and also occurs as irregular concordant tongues intruding the poorly bedded contact rock.

Buchite compositions reflect initial differences in bulk composition of the Old Red Sandstone precursor lithologies. *Glassy buchites* are thought to have developed from quartz-rich rocks and devitrified or *lithic buchites* from quartz-poor sediments that contain abundant feldspathic rock fragments. At one locality, about 10 m from the contact, glassy buchite consists of alternating 4 cm thick layers of different composition; one with orthopyroxene, clinopyroxene, plagioclase, skeletal K-feldspar, magnetite, ilmenite, hematite in a clear to light-brown, largely devitrified glass (*pyroxene buchite*); the other with cordierite ($XMg = 0.94$), plagioclase, K-feldspar, magnetite, ilmenite, hematite in abundant colourless glass (*cordierite buchite*). In both assemblages, detrital alkali feldspar has melted to form plagioclase-glass pseudomorphs.

At another locality, *pyroxene buchites* are interlayerd with *mullite buchites* as 15 cm thick bands and are separated from the intrusion by a zone of contaminated dolerite. The pyroxene buchites are similar to those at the former locality but contain a high percentage of glass and have only a minor amount of feldspathic rock fragments. Relic quartz is resorbed and rimmed by pyroxene; K-feldspar is replaced by plagioclase and glass. Clear to light-brown glass contains microlites of the same minerals as at the other locality with additional occasional tridymite. In the mullite buchites quartz relics are surrounded by rims of microlite-free glass whereas detrital K-feldspar is replaced by dense mats of mullite, forming crudely-radiating aggregates that outline the shape of the replaced feldspar. Glass forms a little more than 52 vol.% and contains quench microlites of cordierite (XMg = 0.83), mullite, K-feldspar and opaque oxides.

Chinner and Dixon (1973) report an additional osumilite-bearing buchite from the Tieveragh aureole. The near K, Mg-end member osumilite contains abundant inclusions of magnetite, hematite and glass, and occurs as masses or trains of euhedral and anhedral crystals within a colourless glass that also contains inverted high temperature hexagonal cordierite, Na-rich plagioclase, sanidine and orthopyroxene with inverted tridymite needles growing from relic quartz.

Two-pyroxene thermometry for dolerite and pyroxene buchite indicate overlapping temperatures between ~920–940 °C (Fig. 3.12) suggesting maximum temperatures for buchite formation were within this range. Temperatures of between 910–740 °C derived from magnetite-ilmenite thermometry in mullite buchites (Fig. 3.12) may partly reflect the effects subsolidus cooling, although Kitchen (1984) infers that quenching temperatures of the mullite buchites were between 940–738 °C. The apparent stable coexistence of Mg-rich osumilite and cordierite in one buchite suggests the reaction

osumilite + vapour = cordierite + K-feldspar + orthopyroxene + liquid + vapour

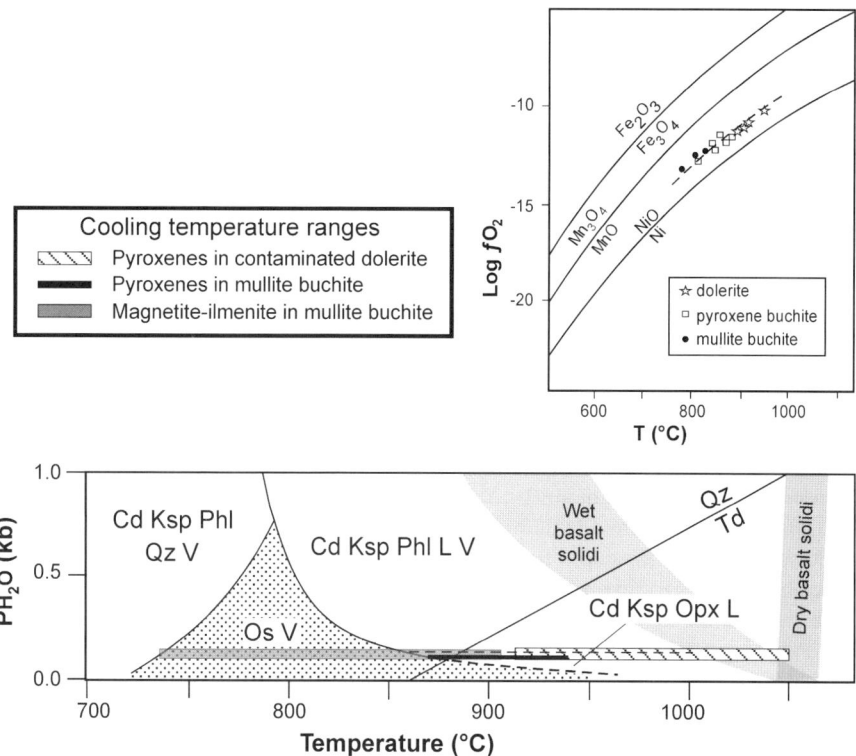

Fig. 3.12. T-P-fO$_2$ data relating to pyrometamorphism at Tieveragh, northern Ireland. Data from Kitchen (1984). See text. The stability fields of osumilite + vapour (*stippled*) and cordierite-bearing assemblages are from Olesch and Seifert (1981).

that is predicted to occur at low $PH_2O \leq 200$ bar (Olesch and Seifert 1981). This implies a maximum temperature of ca. 870 °C for the coexistence of osumilite and cordierite at Tieveragh, i.e. comparable with the lower temperature of the pyroxene cooling interval in mullite buchites and with the presence of inverted tridymite and hexagonal cordierite in the osumilite-bearing buchite (Fig. 3.12). At 200 bar, the temperature of the basalt solidus ranges from ~1070 °C (dry) to 1000 °C (wet).

It would appear that maximum contact temperatures were maintained over a distance of at least 10 m from the magma by convective rather than conductive heat transfer, involving the passage of a single or multiple pulses of hot fluid though the country rock rapidly dissipating heat by condensation, the distance outward from the magma contact being limited by wall rock permeability and gas pressure. Determinations of fO_2-T from coexisting oxides indicate that buchites and dolerites lie along a line between and parallel to the Ni-NiO and MnO-Mn_3O_4 buffer curves (Fig. 3.12) suggesting that oxides in both rocks were in equilibrium and buffered by redox conditions of the dolerite as a consequence of convective heat transfer.

Mull

One of the classic areas from where pyrometamorphosed metasedimentary rocks have been described is the Tertiary minor intrusions of the Isle of Mull, Argyllshire, Scotland. In the southern part of Mull, high-level inclined basalt sheets of the Loch Scridain Complex that intrude Palaeogene lavas and underlying Mesozoic sedimentary and Proterozoic Moine Supergroup metasedimentary rocks, also contain numerous partially melted xenoliths. The type locality of the xenolithic intrusions is a composite sill at Rudh'a'Chromain that consists of a central acidic unit, 6–9 m thick, bounded on each side by sheets of tholeiitic basalt (Thomas 1922; Buist 1961) (Fig. 3.13).

Three main types of xenoliths were first described by Thomas (1922): *siliceous buchites* characterised by partial fusion of quartz and feldspars, the development of tridymite fringes around quartz, and may contain cordierite, orthopyroxene and clinopyroxene in glass; *aluminous sillimanite* (+ mullite [Cameron 1976]) and *cordierite buchites* (Fig. 3.13) with the former having rims of recrystallised, holocrystalline anorthite-corundum-spinel/cordierite-sillimanite-spinel. The xenoliths have been reexamined by Preston et al. (1999) who identified:

- Moine *Quartzites* (abundant) – unmelted and composed of 100% coarse-grained (up to 5 mm), strained anhedral interlocking quartz.
- Moine *Quartzo-feldspathic schist* (minor) – containing highly corroded quartz, together with oligoclase and minor K-feldspar. Melting has occurred along quartz/feldspar grain boundaries as represented by fan spherulites of K-feldspar and devitrified glass. Dark bands (precursor mica-rich layers) consist of hercynite octahedra (1–4 μm) and ilmenite with plates of biotite in a mixture of K-feldspar, andesine and glass. Former garnet is replaced by aggregates of magnetite-ilmenite and spinel.
- Mesozoic *Sandstone* (common) – extensively corroded quartz with fringes of tabular, inverted tridymite. Original feldspar and mica are fused to form pools of cryptocrystalline K-feldspar and quartz together with clinopyroxene.

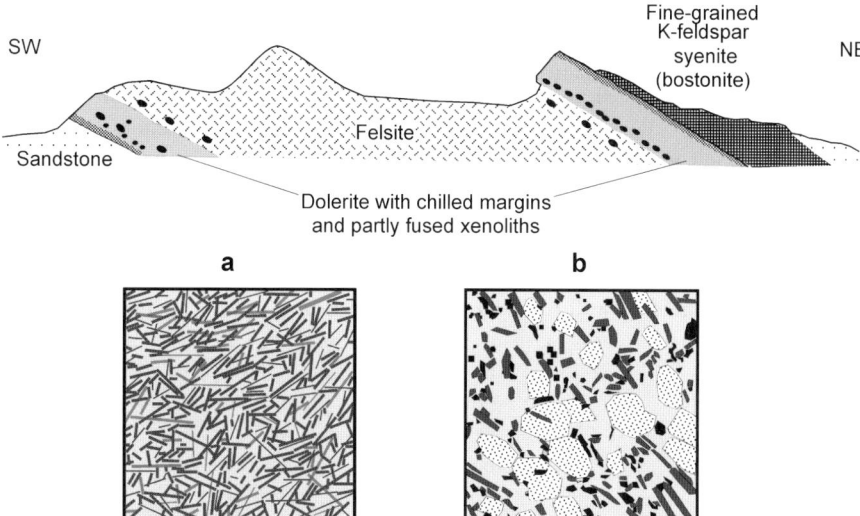

Fig. 3.13. *Above:* Section of the Rudh' a'Chromain xenolithic sill, Isle of Mull, Scotland. Section length is ~9 m (redrawn from Fig. 3 of Thomas 1922). *Below:* Drawings from microphotographs of (**a**) mullite buchite with pale-pink mullite needles in clear, pale lavender glass (×18; Fig. 1, Plate VII of Thomas 1922) and (**b**) mullite-cordierite buchite consisting of well-formed cordierite (*stippled*) with mullite (*grey*) and spinel (*black*) in glass. Textures indicate early crystallisation of spinel and mullite followed by cordierite (×18, Fig. 2, Plate VII of Thomas 1922). A small amount of sillimanite may also be present

Pelitic schist protoliths have produced:

- *Cordierite buchites* – Black, vitreous xenoliths up to 1 m in length. These consist of a dense mat of mullite needles (~25 vol.%) and tiny (10–20 μm) crystals of cordierite (~14 vol.%), rare tridymite anhedra and ragged grains of magnetite (<30 μm) within a clear to red-brown glass.
- *Mullite buchites* – dark grey, through lilac to almost white xenoliths with between ~60–75 vol.% glass varying in size from 10 cm across to large rafts several metres in length. The glass contains a mass of mullite needles, occasionally associated with sillimanite, small octahedra of magnetite and ilmenite, and rare corundum.
- *Plagioclase-rimmed mullite buchites* – many smaller mullite buchite xenoliths (up to 80 cm diameter) have a thick rim of bytownite-labradorite (often forming > 80 vol.% of the xenolith) that contains numerous inclusions of mullite together with lesser amounts of corundum and spinel, the latter increasing in amount towards the basalt contact as found by Thomas (1922) (Fig. 3.14). The plagioclase rims display a variety of textures that indicate a complex growth history. The inner part of the rims consist of large plates of plagioclase typically elongated normal to the buchite core. Closer to the magma contact the plagioclase is finer grained and forms a mosaic of randomly orientated crystals. Crystals in contact with basalt have well developed oscillatory zoning and many examples show evidence of recrystallisation near the basalt contact. Fine grained bytownite rims spinel and corundum suggesting that it formed by reaction of these minerals and an Ca-Al-

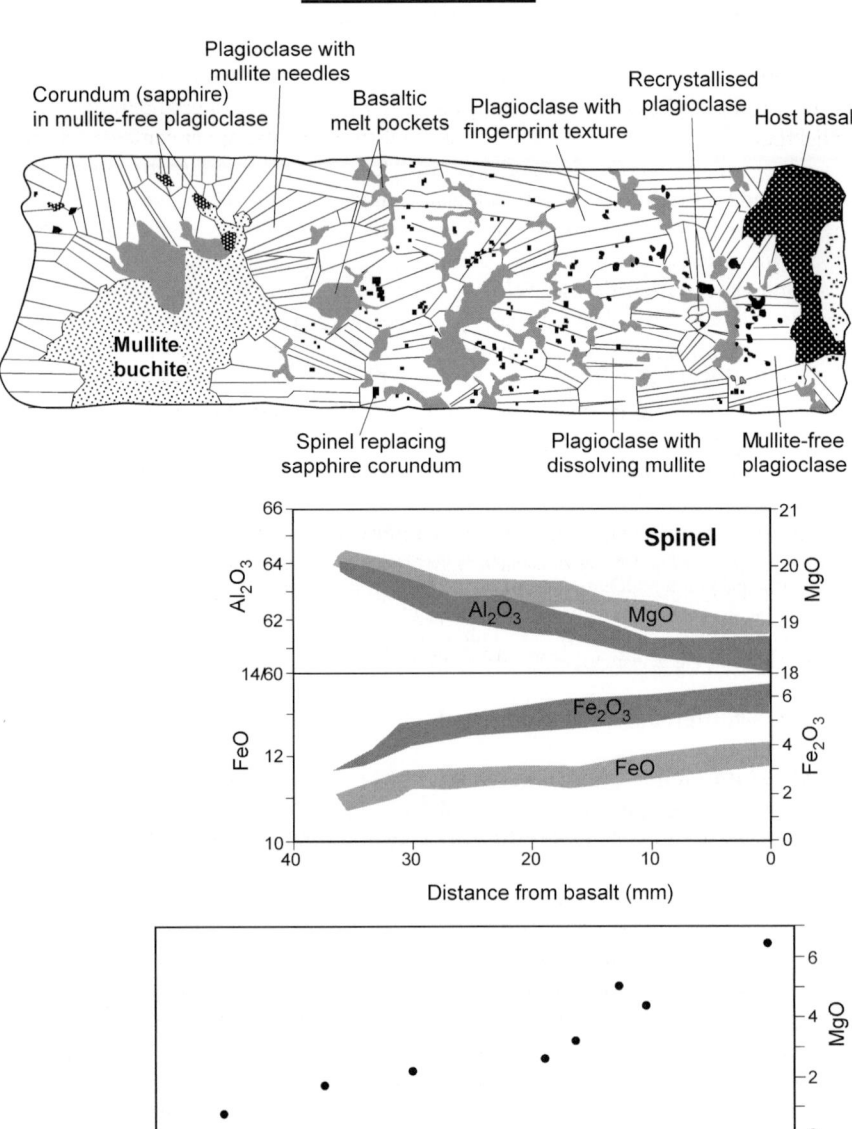

Fig. 3.14. Sketch of major textural relationships across a plagioclase-rimmed mullite buchite within basaltic andesite, Isle of Mull, Scotland. Redrawn and slightly simplified from Fig. 5 of Preston et al. (1999). *Below* are compositional trends of spinel across the plagioclase rim (same distance scale as section) (after Fig. 9 of Preston et al. 1999), and a plot of wt.% SiO_2 versus MgO in basaltic melt pockets within the plagioclase rim with compositions becoming more MgO-rich towards the basaltic andesite (after Fig. 11 of Preston et al. 1999). See text

rich melt in response to penetration of basaltic melt into the xenolith. "Fingerprint" texture indicates reheating after crystallisation.

Except in one case, the basalt sheets have not contact-metamorphosed the country rocks. The exception is a relatively thick (2–6 m), flat lying sheet intruding Moine schist that contains many mullite buchite xenoliths along the upper contact and where there is no chilled margin developed. Locally, the contact rocks have been pyrometamorphosed for up to 40 cm from the basalt contact and Killie et al. (1986) suggest that such occurrences are an indication of areas where there has been turbulent flow in the magma conduit.

Mullite-cordierite-tridymite-Fe-oxide and mullite-Fe-oxide-corundum assemblages in the buchites imply crystallisation temperatures within the range 1205–1210 °C (Fig. 3.4; Table 3.1). An anorthite-spinel-corundum assemblage forms at between 1400–1500 °C in the system $CaO-Al_2O_3-SiO_2$ with 5% MgO (Fig. 3.6). The plagioclase-spinel-corundum rims around mullite buchites have most probably crystallised from a hybrid melt produced by diffusive interaction between aluminous melt in the xenoliths produced from muscovite/biotite breakdown and basaltic magma. This hypothesis is suggested by:

1. Mullite included in plagioclase near the mullite buchite but towards the basalt contact it shows evidence of dissolution and finally disappears in plagioclase adjacent the basalt and basalt melt pockets within the plagioclase rim. If the basalt magma was saturated with clinopyroxene at the time of crystallisation, the decrease and eventual disappearance of mullite may indicate the reaction

    clinopyroxene + mullite (or Al-rich melt) = plagioclase + spinel

2. Corundum inclusions within mullite-free plagioclase suggests the reaction

    mullite (or Al-rich melt) + basaltic melt = corundum + more Si-rich liquid

    Bulk compositions of melt pockets within the plagioclase rim become more Si-rich and Mg-poor towards the buchite (Fig. 3.14) indicating progressive modification of the basalt composition, possible as a result of this reaction.

3. Spinel is concentrated towards the basalt contact and also replaces corundum near basalt. It becomes Al-Mg-poor and Fe-rich towards the basalt (Fig. 3.14) and Preston et al. (1999) consider that this to be the result of post-crystallisation subsolidus equilibration rather than an initial crystallisation trend. However, the change in bulk composition of melt pockets within the plagioclase rims towards the buchite also implies that the change in spinel compositions may be primary and the result of diffusive interaction between xenoliths and basalt magma.

Auckland

The pyrometamorphic effects of alkali olivine basalt of the Auckland volcanic field in northern New Zealand on argillaceous sediment to produce mullite, cordierite and pyroxene-sanidine buchites are documented by Searle (1962). In one example, a zone of yellow-green glass up to 1.3 cm thick is developed between basalt and porcelanite. Against the chilled basalt, devitrified glass is characterised by a felted mass of plumose to radiating clusters of sanidine and sulphur-yellow clinopyroxene needles (Fig. 3.15). The composition of the clinopyroxene is not determined but the distinctive yellow colour suggests that it may be Fe^{3+}-rich and possibly approaching an essenite composition. Further from the contact the glass becomes almost colourless with the clinopyroxene occurring as randomly orientated needles or radiating aggregates and prisms associated with subordinate sanidine. Near the porcelanite a mullite-cordierite rock is developed that contains relic of quartz and feldspar. The porcelanite itself is reddish-brown, intensely indurated and brittle exhibiting columnar joining (ca. 1 cm diameter) and petrographically shows little obvious mineral transformation.

Disco

Unusual buchitic shale xenoliths up to 20 cm in size and containing native iron and graphite, occur in basalt that forms the lower 6 m of a 40–50 m thick sequence of andesite lava and breccia at Asuk in northern Disco, central west Greenland (Pederson 1978, 1979). The xenoliths were derived from carbonaceous shales and sandstones intruded by tholeiitic basalt and show various degrees of melting and reaction with the magma. The andesitic part of the sequence also contains shale xenoliths that are mostly strongly modified by reaction with the magma to form plagioclase ± cordierite-Mg-spinel-corundum-graphite rocks.

One of the buchite shales largely consists of grains of resorbed clastic quartz, spongy feldspar in an advanced stage of fusion, and zircon in a fine-grained partially melted matrix. The glassy matrix contains flakes of graphite and diffuse bands of troilite and native iron that parallel original sedimentary bedding. Troilite often mantles a core of iron (rarely armalcolite), and grains of iron may contain exsolved cohenite ($[Fe,Ni,Co]_3C$). Aggregates of rutile and Al-armalcolite have replaced clastic ilmenite, rutile and fine granulated TiO_2-pseudomorphs.

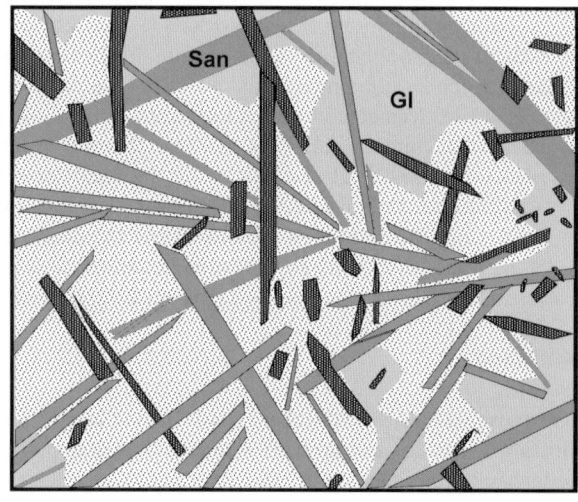

Fig. 3.15.
Drawing from microphotograph (× 43 magnification) of a clinopyroxene-sanidine buchite xenolith in olivine basalt, Auckland volcanic field, New Zealand (after Fig. 8 of Searle 1962). *Dark stippled crystals* = clinopyroxene; *light stippled area* = felted mass of sanidine

Cracks in the shale are occupied by colourless glass that contains cordierite, low Ca-clinopyroxene, plagioclase, ilmenite, rutile and extremely rare armalcolite. Metal grains consist of troilite and iron that is partly mantled by sulphide that contains minor exsolved cohenite and rare schreibersite ($[Ni,Fe]_3P$). Patches within the shale composed of titanite, Ti-diopside, anorthite, and troilite are inferred to have formed by decarbonisation of carbonate-rich material.

The pyrometamorphic assemblages emphasise the role of carbon reduction in carbonaceous sediments when heated to basaltic temperatures of 1150–1200 °C at low pressure. Oxygen fugacities were variable, ranging from slightly above 10^{-13} (at 1180 °C) for melt assemblages in cracks to well below $10^{-13.5}$ (at 1180 °C) for the formation of rutile and Al-armalcolite that result from the reaction of clastic Fe-Ti oxides in the graphite-bearing interior of the shale xenoliths.

Sithean Sluaigh

Pyrometamorphism of greenschist facies pelitic, semipelitic and psammitic phyllites by a dolerite plug at Sithean Sluaigh, Argyllshire, is described in detail by (Smith 1965, 1969). Texture and composition has been significant during their thermal metamorphism, the most important being the relative amounts and arrangement of phyllosilicates relative to quartz and albite (both detrital and neometamorphic). Additional minerals present in the unaltered phyllites include biotite, garnet, epidote. rare microcline, carbonates (calcite, siderite, dolomite), apatite, tourmaline, zircon and rutile. Both massive layers and interlayered (semi-schist) quartz-plagioclase and muscovite-chlorite-rich rocks occur with every gradation in between. Petrographic evidence of mineral breakdown with increasing temperature and coarsening of neometamorphic mineral grain size towards the contact is described by Smith (1969) in several sections at varying distances and localities from the dolerite contact. The mineral replacements observed are summarised below:

- Phengitic muscovite initially reacts to a meshwork of fine mullite/sillimanite needles and glass and at higher temperature to spinel, corundum, cordierite and sanidine.
- Chlorite reacts to spinel, orthopyroxene, cordierite, sometimes pseudobrookite, with additional sanidine in chlorite-muscovite intergrowths.
- Garnet breaks down to orthopyroxene, cordierite and spinels.
- Diopsidic pyroxene forms by reaction between dolomite and quartz.
- Melting of quartz-albite-rich layers produces granophyre that contains quartz paramorphs after tridymite and hopper crystals of sanidine that may act as nuclei for the development of spherulites where glass is still present.
- Near the dolerite contact, quartz segregations in the phyllite may be largely converted to cordierite by metasomatic introduction of Al, Fe, Mg (and alkalis) from adjacent micaceous bands.
- Ilmenite is replaced by pseudobrookite.

In one area, emery-like rocks have formed within 3 m of the dolerite contact as a result of melting and desilicification at near magmatic temperature to produce the following assemblages:

- mullite-spinel-corundum
- corundum-magnetite-spinel-pseudobrookite-sanidine (Fig. 3.16)
- mullite-spinel-pseudobrookite-(ilmenite-hematite)-glass
- spinel-calcic plagioclase-(corundum)
- cordierite-spinel-magnetite-pseudobrookite (black glassy vein)
- cordierite-spinel-mullite-ilmenite-hematite-devitrified glass

Bulk compositions of emery rocks are plotted in Fig. 3.17 in terms of $(Si,Ti)O_2$-$(Al,Fe^{3+})_2O_3$-$(Fe,Mg,Mn)O$ and show a good correlation with the predicted high temperature minerals, taking into account the variability of rock oxidation ratios. With

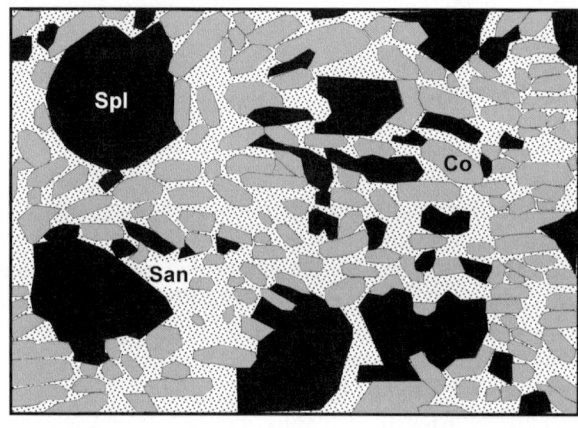

Fig. 3.16.
Drawing from a microphotograph (× 45 magnification) of a corundum-spinel-magnetite-sanidine rock (emery), Sithean Sluaigh. Argyllshire, Scotland (after Fig. A, Plate 6 of Smith 1969). Note the orientation of corundum that probably reflects orientation of original muscovite in the phyllite protolith

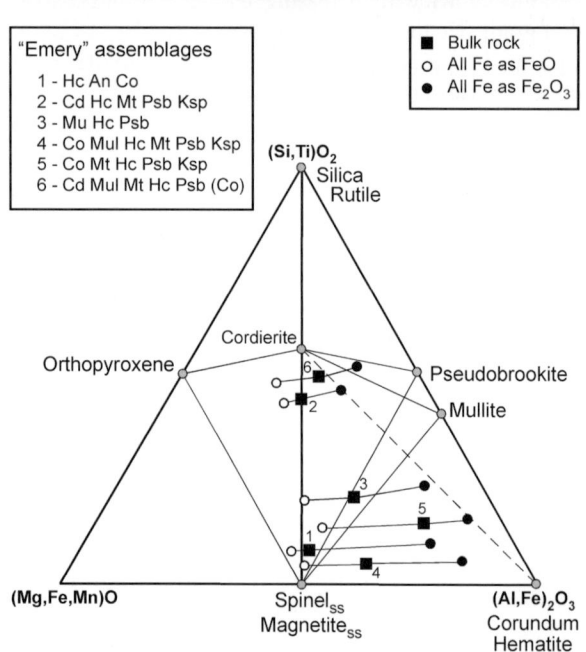

Fig. 3.17.
Emery rocks and mineral components from the contact aureole at Sithean Sluaigh, Argyllshire, Scotland, plotted in terms of mol% $(Si,Ti)O_2$-$(Al,Fe)_2O_3$-$(Fe,Mg,Mn)O$ (after Fig. 8 of Smith 1965)

decreasing temperature away from the contact, wt.% Fe_2O_3 in corundum and the Fe^{3+}/Fe^{2+} ratio of spinel increase with increasing rock oxidation ratio (Fig. 3.18). A decrease in wt.% Fe_2O_3 in corundum 2.7 m away from the igneous contact may indicate that the temperature was too low for oxidation reactions to occur or possibly due to preferential partitioning of Fe^{3+} into coexisting pseudobrookite in this rock.

The emery-like rocks represent highly desilicated Al, Fe, Mg residues of melting and extraction of rheomorphic granophyre as veins from pelitic phyllites. Bands and lenses of granophyre are also typically highly contorted indicating that the rocks were partly molten and plastic. The restricted occurrence of the emeries within the contact aureole could be due to high modal phyllosilicate contents of the emery protolith that resulted in an increased amount of H_2O being available on dehydration to enhance heat conduction.

Bushveld

Symmetrically layered highly aluminous corundum-sillimanite-bearing xenoliths (66.3 wt% Al_2O_3; 29.6 wt.% SiO_2) occur in anorthosite of the Critical Zone of the Bushveld Intrusion, South Africa (Willemse and Viljoen 1970, with additional data by

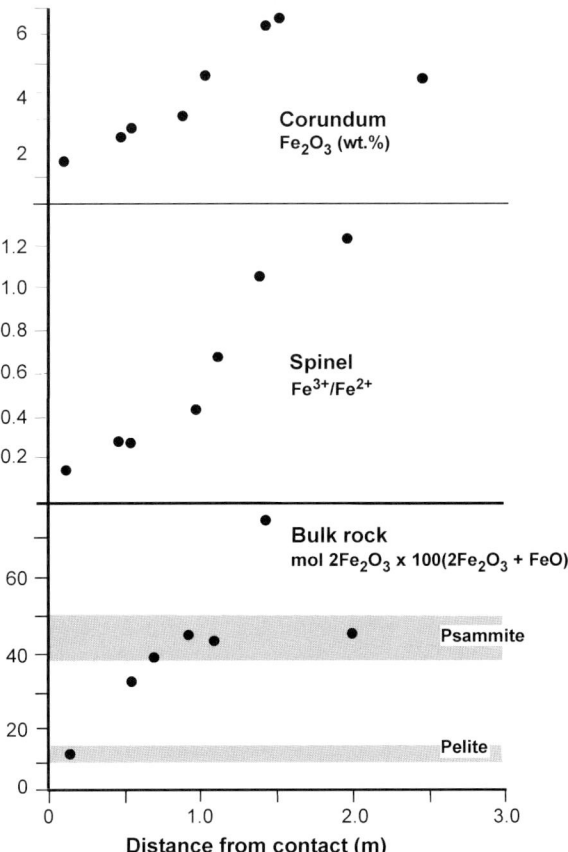

Fig. 3.18. Plots of bulk rock (emery) oxidation ratio, spinel Fe^{3+}/Fe^{2+} and wt.% Fe_2O_3 in corundum versus distance from dolerite contact at Sithean Sluaigh, Argyllshire, Scotland. *Horizontal shaded strips* are the range of rock oxidation ratios in psammitic and pelitic aureole rocks (data from Smith 1965, 1969)

Cameron (1976). One example (at Thorncliff) is ~0.3 m thick and zoned inwards from the anorthosite contact as follows:

- *Spinel zone.* Pleonaste spinel-plagioclase (An_{90-55})-sillimanite-sapphirine.
- *Plagioclase-sillimanite-mullite zone.* Plagioclase (An_{90-95} and An_{75-60})-sillimanite-mullite.
- *Corundum zone.* Corundum with matrix sillimanite-mullite and plagioclase.
- *Main sillimanite-mullite zone.* Sillimanite-mullite matrix with corundum and accessory rutile and sapphirine.

This example is one of the few recorded occurrences of sapphirine in low pressure metamorphic rocks. In the system $MgO-Al_2O_3-SiO_2$, a small sapphirine field coexists with spinel, cordierite and mullite at high temperatures between 1460 °C and 1453 °C (Fig. 3.4) and in the hydrous system, the reaction

cordierite + corundum = sapphirine + mullite

occurs between ~1270 °C (atmospheric pressure) and 1255 °C (2 kb) (Seifert 1974). Addition of iron and K_2O in the corundum-bearing, quartz-absent KFMASH system produces a sapphirine-K-feldspar assemblage at 865 °C with the reaction curve converging to an invariant point ("I" in Fig. 3.19) at ~900 °C/1.6 kb were the reaction

Bt + Opx + Cd + Co = Sa + Sil + L

occurs (Kelsey et al. 2005). This suggests that corundum in the main sillimanite-mullite zone of the Bushveld xenolith could be relic due to the reaction Sil + Co = Mul, as also implied by the bulk rock composition that lies within the sillimanite-mullite-sapphirine triangle shown in Fig. 3.19. Cordierite does not occur with sapphirine, but is found in a "mottled" zone associated with spinel that is developed along the contact with anorthosite and where sapphirine may be replaced by spinel. In another xenolith, cordierite occurs with sillimanite, mullite, spinel ± minor corundum without sapphirine (Fig. 3.19). Elongated quartz nodules 3 cm or more in length in the spinel zone extend into and deform the corundum zone and may have formed from silica produced by the reaction Sill = Mul + SiO_2 (Fig. 3.19). Assuming that the formation of sapphirine occurred via the Cd Co = Spr Mul reaction, minimum P-T conditions of 2.9 kb and 1250 °C are indicated from intersection of the two reaction curves (Fig. 3.19). The temperature of the Bushveld magmas is estimated at between 1160–1300 °C by Cawthorn and Walraven (1998).

The marginal zonation of the xenoliths is best explained by diffusive interaction with the Bushveld magma that involved diffusion of Ca, Mg, Fe, Si and Na into the xenoliths. For example, the outer *spinel zone* of the Thorncliffe xenolith is similar to that developed around mullite xenoliths from Mull (see Fig. 3.14) except for the presence of sapphirine that may have formed by the reaction

$2CaMgSi_2O_6 + 4Al_2SiO_5 = 2CaAl_2Si_2O_8 + MgAl_2O_4 + 0.5Mg_2Al_4Si_2O_{10} + 3SiO_2 + 0.5O_2$
clinopyroxene   sillimanite   anorthite   spinel   sapphirine   quartz

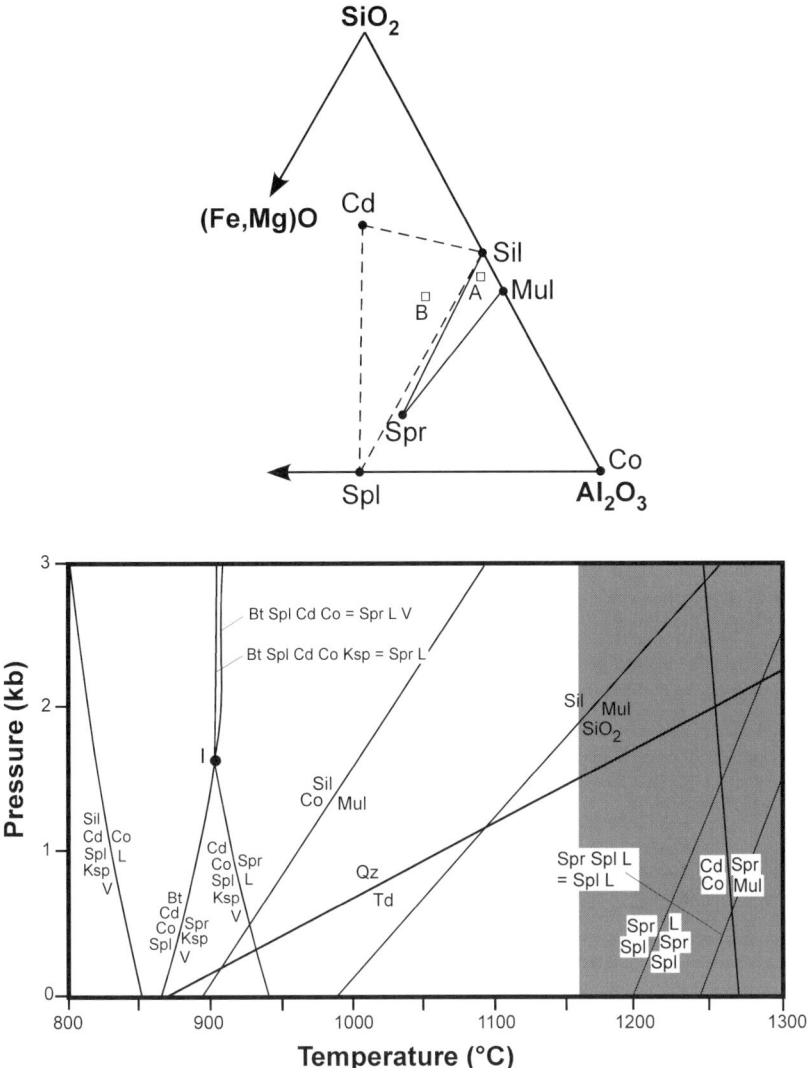

Fig. 3.19. *Above:* Compatibility relations in terms of mol% (Fe,Mg)O – Al_2O_3 – SiO_2 (total iron as FeO) of mineral phases and Al-rich xenolith compositions in the marginal part of the Bushveld Intrusion, South Africa (modified after Fig. 3 of Cameron 1976a). *A* = sillimanite-mullite-sapphirine-corundum assemblage; *B* = cordierite-sillimanite-spinel assemblage with minor mullite and corundum. In both assemblages corundum may be a relic phase. *Below:* P-T diagram of mullite, sillimanite, cordierite, corundum, spinel, sapphirine stabilities relevant to assemblages in Al-xenoliths of the Bushveld Intrusion. Lower temperature equilibria involving Ksp are calculated for sapphirine-bearing, silica-undersaturated metapelitic compositions in the KFMASH system (Kelsey et al. 2005). Higher temperature reactions involving Co, Sil, Mul, Spr are the same as given in Fig. 3.7. High temperature sapphirine breakdown reaction curves were calculated using the average Bushveld sapphirine composition analysed by Cameron (Table 3; 1976a) using the Teriak/Domino software of de Capitani and Brown (1987). *Dark grey shaded area* = temperature range of Bushveld magmas

where Ca, Mg and Si components are represented by a clinopyroxene-saturated Bushveld magma and sillimanite is relic. In the *plagioclase-sillimanite* and *corundum zones*, the presence of more sodic plagioclase, An_{60-75} in the former and An_{25-40} in the later respectively, and zonation from anorthite cores containing sillimanite inclusions (?original restite association) to more sodic, sillimanite-free rims, reflects the relatively greater diffusion distances of Na and Si into the xenolith. The abundance of corundum (~50 vol.%) in a predominantly sillimanite matrix in the *corundum zone* suggests that these phases may represent the restite assemblage from reaction of an original muscovite-rich layer in the xenolith.

Skaergaard

Metapelitic country rocks, most likely Precambrian basement pelitic schist/gneiss with > 20 wt.% Al_2O_3 (Kays et al. 1981, 1989), have been pyrometamorphosed at ~650 bar by basaltic magma of the Skaergaard intrusion, Greenland. The xenoliths occur within the marginal part of the intrusion and range from small rounded examples a few cm to several decimeters in size to larger ones that are metres to tens of metres in size. Examples from four localities have been studied in detail by Markl (2005). The xenoliths are either layered between blue to purplish-blue cordierite-rich layers to darker nearly black spinel-rich layers or zoned examples (smaller xenoliths) with cores of corundum, mullite, sillimanite, spinel and rutile within a plagioclase matrix. Associated plagioclase-rich schlieren and stringers in the layered xenoliths are considered to respresent anatectic melts. Textures of the xenoliths are extremely complex because of variable grain size, the presence of at least two generations of high temperature mineral assemblages, and diffusive interaction with ferro-basaltic magma and possibly also a granophyric melt.

In the layered xenoliths, cordierite and spinel are abundant and in places constitute > 80% of the rock. Texturally, two generations of cordierite and spinel occur. Early cordierite (up to 5 mm) contains needles of mullite and is in apparent equilibrium with spinel, plagioclase (An_{90-30}), ilmenite, ± corundum, ± rutile. Later cordierite forms small grains in fine-grained intergrowths with spinel, plagioclase, and minor ilmenite, K-feldspar and tridymite. Relics of early corundum and mullite remain as strongly resorbed crystals in plagioclase.

The zoned xenoliths are characterised by cores that consist of euhedral (up to 5 mm) crystals of corundum, euhedral rutile, abundant mullite and sillimanite needles/blocky crystals that are sometimes in felted intergrowth with a silica phase ± spinel. Large crystals of zoned plagioclase (An_{90-50}) form the matrix of all these phases (Fig. 3.20). At the contact with gabbro, the above assemblage has been modified (Fig. 3.21). A 5 mm–1 cm thick plagioclase (An_{90-60}) rim is developed around the xenoliths that changes inward to a fine-grained intergrowth of spinel and plagioclase (An_{30-15}) (~5 mm thick) where spinel replaces partly resorbed mullite of the early assemblage. Some of the corundum and rutile are also partly replaced by spinel. Thin seams of plagioclase, ± K-feldspar, ± ?tridymite surrounding corundum, corundum/spinel and spinel aggregates may represent crystallised melt. Only minor cordierite is present and is associated with fine-grained later plagioclase.

3.2 · Contact Aureoles and Xenoliths   71

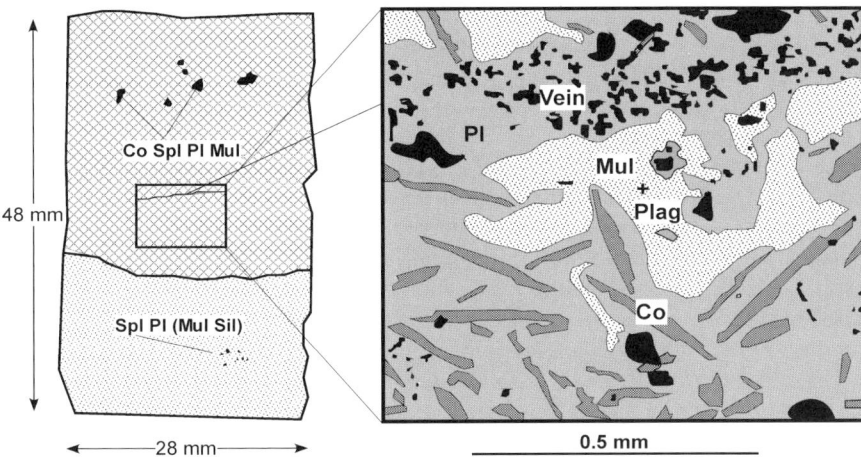

Fig. 3.20. Drawing from a thin section photomicrograph of a zoned aluminous xenolith from the Skaergaard Intrusion, Greenland, together with a diagram showing textural relationships in one area of the thin section (after Fig. 3A of Markl 2005). Spinel grain sizes are indicated in the thin section. The spinel-plagioclase vein is interpreted to be a late stage melt that entered the xenolith along a crack. See text

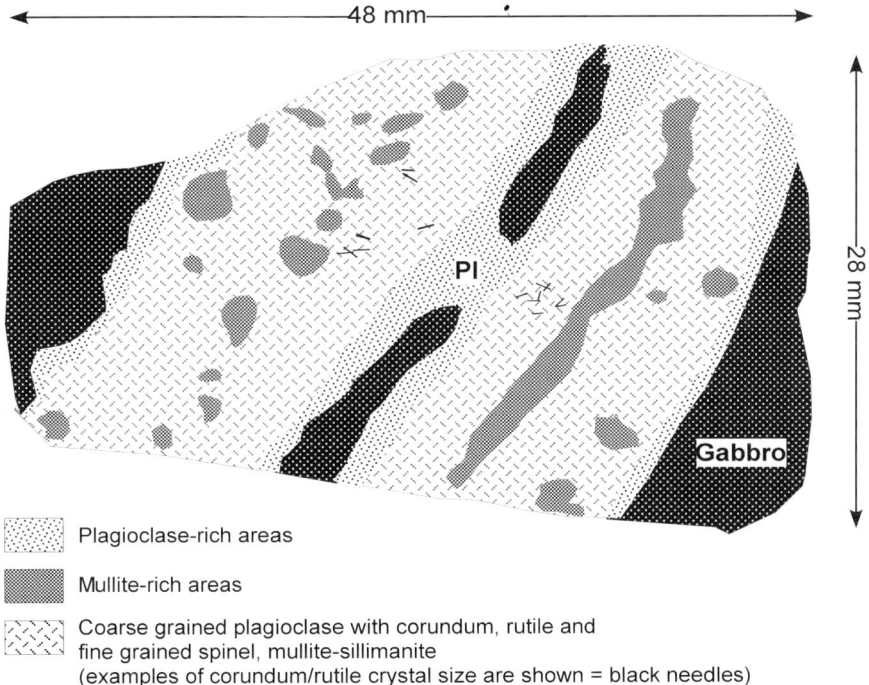

Plagioclase-rich areas

Mullite-rich areas

Coarse grained plagioclase with corundum, rutile and fine grained spinel, mullite-sillimanite
(examples of corundum/rutile crystal size are shown = black needles)

Fig. 3.21. Drawing from a thin section photomicrograph of a banded aluminous xenolith from the Skaergaard Intrusion, Greenland (after Fig. 5B of Markl 2005). See text

The mineral assemblages in the Skaergaard xenoliths are similar to those developed in emeries (e.g. Slithean Sluaigh, Bushveld) that involve extreme desilicification (removal of granitic melt) and diffusive exchange with magma. In the system $FeO-Al_2O_3-SiO_2$, assemblages of mullite-cordierite-hercynite and mullite-corundum-hercynite crystallise at 1205 °C and 1380 °C respectively (Fig. 3.4; Table 3.1) and it would need prohibitively high temperatures, i.e. higher than that of the Skaergaard magma temperatures of between 1170–1200 °C at atmospheric pressure (Hoover 1977), to melt such mineral assemblages as proposed by Markl (2005). Although euhedral and occasional skeletal habit of corundum and early spinel, mullite needles in plagioclase, and optically continuous plagioclase crystals are cited as evidence of crystallisation from a melt, such features are also typical of restite phases (mullite-sillimanite, corundum, spinels, rutile) formed from melt-producing breakdown reactions involving biotite, cordierite, sillimanite and garnet that occur in the metapelitic rocks exposed along the western contact of the Skaergaard Intrusion. The restite phases may be enclosed by subsequent growth of plagioclase from associated peraluminous melt compositionally modified by diffusion (e.g. introduction of Ca) from the basaltic magma, with restite crystal coarsening occurring during high temperature annealing. The plagioclase-rimmed parts of the xenolith in contact with gabbro shown in Fig. 3.21, and probably the lower part of the xenolith shown in Fig. 3.20, clearly indicate the effects of diffusive interaction with the Skaergaard magma.

Ngauruhoe

Buchitic xenoliths, a few centimeters to several decimeters diameter described by Steiner (1958) and Graham et al. (1988) occur in andesite flows and glowing avalanche deposits erupted from Mount Ngauruhoe, New Zealand, in 1954. The xenoliths were derived from schistose greywacke sandstone and argillite basement rocks underlying the central North Island volcanic area. Contacts between xenoliths and andesite host are sharp, although often fragmented, and there is no microscopic evidence of a reaction zone. Most xenoliths are characterised by fine (mm-scale) layering that reflects original quartz-plagioclase-rich and quartz-poor (mica-rich) compositions. These layers are either parallel, contorted and/or discontinuous over a few centimeters, suggestive of viscous flow in a partially molten state.

The xenoliths contain up to 80% glass with quartz, zircon and apatite remaining as unmelted relics. Newly-formed phases are dominated by cordierite (mainly in quartz-poor layers), orthopyroxene (mainly in quartz-rich layers), and spinel (usually associated with cordierite), which are associated with lesser amounts of Mg-rich ilmenite, rutile, pyrrhotite and rare V-Cr-Ti spinel. Quartz-carbonate veins in the xenoliths are converted to quartz, wollastonite and calcic plagioclase. Phase relations from disequilibrium melting experiments of the basement greywacke (Kifle 1992) suggests a minimum temperature of 750 °C and maximum PH_2O of 1.3 kb for fusion of the xenoliths determined by positions of the biotite-out and cordierite-out curves in Fig. 3.22.

Traigh Bhàn na Sgùrra

Pyrometamorphic changes in garnet-grade pelitic and psammitic gneiss occur within 300 cm of the basal contact of a 6 m thick gabbroic sill at Traigh Bhàn na Sgùrra on

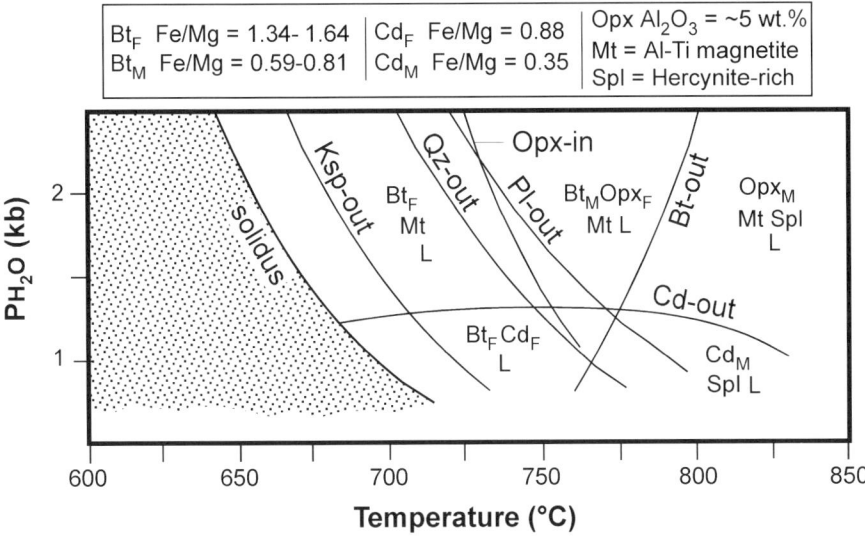

Fig. 3.22. Phase relations of disequilibrium melting of a greywacke composition, Taupo Volcanic Zone, New Zealand (after Fig. 7 of Graham et al. 1988)

the Ross of Mull, Scotland (Holness and Watt 2002). Mineralogical changes toward the gabbro contact are summarised as follows:

- *Psammite*
  - *300 cm*. Muscovite is reacted along cleavage planes and at elongate grain ends contacting K-feldspar to a submicroscopic intergrowth of biotite, sanidine, spinel ± corundum ± mullite. The presence of clear feldspar along some quartz-feldspar contacts indicates the beginning of melting.
  - *263 cm*. Biotite is reacted to ilmenite (oriented plates at grain margins) and K-feldspar.
  - *173 cm*. Reacted muscovite grains in contact with quartz are surrounded by a 10 micron-thick rim of polycrystalline K-feldspar (Ab_{37-56} Or_{61-38} An_{2-6}). Development of clear, optically continuous cuspate margins on feldspar adjacent quartz and extensions along quartz-quartz boundaries indicate the formation of an initial melt.
  - *80 cm*. Sites of reacted muscovite are represented by smoothly rounded lozenges containing clusters of euhedral cordierite grains and fine radiating granophyric intergrowths rarely associated with biotite.
  - *74 cm*. Feldspars are surrounded by granophyric rims of ~40 microns thickness. The granophyric intergrowths contain polycrystalline quartz paramorphs after tridymite.
- *Pelite*
  - *120–100 cm*. Biotite is reacted to new strongly pleochroic fox-red biotite, together with K-feldspar, hercynitic spinel and magnetite.
  - *90 cm*. Muscovite is reacted to mullite and K-feldspar is surrounded by a thick granophyric rim when in contact with quartz.

- *60–70 cm.* Garnet is partly replaced by an aggregate of euhedral hercynitic spinel, pyroxene (identified only from shape as it is replaced by chlorite) and quartz, together with plagioclase ($Ab_{46}An_{50}Or_{4}$), which formed from a melt rather than as a reaction product of the garnet. The mullite reaction product of former muscovite begins to be replaced by fine grained aggregates of spinel.
- *57 cm.* A marked increase in the amount of melt (identified as fine-grained aggregates of euhedral feldspars [An_{35-40} or Or_{60-65}] in a granophyric matrix). Relic plagioclase (An_{35-55}) has a sieve-like appearance due to melting and has overgrowths of Or_{60-65}. Biotite is replaced by elongate aggregates of spinel aligned parallel to the (001) cleavage associated with ilmenite and magnetite, that are enclosed by cordierite (pinitised), plagioclase (~Ab_{56}) and K-feldspar.
- *35 cm.* Garnet is completely replaced. Relic quartz is enclosed in a matrix of euhedral to swallow tail feldspar-bearing granophyre. Mullite after muscovite is replaced by spinel.
- *10 cm.* Trails of spinel lose their coherence as a result of movement in the melt phase that forms ~70% of the rock. Minor polycrystalline aggregates of quartz remain. Inverted tridymite is rare.

A temperature profile for the lower contact of the gabbro sill is shown in Fig. 3.23 assuming a simple, two-stage thermal model (maintaining magmatic temperature of

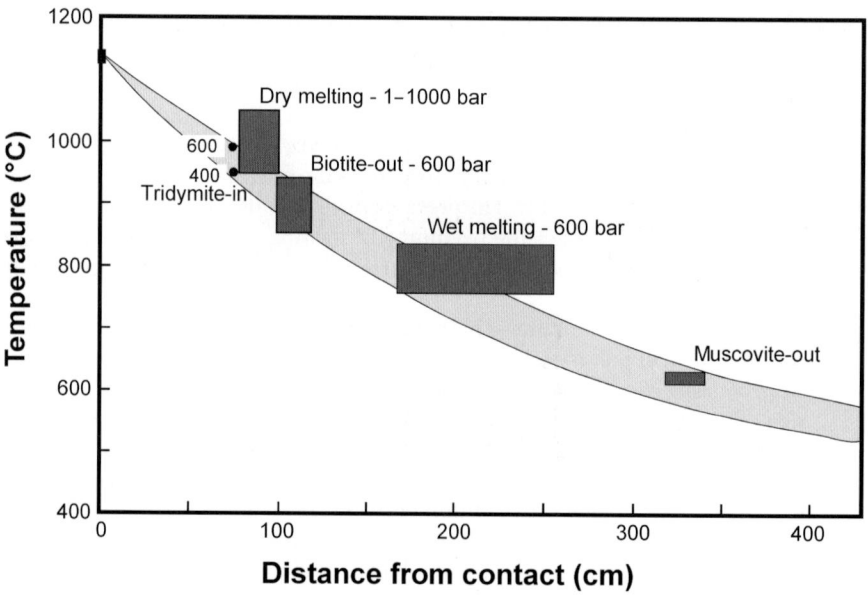

Fig. 3.23. Thermal profile (*shaded strip*) at 600 bar developed at the lower contact of the Traigh Bhán na Sgúrra gabbro sill, Isle of Mull, Scotland (after Fig. 12 of Holness and Watt 2002). See text. The higher temperature boundary of the thermal profile is for a thermal diffusivity of 10^{-6} m^2 s^{-1} and for the lower temperature boundary the thermal diffusivity is 6×10^{-7} m^2 s^{-1}. Magma temperature is estimated at 1130 °C. Bulk of the wet melting reaction occurs at between 200–220 cm from the contact. Muscovite-out reaction = metastable melting reaction. Quartz-tridymite inversion temperatures are given for 600 and 400 bar

1130 °C constant at contact of the sill during which time magma was flowing, followed by cooling of the sill and country rock) fitted to the maximum temperature profile, and suggesting that pyrometamorphism occurred over a period of about five months.

Eifel

One of the classic areas from which pyrometamorphosed xenoliths were first described by Brauns (1912a) is the east Eifel area, Germany. More recent studies by Worner et al. (1982) and Grapes (1986) concentrate on the latest trachyitic pyroclastic unit of the Wehr maar-volcano that contains a large number of xenoliths usually a few cm in size derived from the upper part of amphibolite facies basement schist (Fig. 3.24).

Typical textures of recrystallised mica-rich layers of the schist xenoliths are shown in Figs. 3.25 and 3.26 and their bulk and respective layer compositions are plotted in terms of SiO_2-Al_2O_3-$(Na,K)_2O$ and SiO_2-Al_2O_3-$(Fe,Mg)O$ in Fig. 3.27. Schist compositions plot within the field of pelitic metamorphic rocks and in comparison with desilicated aureole rocks such as at Sithean Sluaigh or aluminous xenoliths in the Skaergarrd and Bushveld intrusions, the pelitic xenoliths did not suffer extreme silica-depletion in that primary compositions of the mica-rich layers are chemically unmodified either by removal of a "granitic" melt or from diffusive interaction with surrounding magma.

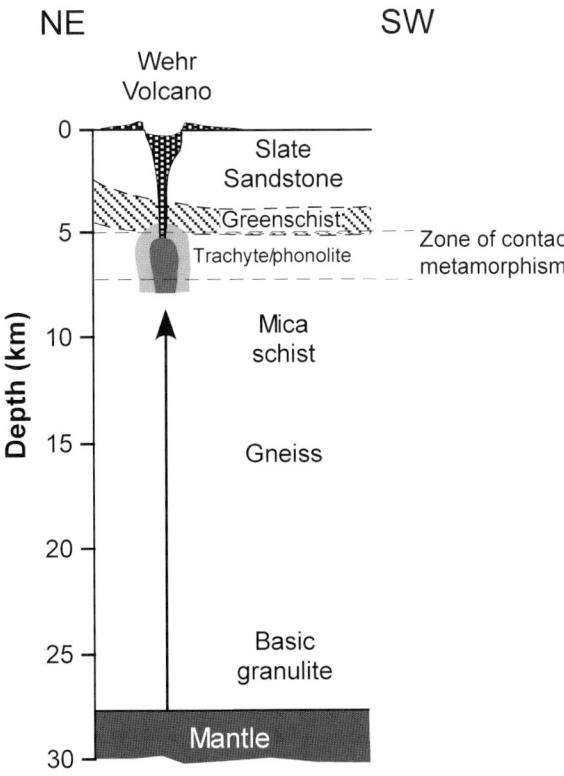

Fig. 3.24.
Hypothetical crustal section for the East Eifel area, Germany, showing the position of the Wehr phonolite-trachyte magma chamber within the upper part of basement schist indicating a minimum depth of contact metamorphism of ~5 km (modified from Fig. 13 of Wörner et al. 1982)

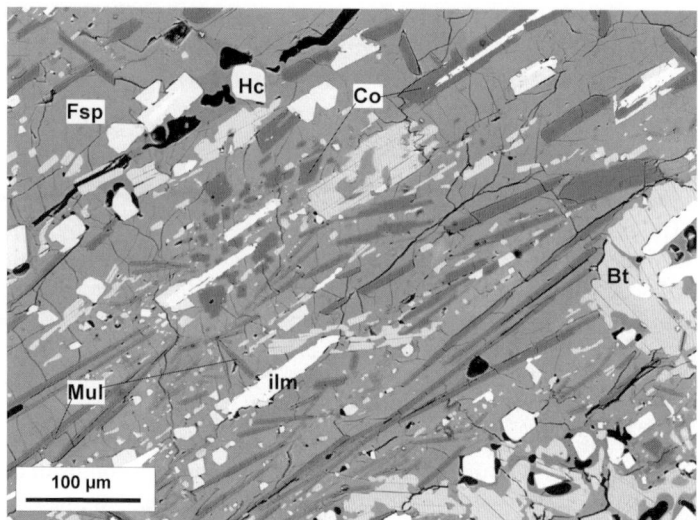

Fig. 3.25. BSE image photograph showing texture of a pyrometamorphosed mica schist xenolith, Wehr volcano, east Eifel, Germany. The preferred orientation of restite mullite and corundum crystals reflects original muscovite fabric of the schist. Biotite is a also a restite mineral and is Ti-rich (up to 6 wt.% TiO_2). The feldspar matrix (*Fsp*) is Ca-plagioclase and K-sanidine

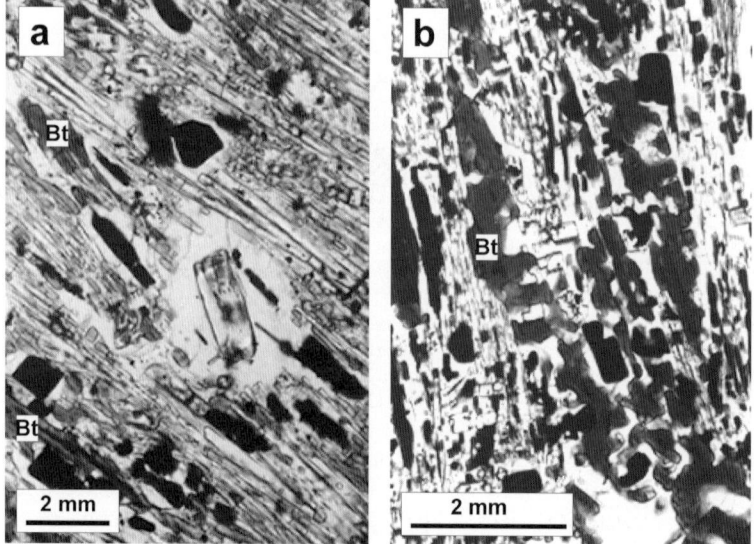

Fig. 3.26. Photomicrographs showing textural detail in reconstituted micaceous layers in pyrometamorphosed schist xenoliths, Wehr volcano. East Eifel, Germany. **a** Large corundum crystal (*centre*) separated from needles and rhombs of mullite-sillimanite within a matrix of K-sanidine. Elongate black crystals are ilmenite; the others are hercynitic spinel. **b** Newly-formed skeletal crystals of dark redbrown Ti-rich biotite associated with needles and rhombs of mullite-sillimanite, hercynitic spinel and ilmenite (*black*). The matrix is K-sanidine

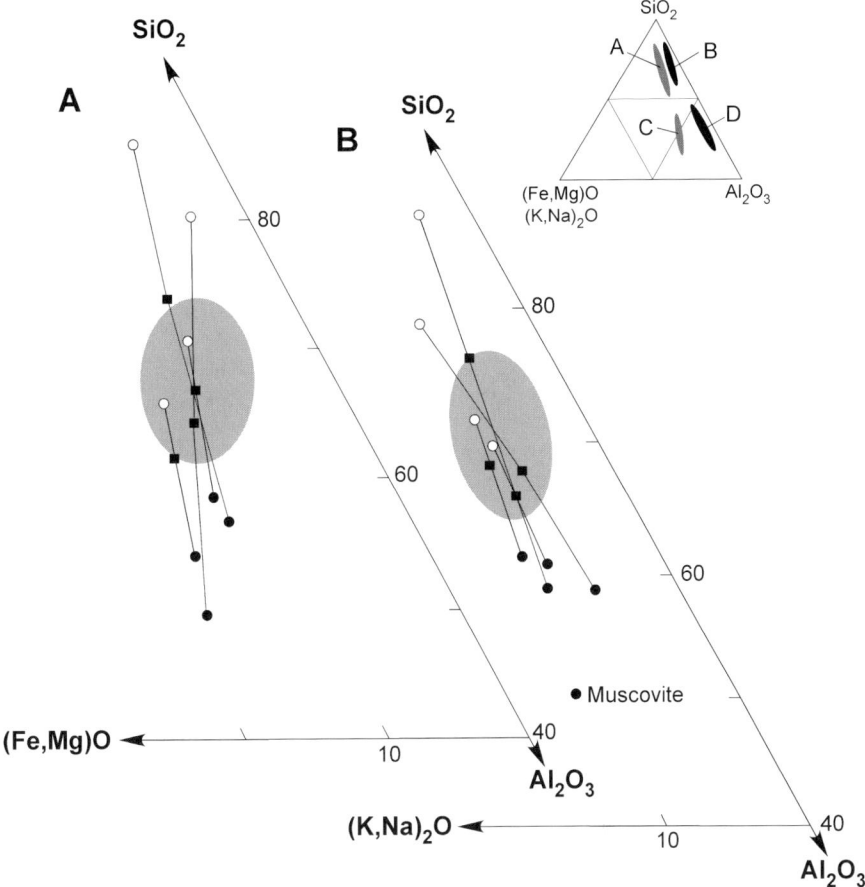

Fig. 3.27. Bulk compositions of pyrometamorphosed schist xenoliths, Wehr volcano, East Eifel, Germany, plotted in terms of wt.% $SiO_2 - Al_2O_3 - (Fe,Mg)O$ (**a**) and $SiO_2 - Al_2O_3 - (Na,K)_2O$ (**b**) and (after Fig. 3 of Grapes 1986). The *tie lines* join bulk rock (*filled squares*) to quartz-plagioclase-rich layers/buchite derivatives (*open circles*) and mica-rich layers/restite equivalents (*filled circles*). *Grey-shaded field* = field of metamorphic pelitic rocks (Mehnert 1969; Wörner et al. 1982). *Inset diagram* shows composition fields of Wehr xenoliths (*A, B*) and aluminous xenoliths and hornfels (*C, D*) after Leake and Skirrow (1960) and Smith (1965, 1969) where $A, C = SiO_2 - Al_2O_3 - (Fe,Mg)O$; $B, D = SiO_2 - Al_2O_3 - (Na,K)_2O$

Quartz-plagioclase-rich layers have undergone melting leading to the formation of siliceous buchites from which new, high temperature phases have crystallised such as β-quartz, cordierite ($XMg = 65$), Ca-rich plagioclase, with minor mullite-sillimanite, magnetite and ilmenite. In contrast, the micaceous layers were reconstituted through dehydration reactions, melting and crystallisation to a more refractory (restite) assemblage of Ca-plagioclase, sanidine ($Or_{61-83}\ Ab_{37-16}\ An_{1-2}$), Fe-rich mullite-sillimanite, Fe-rich corundum, Ti-rich biotite, pleonaste spinel, magnetite, ilmenite, with cordierite ($XMg = 69$) typically developed along contacts with largely melted quartz-plagioclase-rich layers (Fig. 3.28). Melting between quartz, oligoclase, muscovite and biotite led to

Fig. 3.28.
Microphotograph (plane polarised light) showing texture of reconstituted micaceous layer (*right*) overgrown by cordierite along the contact with an adjacent largely melted quartz-plagioclase-rich layer (*left*). The reconstituted micaceous layer consists of an assemblage of mullite-sillimanite, corundum, Ti-rich biotite, hercynitic spinel, ilmenite in a sanidine matrix (see also Figs. 3.25 and 3.36). The schistose metamorphic fabric is preserved in the preferred orientation of the restite minerals. Black areas within the glass-rich quartz-plagioclase layer are holes caused by vesiculation of the melt on eruption. Width of photo = 10 mm

the production of peraluminous melts (A/CNK = 1.2–2.1), their compositions depending on what minerals and their proportions were involved in the melting process.

Breakdown relations of plagioclase, muscovite and biotite that occur in the Eifel xenoliths and other pyrometamorphosed rocks are described in detail in Chapter 7 and characteristics of the ubiquitous spinels and ilmenite in the xenoliths are described here. Spinels of the pleonaste-magnetite series occur in all the Eifel xenoliths and have a wide compositional range within the $FeAl_2O_4$-$MgAl_2O_4$-Fe_3O_4 volume depending on whether they result from muscovite and biotite breakdown reactions or crystallise from peraluminous melt (Fig. 3.29). Because there are extensive regions of immiscibility between $FeAl_2O_4$ and Fe_2TiO_4 and $MgAl_2O_4$ and Mg_2TiO_4 at high temperature (Muan et al. 1972) and between $FeAl_2O_4$ and $MgAl_2O_4$ below ~800 °C (Turnock and Eugster 1962), the field of stable spinel compositions will be complex. Increasing fO_2 tends to stabilise the $MgFe_2O_4$ an Fe_3O_4 components with respect to $FeAl_2O_4$. In the Eifel xenoliths, pleonaste spinels have up to 28 mol% Fe_3O_4 and magnetite up to ~25 mol% $FeAl_2O_3$ that suggests a possible miscibility gap shown in Fig. 3.29.

Redox conditions in the xenoliths derived from coexisting magnetite-ilmenite lie between the Ni-NiO and Mt-Hm buffer curves and 610–1050 °C (Fig. 3.30) that presumably reflect temperatures of crystallisation and subsolidus composition adjust-

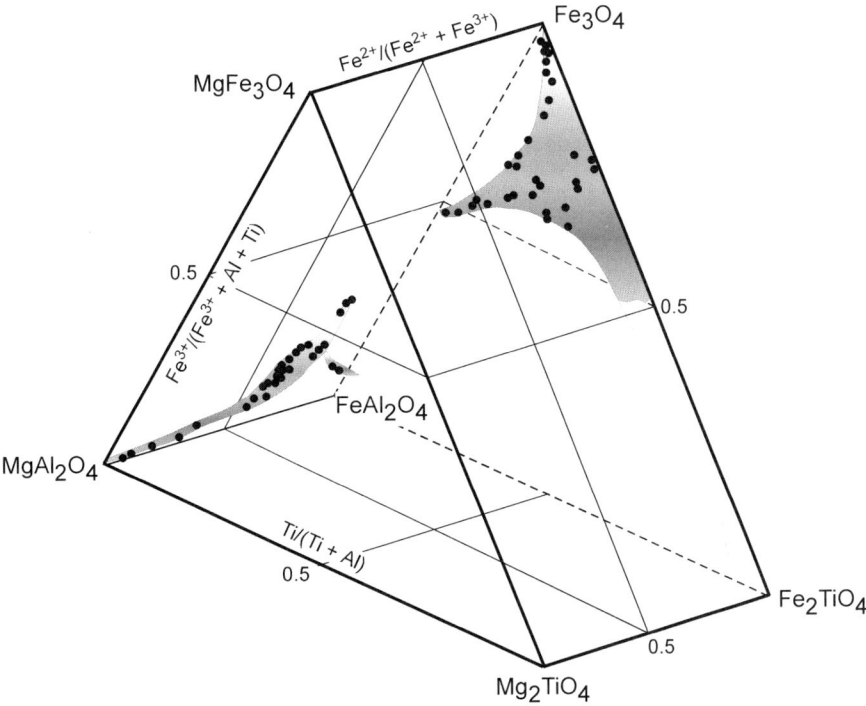

Fig. 3.29. Compositions of pleonaste spinels and magnetites in pyrometamorphosed xenoliths from the Wehr volcano, East Eifel, Germany, plotted in terms of a modified ($MgCr_2O_4$-$FeCr_2O_4$-absent) multicomponent spinel prism (after Fig. 27 of Grapes 1986). Spinel compositions from Sithean Sluaigh (Smith 1965), Mull (Thomas 1922; Buist 1961; Preston et al. 1999), and Disco (non-Cr_2O_3-bearing spinels; Pederson 1978) and also plotted

ments and/or oxidation with cooling, i.e. the heating-cooling temperature range of the contact aureole rocks and xenoliths. Highest temperatures of ~1050 °C are recorded by oxide pairs in buchitic xenoliths; lowest temperatures of ~620 °C are recorded by oxide pairs in partially melted xenoliths where compositional profiling indicates intergrain cation diffusion and oxidation that may be related to dehydroxylation of associated micas. Core and rim compositions of coexisting magnetite-ilmenite in the host trachyte give subsolidus T-log fO_2 quenching values of 781/ −13.9 and 743/ −14.6 respectively.

Glass within the xenoliths is highly vesicular (Fig. 3.31) reflecting the sensitivity of H_2O-saturated melts to a rapid decrease in confining pressure caused by failure of the cover rocks overlying the trachyte magma chamber (Fig. 3.24). When this occurred melts in the variously molten xenoliths became vesiculated by exsolution of H_2O resulting in an increase in volume causing fragmentation and disorientation of more solid, restite micaceous layers. As a result of volume change and PH_2O increase during vapour exsolution both within the xenoliths and magma, the whole mass moved upward as an entrained fluidised system to be explosively erupted.

Fig. 3.30.
T-fO_2 plot from magnetite-ilmenite thermometry, pyrometamorphosed schist xenoliths from the Wehr volcano, East Eifel, Germany (after Fig. 28 of Grapes 1986). *C* and *R* refer to core and rim values respectively of ilmenite/magnetite microphenocrysts in a trachyte selvage around one of the xenoliths. *Grey-shaded field* = acid extrusive suites (Haggerty 1967). See text

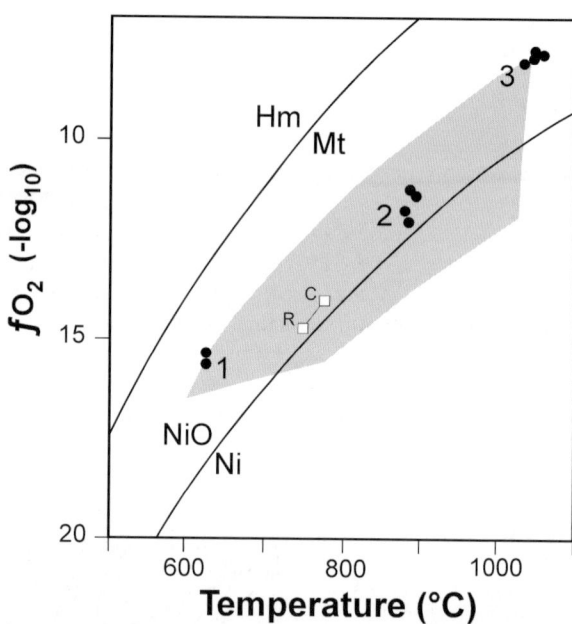

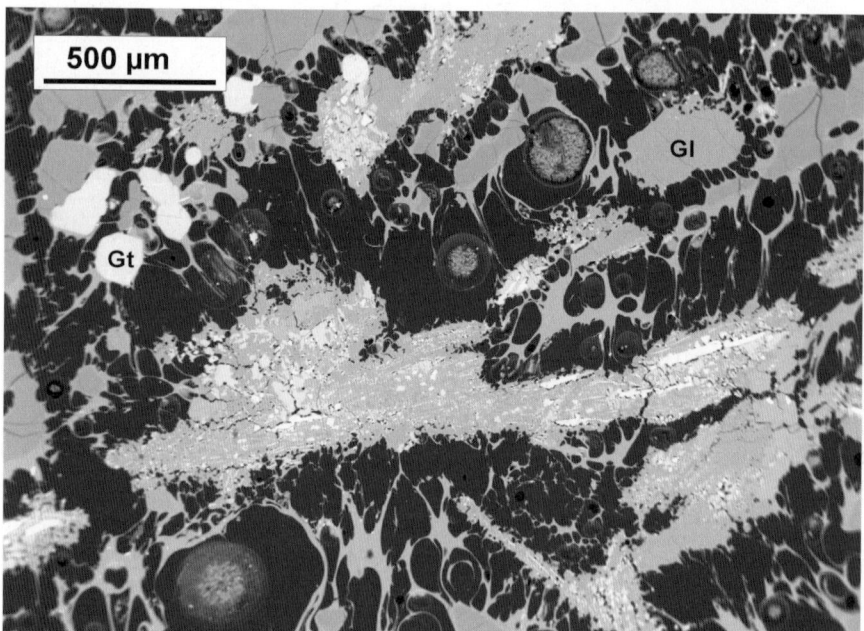

Fig. 3.31. BSE image photo showing vesicular glass in a buchitic xenolith from the Wehr volcano, East Eifel Germany. *Black areas* = holes resulting from H_2O exsolution during eruption. Note the web-like texture of the glass (*Gl*) implying expansion and consequent separation of different crystal-melt areas of the buchite

3.2.2
Sanidinite

Eifel

Sanidine-rich xenoliths of different origin were first described in volcanics of the Eifel area, Germany, by Brauns (1912a,b), and later by Kalb (1935, 1936) and Frechen (1947) who distinguish sanidine-bearing regional metamorphic rocks from sanidine-rich rocks of igneous origin. In the former, the sanidine is considered to have resulted from alkali metasomatism of country rock by phonolite/trachyite magma that has also produced aegerine-augite, nosean, nepheline, cancrinite, scapolite and hauyne-bearing sanidinites (fenites) on the one hand, and biotite sanidinite (termed *laachite*) with relic garnet ± andalusite on the other. In the biotite sanidinites, which preserve regional metamorphic fabrics and are the type example of sanidinite facies rocks, sanidine typically forms large crystals that poikilitically enclose minerals such as corundum, mullite-sillimanite, Ti-rich biotite, hercynitic spinel and ilmenite (Fig. 3.32) and an evolutionary scheme

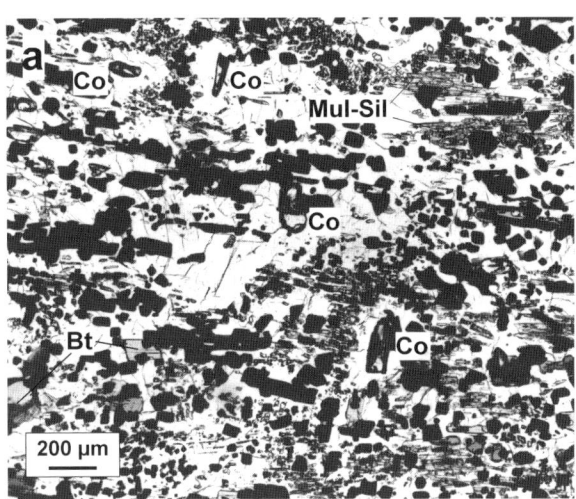

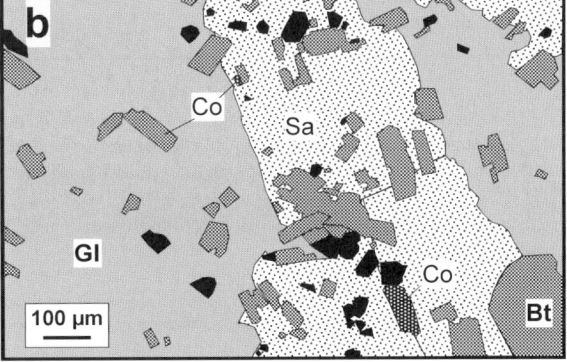

Fig. 3.32.
a Photomicrograph of a sanidinite, Wehr volcano, East Eifel, Germany (after Fig. 1B2 of Grapes 1991). *White areas =* Na-sanidine. Four crystals of corundum are indicated. Black crystals are hercynitic spinel and ilmenite (elongated crystals). Note the similarity with reconstituted micaceous layers shown in Figs. 3.25, 3.26 and 3.28. **b** Drawing from photomicrograph of a partly recrystallised biotite sanidinite xenolith, Laacher See, East Eifel, Germany (after Fig. 1A2 of Grapes 1991). Large plates of Na-sanidine are in the process of crystallising from a K-Na-rich peraluminous melt to enclose restite biotite, hercynitic spinel and minor corundum

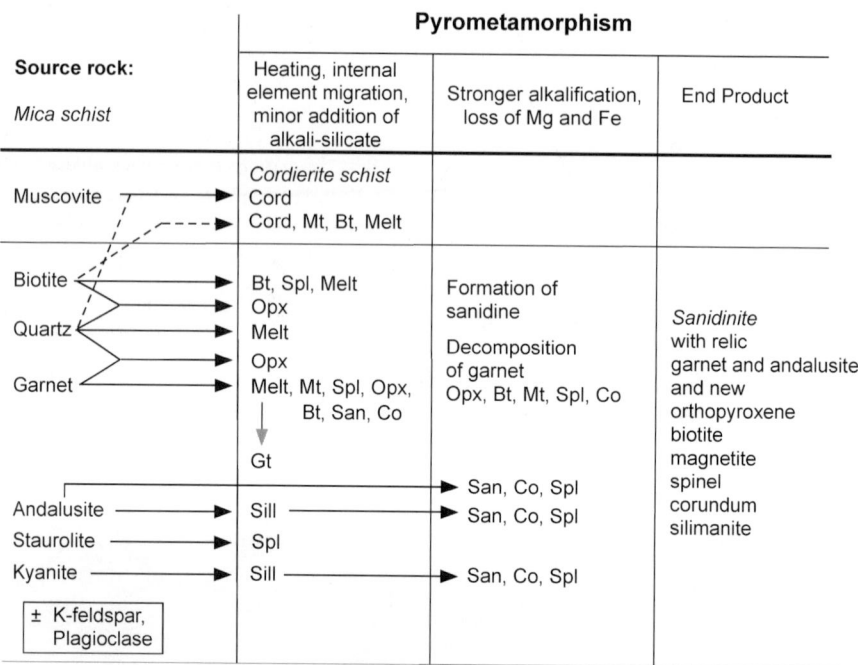

Fig. 3.33. Evolutionary flow diagram to illustrate the formation of sanidinite from mica schist during pyrometamorphism based on evidence in xenoliths from the Laacher See area, Eifel volcanic field, Germany (after Table 2 of Frechen 1947)

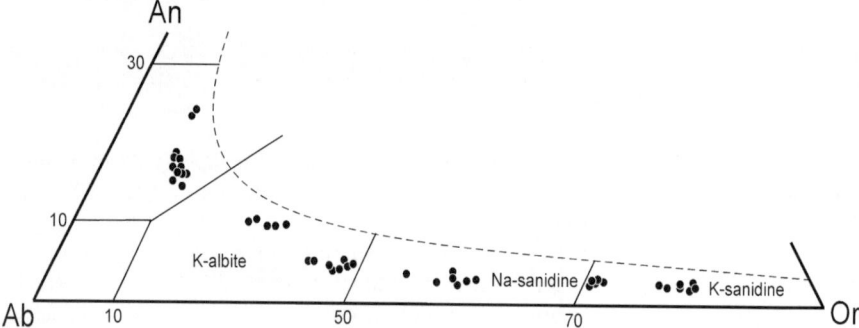

Fig. 3.34. An-Ab-Or plot showing the range of feldspar compositions in sanidinite xenoliths, East Eifel area, Germany (after Fig. 2 of Grapes 1991). The feldspars occur as intergrowths, overgrowths and replacement rims. Compositions within the K-sanidine field are from buchite xenoliths from the Wehr volcano

to illustrate the change from mica schist to sanidinite proposed by Frechen (1947) is shown in Fig. 3.33.

The "sanidine" in biotite sanidinites from the Eifel that lack relic garnet is a mixture of Na-sanidine, K-albite, ± K-oligoclase as patchy intergrowths, overgrowths/ replacement rims (Fig. 3.34). In terms of wt.% $SiO_2 - Al_2O_3 - [(K,Na)_2O + CaO]$ and

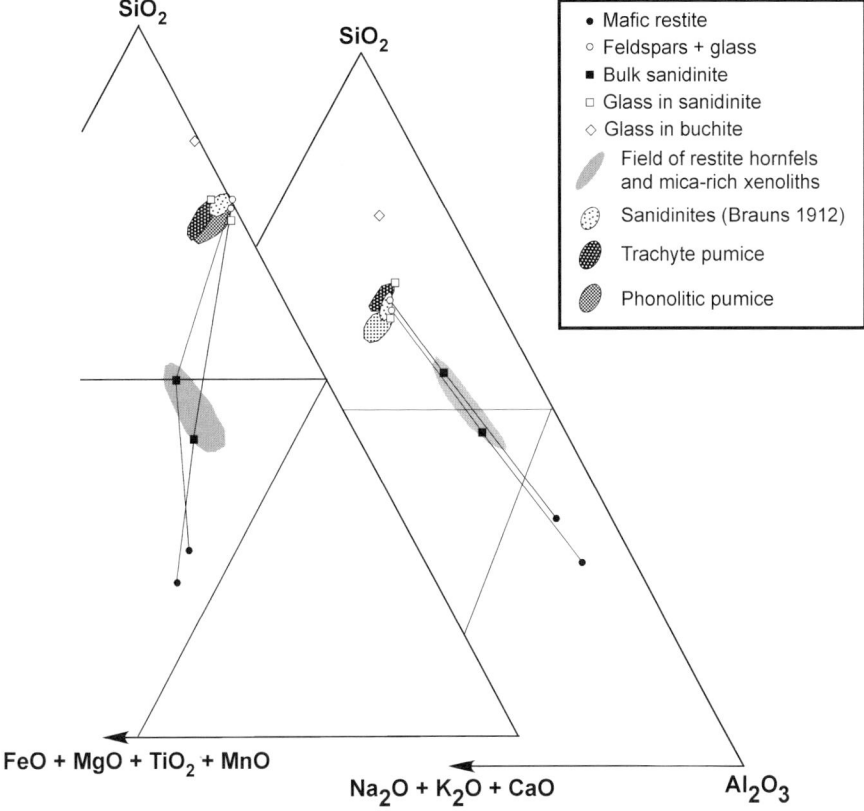

Fig. 3.35. Plots of wt.% $SiO_2 - Al_2O_3 - [(Fe,Mg,Mn)O + TiO_2]$ and $SiO_2 - Al_2O_3 - [(Na,K)_2O + CaO]$ of two sanidinite xenoliths, East Eifel volcanic area, Germany (after Fig. 3. of Grapes 1991). *Tie lines* join bulk sanidinite composition with respective mafic and felsic + glass components. See text

$SiO_2 - Al_2O_3 - (Fe,Mg)O$, bulk compositions of typical sanidinites plot within the field of mica-rich layers and restitic hornfels of basement schist underlying the Eifel area (Fig. 3.35). Feldspar + glass-rich parts of two biotite sanidinite xenoliths have compositions almost identical with those analysed by Brauns (1912b) and also plot within the composition fields of phonolitic/trachytic pumice from the Eifel area. The aluminous mafic fraction of the xenoliths that is poikilitically enclosed by feldspar consists of oriented crystals biotite, mullite-sillimanite, corundum, hercynite, ilmenite that reflect the metamorphic fabric of the schist protolith as shown in Figs. 3.26 and 3.28, and compositions are those of desilicated xenoliths and hornfels, i.e. emeries. These relationships indicate that the aluminous mafic parts of the sanidinite are the high temperature refractory end product (restite) of the melting of mica-rich layers and that the feldspar-rich "matrix" represents the peraluminous melt (buchitic stage) of this melting process (open diamonds in Fig. 3.35) modified by diffusive exchange of with phonolitic/trachytic magma involving Na, to a lesser extent K, enrichment and Si-loss concomitant with Al-enrichment of the sanidinite melt (Wörner et al. 1982;

Grapes 1986, 1991). Mg/Fe ratios of spinel/biotite pairs indicate a final equilibration temperature of about 830 °C for the two sanidinites plotted in Fig. 3.35 with a lower temperature of 750 ± 10 °C at 1 kb from plagioclase-K-feldspar thermometry (Brown and Parsons 1981) for crystallisation of latest-formed Na-sanidine–K-albite (Fig. 3.32b).

Mt. Amiata

Sanidinite xenoliths described by van Bergen and Barton (1984) occur in K-rich rhyodacite at Mount Amiata, Central Italy. During pyrometamorphism, regionally metamorhposed pelitic schist and gneiss have largely recrystallised to form sanidinite consisting of biotite, spinel, andesine-labradorite, andalusite, sillimanite, corundum, Fe-Ti oxides and graphite enclosed in Na-sanidine (Or_{86}). There is very little glass present and regional metamorphic microfoliation is preserved by oriented ilmenite and flakes of graphite. Many of the xenoliths show millimeter-scale mineralogical and textural core to rim zoning (Fig. 3.36). The sanidinite cores are enclosed by a narrow plagioclase-rich zone with subordinate orthopyroxene, biotite and spinel, surrounded by a brown-vesicular glass zone containing crystals of biotite, plagioclase and/or orthopyroxene, with a rarely-developed outermost clinopyroxene-rich rim. These changes are thought to reflect reaction between xenoliths and magma involving melt separation from the xenoliths (brown glass zone) and diffusive exchange gradients to form the plagioclase-rich and clinopyroxene zones as illustrated in Fig. 3.36.

3.2.4
Granitoids

Iceland

A large number of light coloured, dense, sometimes flinty acidic xenoliths, averaging 2–10 cm in diameter within basaltic lava and tephra occur on the island of Surtsey off the south coast of Iceland, and associated with post-glacial cinder cones (Sigurdsson 1968). Highly vesicular acidic xenoliths are also found and during the eruptions that formed Surtsey between late 1963 and mid 1967, light, frothy glassy fragments drifted ashore on its beaches.

The fine grained xenoliths have sharp contacts with the host basalt and lack any reaction phenomena. Despite evidence that the xenoliths were molten, lobes and folds intertwining basalt and xenolith suggest immiscibility between the two. Glass is rare or absent in most of the fine grained xenoliths and they have a sub-spherulitic texture largely composed of calcic plagioclase (An_{85-95}) associated with aggregates of wedge-shaped twinned tridymite, and occasional relic quartz grains in optical continuity with the tridymite. Rare orthopyroxene, cordierite and mullite are also present. Small patches of brown glass containing scarce pyroxene microlites are thought to represent areas of former biotite or possibly hornblende. Increasing fusion is accompanied by increasing vesiculation so that some xenoliths are characterised by abundant vesicular or frothy, almost transparent glass with occasional relic quartz and K-feldspar preserving a granitic texture.

3.2 · Contact Aureoles and Xenoliths 85

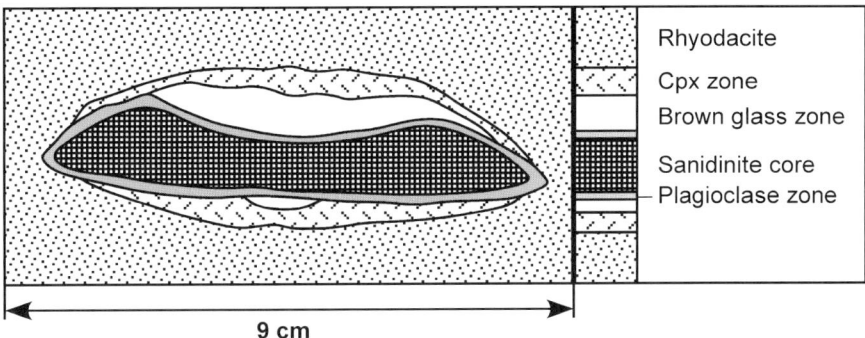

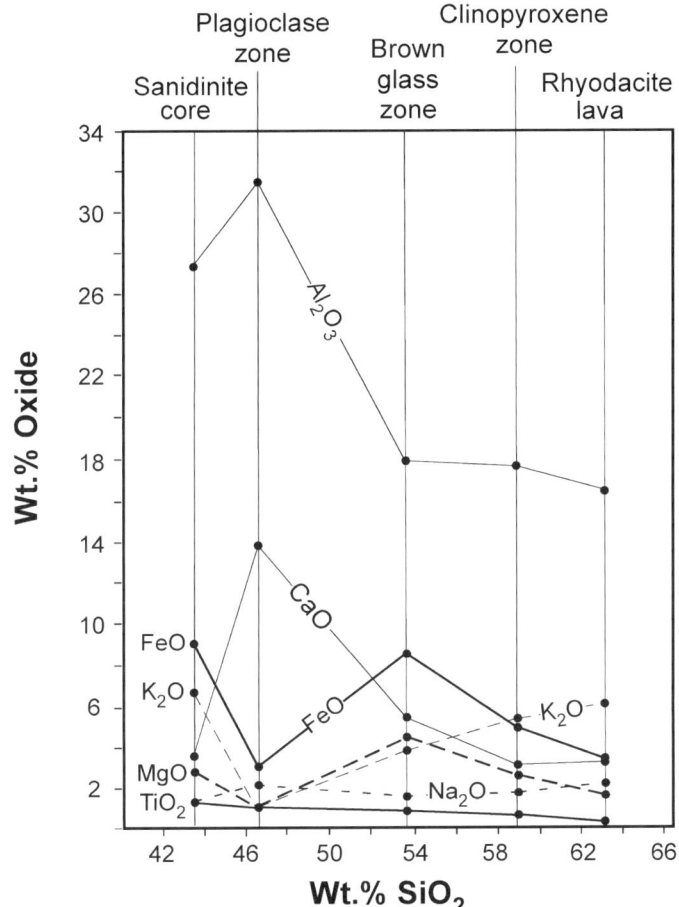

Fig. 3.36. *Above:* Section of a zoned sanidinite xenolith in siliceous lava, Mt. Amiata, central Italy (after Fig. 1 of van Bergen and Barton 1984). *Below:* Major oxide variation as a function of SiO_2 content from core to lava-envelope of the sanidinite xenolith (after Fig. 7 of van Bergen and Barton 1984). See text

Leucocratic xenoliths of 1 to 30 cm diameter also occur in rhyolite erupted in 1875 from the Askja central volcano (Sigurdsson and Sparks 1981). They range from partially fused trondhjemitic, granodiorite to dense, highly siliceous types (quartz trondhjemite) and are comparable to those from Surtsey and western Iceland described above. Primary minerals in the trondhjemite are quartz and normal zoned andesine (An_{43-37} to An_{40-32} and sometimes anorthoclase), with minor amounts of biotite. The beginning of melting is indicated by the presence of thin films of glass around quartz in granular aggregates with plagioclase. With more advanced melting the quartz becomes scalloped and laths of tridymite occur in the surrounding glass. In extensively fused xenoliths, tridymite forms the dominant phase. Rims of plagioclase become granulated with individual granules of An_{35-37} surrounded by glass to produce a sieve or characteristic fingerprint texture (see Chapter 7). The granules are optically continuous and typically preserve the original outline of the plagioclase crystal. Fine-grained aggregates of magnetite and orthopyroxene ($Wo_{1.3} En_{75} Fs_{24}$) occur after biotite (see Chapter 7) and extensively fused siliceous xenoliths contain cordierite and mullite.

Plagioclase thermometry for the rhyolite tephra that contains the xenoliths gives crystallisation temperatures of 1043 to 1091 °C under dry conditions and 990–1010 °C under PH_2O = 0.5 kb. The lower temperature for wet melting is consistent with a minimum temperature derived from intersection of the quartz-tridymite inversion and the dry Qz Ab Or melting curve at 970 °C/550 bar but would imply overstepping of the wet melting curve by ~200 °C (Fig. 3.37). A similar temperature overstepping is also indi-

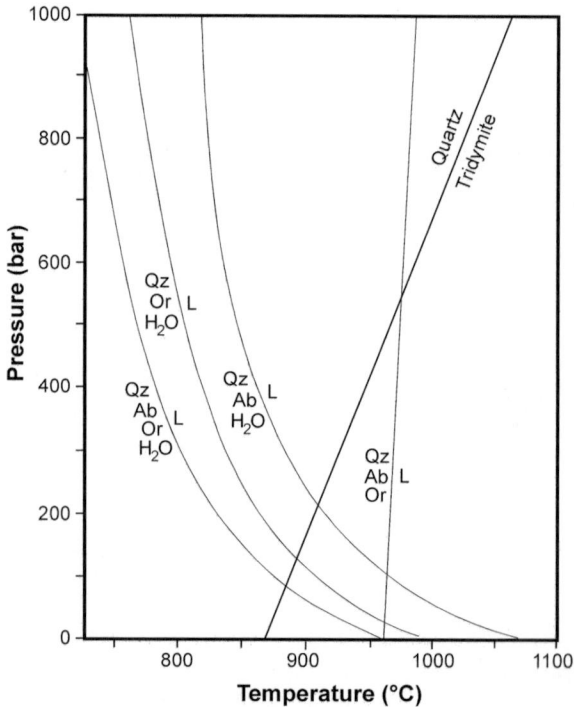

Fig. 3.37. Phase diagram for Qz Ab Or ± H_2O melting in relation to the Qz/Td inversion at pressures < 1000 bar (solidus curves after Tuttle and Bowen 1958; Shaw 1963). The Qz/Td inversion occurs at 867 °C at 1 bar and has a slope of 200 °C kb^{-1}

cated from experimentally determined biotite-out curves in granodiorite and peraluminous granite compositions (Fig. 3.38). With a plagioclase composition of An_{40} as in the partly fused Icelandic tronjhemites, melting temperatures would be ~10 °C higher than for pure albite (e.g. Fig. 6.1 of Johannes and Holz 1996).

Sierra Nevada

At *Owens Valley*, southern Sierra Nevada, California, a 61 m long by 8 m high exposure of columnar jointed granodiorite has been partially fused by olivine basalt (Knopf 1938). The granodiorite shows evidence of disintegration at the contact with the basalt which contains innumerable feldspar and quartz fragments as well as blocks of granodiorite. Feldspar xenocrysts are honeycombed with brown glass around their peripheries, along cleavage planes and as circular patches inside the crystals. Tridymite is abundant in the surrounding glass.

The granodiorite is principally composed of quartz, plagiolcase (An_{45}), orthoclase and biotite. In the partially melted rock, biotite is pseudomorphed by a mixture of magnetite, spinel and newly-formed red biotite, quartz is conspicuously resorbed and cracked although orthoclase has not been converted to sanidine. The more glass-rich parts of the granodiorite give it the appearance of a porphyry. The glass is colourless, has an inferred "silicic rhyolite" composition, and is typically developed along quartz and K-feldspar contacts. It contains crystals of orthopyroxene that are especially abundant near altered biotite. Perlitic cracking in the glass is common, especially near crystals of unreacted zircon.

Comparison with extrapolated experimental melting data for a granodiorite composition in Fig. 3.38a indicates that between 100 and 500 bar, the presence of tridymite in the more fused granodiorite implies minimum temperatures of between 870–970 °C

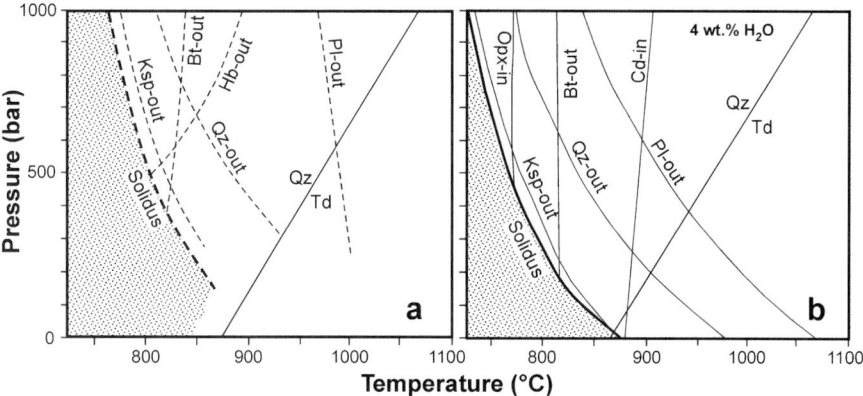

Fig. 3.38. Phase diagrams for melting of granodiorite (**a**) and peraluminous granite (**b**) compositions in relation to the Qz/Td inversion at pressures < 1000 bar. In (**a**), solidus and mineral-out curves with excess H_2O are extrapolated below 1000 bar from experimental data given in Fig. 2 of Piwinskii (1968). In (**b**), solidus, mineral-out curves and Cd-in curve with 4 wt.% H_2O are from experimental data given in Fig. 4 of Clemens and Wall (1981). The Cd-in curve is coincidental with the Opx-out curve

respectively, i.e. ca. 180 °C above the granodiorite solidus at 500 bar. At this pressure biotite stability is exceeded at ~810 °C and in the Owens Valley granodiorite it has reacted to form magnetite, orthopyroxene and new biotite (see Chapter 7).

An elliptical (~1.6 km in its longest dimension) plug of andesite containing an "egg-shaped" inclusion of granodiorite some 12 m wide and 15 m long, occurs near *Carlsbad*, San Diego County, California (Larsen and Switzer 1939). In comparison with the unmelted rock, the granodiorite, with an original mineral composition of quartz, plagioclase (An_{34-20}), perthite, brown biotite, green hornblende, with accessory magnetite, apatite and zircon, shows evidence of extensive fusion. Perthite has completely melted; biotite and hornblende are replaced by pyroxene and Fe-oxide. Quartz and plagioclase have undergone partial melting as indicated by quartz having sharp, embayed boundaries against glass. Brown glass makes up 40-50% of the rock and contains abundant, needle-like crystals of plagioclase (An_{15}) with a distinctive "swallow-tail" habit, pyroxene and very fine magnetite. The pyroxene (identified as augite) is typically concentrated around resorbed grains of quartz. The absence of tridymite implies that fusion temperatures were lower than the Owens Valley granodiorite. Complete melting of K-feldspar, reaction of biotite and hornblende, and the presence of quartz suggests fluid pressures of between 500–700 bar over a temperature range of ~810–840 °C (Fig. 3.38a).

Al-Rawi and Carmichael (1967) document fusion of biotite granite wall rock by a small Tertiary trachyandesitic plug, Sierra Nevada batholith, some 61 m in diameter that acted as a feeder to basaltic-andesitic lavas. An extensive amount of black glass is developed along the andesite-granite contact and the principal constituents of the granite, –quartz, plagioclase, K-feldspar, biotite, show progressive changes in response to heating towards the andesite (Fig. 3.39). Biotite changes from greenish-brown to reddish-brown prior to initial breakdown to fine grained Fe-oxide and then to a fine intergrowth of Ti-magnetite, orthopyroxene and Na-plagioclase that persists through the zone a melting, i.e. within 8 m of the andesite contact. The optic axial angles of

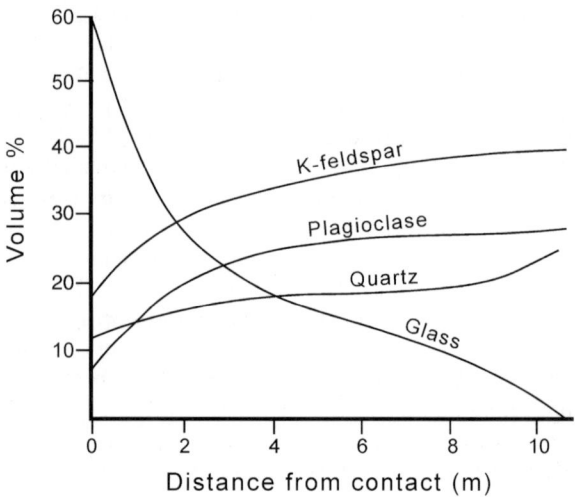

Fig. 3.39. Variation of volume % minerals and glass in a partially fused granite with distance from a trachyandesite plug, near Mono Lake, Sierra Nevada batholith, California (after Fig. 1 of Al-Rawi and Carmichael 1967)

both plagioclase and orthoclase (converts to sanidine) decrease towards the contact. Hornblende is replaced by Fe-oxide, the grain size of which coarsens as the contact is approached and where it is difficult to distinguish replaced hornblende from biotite. Titanite becomes rimed by Fe-oxides.

Melting and mobilisation of granite by another small 60×40 m diameter trachyandesite plug intruding the same Cretaceous granite as that reported by Ali-Rawi and Carmichael (1967) has also been examined by Kaczor et al. (1988) and Tommasini and Davies (1997) (Fig. 3.40). Field relations indicate a complex interplay between the intrusive and intruded rocks. The partially melted granite consists of numerous inclusions of unmelted granite (1–60 cm diameter), partially melted granite as well as trachyandesite, suggesting that wall rock fusion occurred at deeper levels. The density difference between the two rocks resulted in partially melted granite intruding the trachyandesite after cooling to a point where jointing was developed. Later, the formally mobile, partially melted granite was in turn intruded by trachyandesite.

The unmelted peraluminous granite (alumina saturation index $[Al_2O_3/(CaO + Na_2O + K_2O)] = 1.02$) is coarse grained, consisting of quartz, plagioclase, microcline, minor biotite, and magnetite, with accessory apatite, allanite and zircon. When partially melted, the granite develops a vitrophyric texture and crystal size decreases as the volume of glass increases. Relic phases are quartz, plagioclase, sanidine (recrystallised microcline), magnetite, minor ilmenite, apatite and zircon. In one sample with 25% glass, additional newly-formed cordierite, sillimanite and rutile occur. Glass is intergranular, occurring along fractures and rims inclusions within grains. Where the amount of glass is small it is typically clear, but with increased melting it becomes brown. The two types of glass shown in Fig. 3.41 are compositionally different. Overall, the less siliceous brown glass formed by a higher degree of melting contains less K_2O and more $(Ca,Mg,Fe)O$ and approaches the granite composition as expected. In detail, however, along contacts with residual minerals the composition of the brown glass is heterogeneous on a micron scale implying diffusion-controlled dissolution across the boundary layer at the solid-liquid interface. On devitrification, fan spherulites are developed in the brown glass. Glass-mineral associations and the dominance of brown glass with increased fusion, implies that the glasses represent heterogenous melts due to insufficient time for complete mixing rather than immiscible liquids. On the basis of Na, K, Rb and Ca, Sr diffusion calculations, Kaczor et al. (1988) suggest that the period of melting and quenching was between $> 1 - < 6$ months. The two types of glass in the partially melted granite contain different forms or associations of quench crystals. Rectangular or H-shaped hollow *microlites* are common in the clear glass whereas microlites together with rod-like to globular *crystallites* occur in the brown glass.

Glass first appears along quartz–K-feldspar contacts and gradually increases in amount towards the trachyandesite as shown by Al-Rawi and Carmichael (1967), becoming brownish in colour due to diffusion of salic components from biotite and progressively pervades the whole rock as veins and pools. Alkali feldspar appears to have undergone melting along cleavages and perthite lamellae interfaces that develops into a typical fingerprint texture through the whole crystal with increasing fusion. The new-high temperature K-feldspar resulting from the melting of sanidine after microcline is sodic sanidine (Fig. 3.41). Plagioclase remains unaffected until the

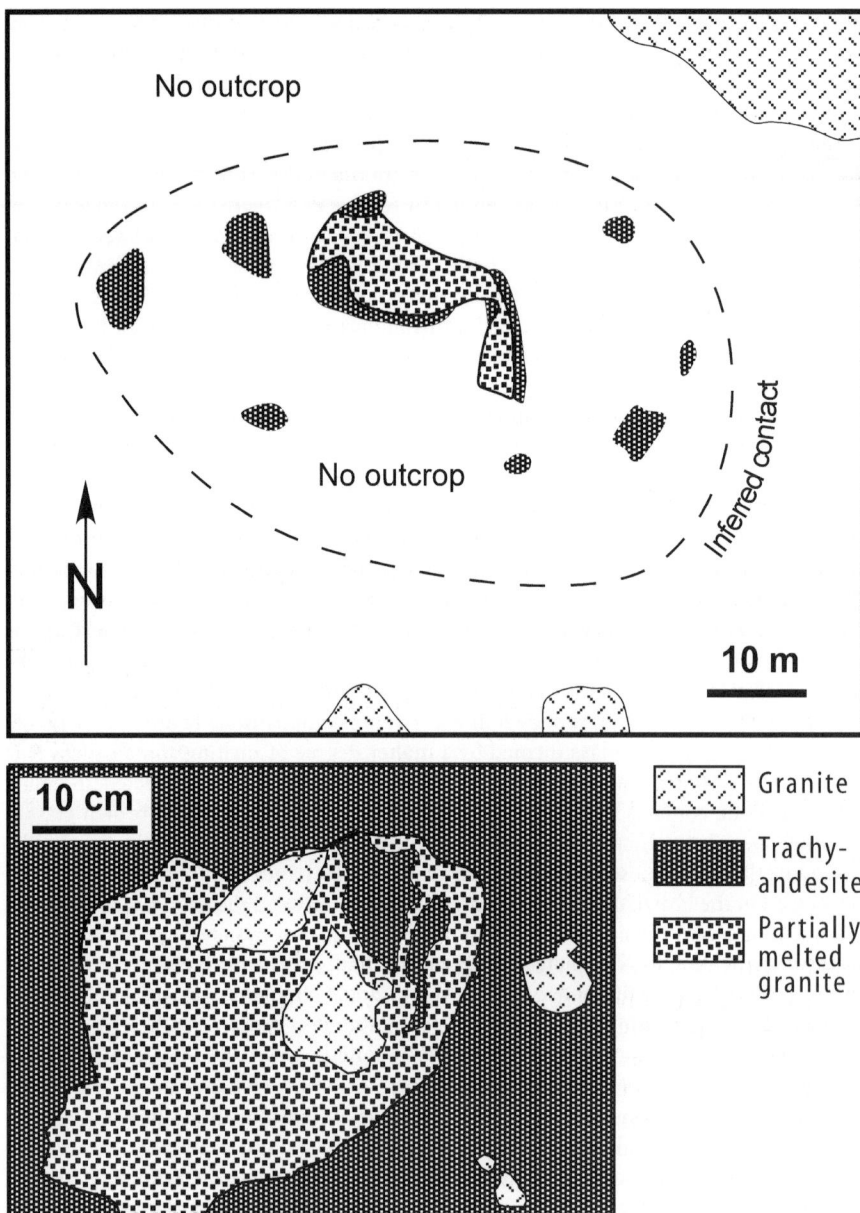

Fig. 3.40. *Above:* Geologic map showing the relationship between unmelted granite (Rattlesnake Gulch, Sierra Nevada batholith, California, USA), trachyandesite and partially melted granite (after Fig. 3. of Kaczor et al. 1988). *Below:* Xenolith of a partially melted granite xenolith exhibiting strong flow structure at the trachyandesite contact. The xenolith includes fragments of granite and trachyandesite (after Fig. 4B of Kaczor et al. 1988). See text

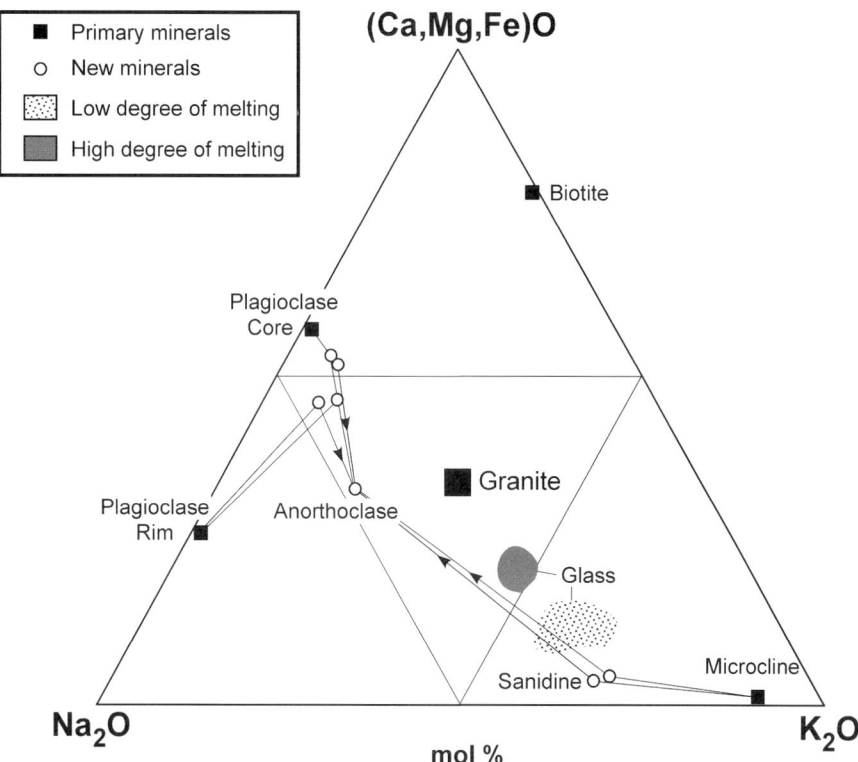

Fig. 3.41. Mol% Na_2O-K_2O-(Ca,Mg,Fe)O diagram for granite, relic and new minerals and glass (Rattlesnake Gulch, Sierra Nevada batholith, California, USA), (after Fig. 1 of Tommasini and Davies 1997). *Tie lines with arrows* indicate compositional changes in feldspars with increasing degrees of melting of the granite. See text

trachyandesite contact is reached at which point it also becomes "spongy" in appearance indicating melting, with rim compositions becoming notably more Ca-rich (see Chapter 7) (Fig. 3.41). Anorthoclase is the end product of both plagioclase and sanidine melting reactions (Fig. 3.41). In contrast, quartz remains clean and has sharp but irregular (resorbed) junctions with the glass. At the trachyandesite contact, dark brown to black glass forms more than 60% of the rock. It contains microlites of orthopyroxene, as radiating clusters and as fringes around residual quartz. Fine grained spherulitic intergrowths of quartz and feldspar form when the glass devitrifies.

Silicate, oxide and apatite geothermometry of the trachyandesite yield temperatures within the range of ~900–1150 °C (Kaczor et al. 1988). Thermal modeling by Tommasini and Davies (1997) assuming an intrusion temperature of 1100 °C indicates that the granite xenolith exceeded its solidus within 3 months of trachyandesite intrusion and attained a maximum temperature of ~1000 °C after ca. 1.5 years (see Chapter 2). At this temperature, the absence of tridymite suggests a minimum pressure for pyrometamorphism of ~680 bar (Fig. 3.37).

3.3
Combustion Metamorphism

3.3.1
Sandstone, Siltstone, Shale, Diatomite

India

In 1914, Fermor described a "very puzzling series of 'volcanic rocks'" from the *Bokaro coalfield*, Madhya Pradesh, India. The rocks are vesicular, sometimes pumiceous, and consist of laths of labradorite and skeletal magnetite, but lacking augite, in a black glassy matrix. Subsequent investigation revealed that the 'scoriaceous rocks' are in fact the products from the burning of coal seams above the water table (Fermor 1918) and they were later described as 'para-lavas' (Hayden 1919; Fermor 1924) consisting of Fe-cordierite, sillimanite (?mullite), orthopyroxene, plagioclase and spinel in a glassy matrix.

One burnt section described by Fermor (1918) is associated with a 13 m thick coal seam. The upper part of the variably burnt seam is marked by some 0.5 m of vesicular 'semi-coke' overlain by 1.4 m white coal ash the top of which defines the upper surface of the coal seam that is overlain by white, bleached and sometimes reddened shale. Where the seam has been entirely burnt its original thickness is represented by a "few feet of pumiceous scoria" and the overlying shale has been partially fused with the formation of breccias of porcellanitised shale in a fused shale matrix. Fermor also describes a 5 cm vein cutting bleached shale, ascribing it to the intrusion of a 'shale melt'.

Other examples of paralavas occur in the eastern part of the *Jharia coalfield* in the Dhanbad district of Bihar State, NE India (Fig. 3.42) where in 1995 some 96 coal fires were burning at depths between 45–55 m (Saraf et al. 1995). The paralavas are described by Sen Gupta (1957) as very localised, occurring as small bouldery masses on the hanging-wall sides of coal seams and as more or less continuous exposures immediately overlying coal seams. The bouldery masses are attributed to fires that began at coal outcrops and are therefore confined to the surface or to shallow depth; the continuous paralava exposures formed when coal was ignited after quarries were opened and in some cases these could be directly correlated with burning seams. In the overlying sediments, fusion is most marked along bedding planes and joints in shale whereas in fine-grained sandstone loci of fusion are comparatively uniformly distributed and appear to be controlled by pore spaces. Larger bodies of paralava occur in fissures where the shales contain well-developed joints, and especially where two or more joint planes intersect. Here the paralava exhibits ropy, lobate and tongue-like structures similar to those in basaltic lava. It occurs as pendants in joint cavities, has a smooth glazed surface and is hollow being usually very light with a thin glassy skin. Such bodies are characterised by circular or ellipsoidal vesicles, the elongation of which is mostly aligned along bedding planes within the shale confirming the direction of flow along such planes.

The glass-rich nature of the paralavas and characteristic igneous textures (vitrophyric, hyalopilitic, intersertal), the presence of metastable structures such as spherulites and idiomorphic habit of constituent minerals indicates complete melting and crystallisation from a molten state. Incompletely fused rocks retain relic minerals such

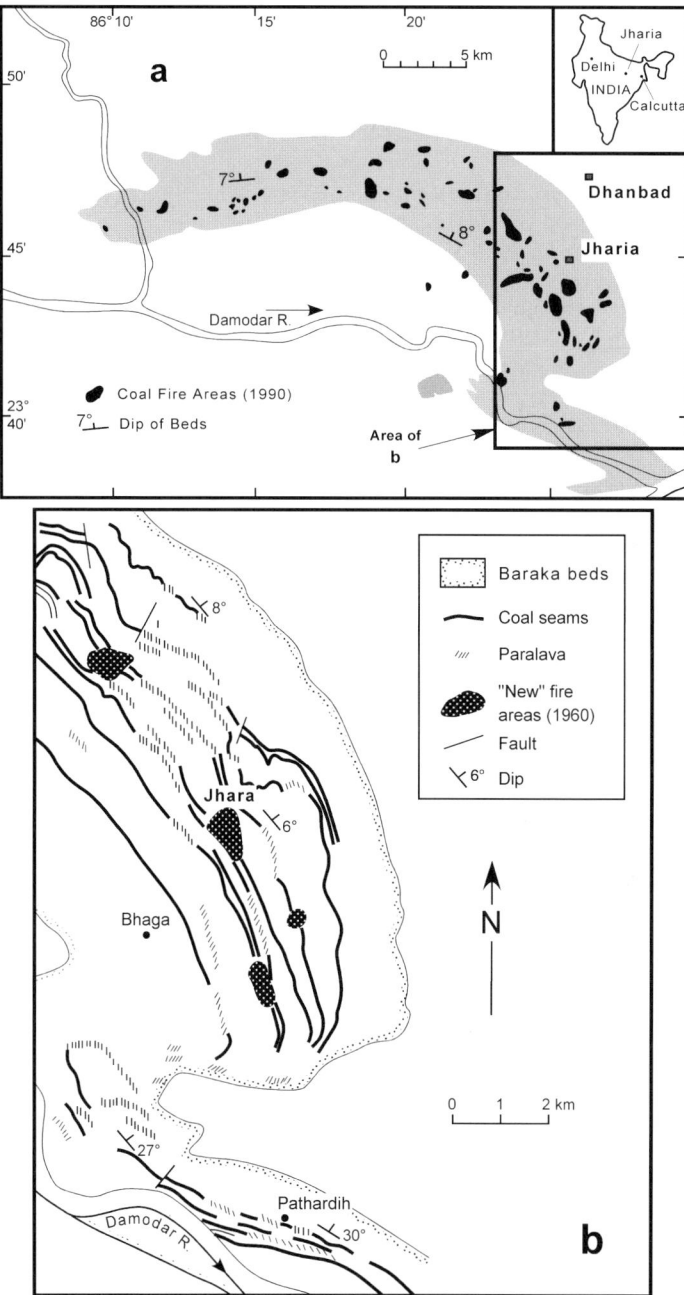

Fig. 3.42. Maps of the Jhara coalfield, India, showing areas of coal fires in 1995 (**a**) (after Fig. 6 of Saraf et al. 1995) and distribution of paralavas and burning areas associated with paralavas in the eastern part of the Jhara coalfield (**b**) (after Fig. 13 of Sen Gupta 1957)

as quartz, preserve traces of bedding planes. The following paralava types are identified by Chatterjee and Ray (1946) and Sen Gupta (1957):

- *cordierite-tridymite buchites* (vesicular, aphanitic, ash-grey)
- *fayalite-cordierite-tridymite buchites* (extremely rare, hard, heavy, aphanitic and jet black)
- *mullite-tridymite buchites* (most common, aphanitic, highly vesicular, ash-grey – lilac and form the largest paralava bodies)
- *hypersthene-cordierite-tridymite buchites* (hard, aphanitic, vesicular resembling basalt, associated with medium-grained sandstones)
- *mullite glass-rich buchites* (dark brown – black, resembling obsidian; banded ash-grey/black; lilac)
- *plagioclase-pyroxene buchites (referred to as para-basalts)* (hard, minutely vesicular, aphanitic, ash-grey to black, hard, occurring as small bouldery masses)

The high temperature hexagonal form of cordierite known as *indialite* was first described from paralavas of the Bokaro coal field, India, by Miyashiro and Iiyama (1954) and Miyashiro et al. (1955) and growth forms (Fig. 3.43) are discussed by Venkatesh (1952). These, and many other examples of high temperature cordierites occurring in pyrometamorphosed rocks have elevated alkali contents (up to 2.74 wt.% K_2O and 1.25 wt.% Na_2O; e.g. Schreyer et al. 1990; Nzali et al. 1999) that reflects the coupled substitution $(K,Na)^+ + (Al,Fe)^{3+} = \Box + Si^{4+}$.

The tridymite-, Fe-cordierite-, mullite-, fayalite-bearing paralavas of the Jharia coalfield have bulk compositions with 93 wt.% SiO_2, Al_2O_3, Fe_2O_3 and FeO (Sen Gupta 1957) and their crystallisation can be discussed in terms of the system $FeO-Al_2O_3-SiO_2$. Textural relations between tridymite, cordierite, mullite and fayalite are shown in Fig. 3.44. Progressive stages of crystallisation of a *tridymite-cordierite-fayalite paralava* are illustrated by a, b and c and the bulk composition is represented by point 1 in Fig. 3.45. Crystallisation begins with tridymite (at ~1450 °C) followed by cooling to 1210 °C where cotectic precipitation of tridymite and cordierite occur (Fig. 3.44a). Any mullite formed would have been dissolved at the peritectic point K with simultaneous crystallisation of cordierite. Cooling continues with coprecipitation of cordierite and tridymite until

Fig. 3.43.
BSE image photo showing the growth habit of cordierite in K-rich (~7.5 wt.% K_2O) siliceous glass in red clinker in contact with parabasalt overlying a burning coal seam, Yellow River, Inner Mongolia, China (sample collected by Professor K. Zang, Department of Earth Sciences, Zhongshan University, Guangzhou, China)

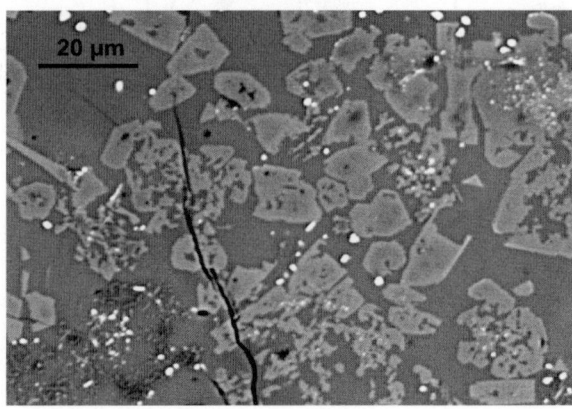

they are joined by fayalite at the ternary eutectic N at 1083 °C (in Fig. 3.44b). Inclusions of tridymite in fayalite and needles of tridymite in the glass (in Fig. 3.44c) indicate that tridymite continued to form over the entire cooling interval of the buchite. The bulk composition of a *tridymite-mullite-cordierite paralava* (point 2 in Fig. 3.45) plots in the mullite field close to the mullite-tridymite cotectic. Crystallisation begins with mullite at about 1310 °C (needle-like crystals in Fig. 3.44d) and is joined by tridymite on the cotectic boundary. Mullite and tridymite continue to crystallise until the ternary invariant point K when tridymite and mullite react with liquid K to produce Fe-cordierite. Further cooling results in continued coprecipitation of tridymite-

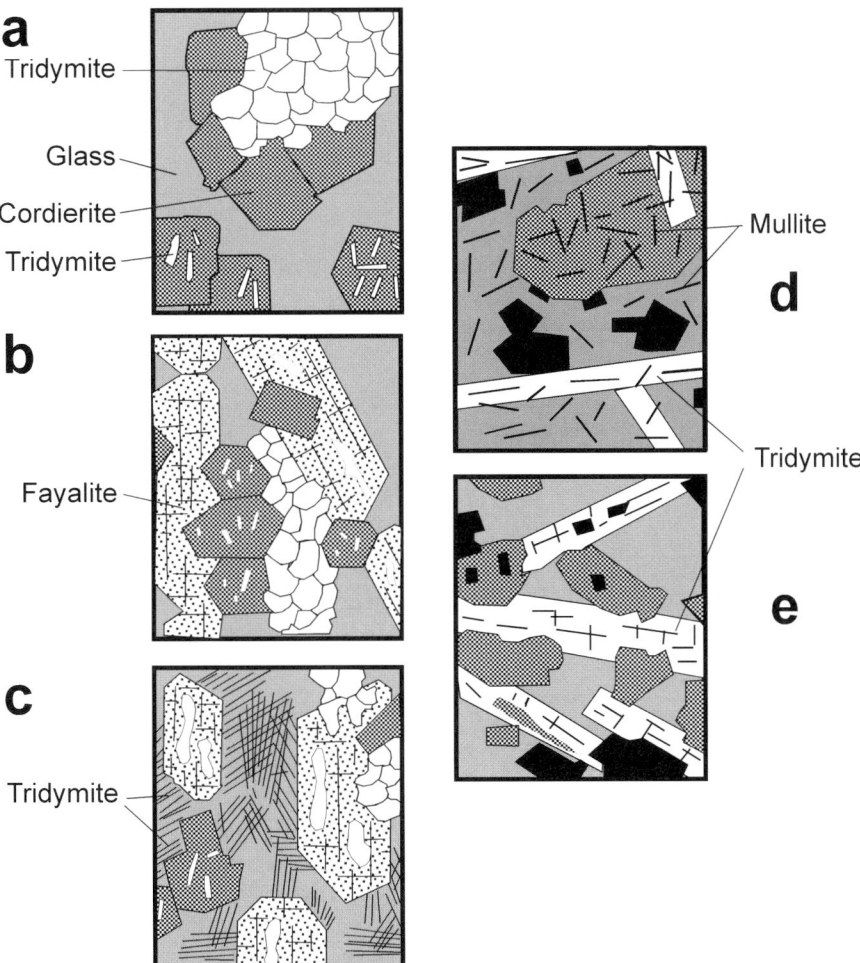

Fig. 3.44. Textural development in tridymite-cordierite-fayalite paralava (**a, b, c**) after Fig. 5 of Sen Gupta (1957), and tridymite-mullite-cordierite paralava (**d, e**) after Fig. 4 of Sen Gupta (1957). The sequence of crystallisation in both examples in discussed in terms of the system $FeO-Al_2O_3-SiO_2$ in Fig. 3.45. See text

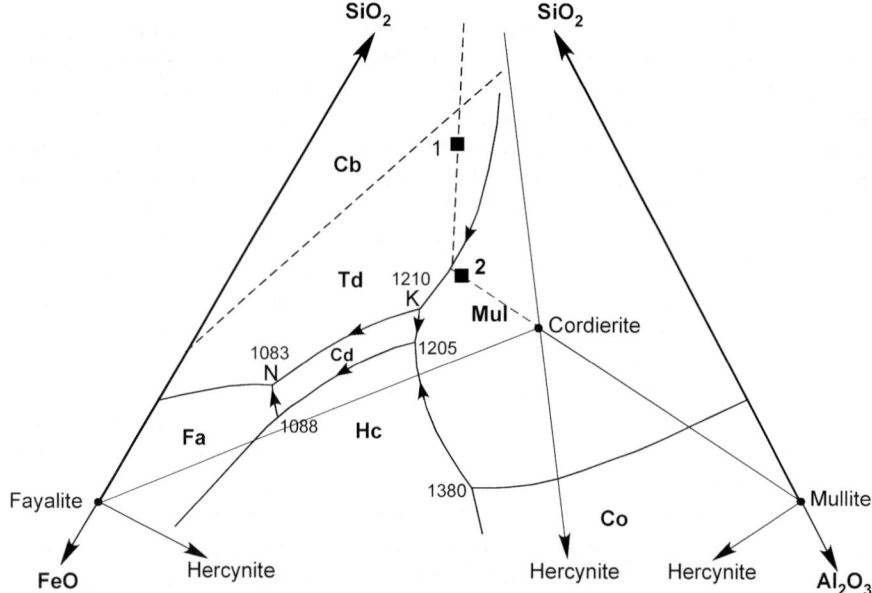

Fig. 3.45. Central part of the system FeO-Al$_2$O$_3$-SiO$_2$ after Schairer and Yagi (1952). Bulk paralava compositions indicated by *1* (tridymite-cordierite-fayalite) and *2* (tridymite-mullite-cordierite). See text

cordierite (in Fig. 3.44e) and the rock is quenched at some point along this cotectic somewhere between 1210 °C and 1083 °C. In both buchites Fe-oxide (magnetite) is an additional phase, forming 18 modal% in the tridymite-cordierite-fayalite paralava (not shown in Fig. 3.44a, b and c) and 34 modal% in the tridymite-mullite-cordierite paralava. It is texturally the earliest phase and is the most common mineral in all the paralavas. The magnetite may have formed during the heating stage from the breakdown of phyllosilicates or possibly the melting of siderite and persisted to higher temperatures of paralava formation.

Western United States

> *"hell with the fires put out"* – description of the effects of combustion metamorphism seen in the badlands of the Little Missouri by General Sully when he crossed it in 1864 (quoted in Allen 1874).

Regional pyrometamorphism of Cretaceous and Tertiary sediments associated with burning lignitic or sub-bituminous coal seams ranging in thickness from 1–70 m has affected an enormous area (> 518 000 km^2) of the great western coal basins of the United States, extending from Texas in the south to Canada in the north. The pyrometamorphic rocks have been described by many writers since the Government-sponsored Great Plains expedition of Lewis and Clark in 1805–1806, who, while traveling in the badlands of North Dakota along the Missouri River, record the occurrence of clinker, pumice and paralava and correctly attributed their origin to the action of burning coal. While navigating the Missouri River William Clark records (March 21, 1805) that they *"saw an emence (sic) of quantity of Pumice Stone on the sides & feet of the hills and emence (sic) beds of Pumice Stone near the Tops of them, with evident marks of the hills having once been on fire. I Collecte Somne (sic) of the different sorts i.e. Stone Pumice & a hard earth, and put them into a furnace, the hard earth melted and glazed the others two and the hard Clay became pumice Stone glazed"*. In his journal entry of 11 April, 1805 he writes that, *"… the hills on either side are from 5 to 7 miles asunder and in maney (sic) places have been burnt, appearing at a distance of a reddish brown choler (sic), containing Pumice Stone & lava, some of which rolin (sic) down to the base of the hills."*, and on the 16 April Merriwether Lewis believed that *"… the stratas of coal seam seen in those hills which causes the fire and burnt appearances frequently met with in this quarter. Where those burnt appearances are to be seen in the face of the river bluffs, the coal is seldom seen, and when you meet with it in the neighborhood of the stratas of burnt earth, the coal appears to be precisely at the same height, and as nearly the same thickness, together with the sand and sulphurous substance which usually accompanies it."* (Thwaites 1969). In 1839, Nicollet in an account of a journey up the Missouri River, relates information about dense smoke rising from hills and issuing from crevices that *"is said to last at the same spot for a long time – say two or three years; indicating at them a large accumulation of combustible materials … these pseudo-volcanic phenomena may be compared with those described as occurring in other parts of the globe, under the name* **terraines ardens**; *although they are not here accompanied by the emission of flames."* Nicolet was of the opinion that the burning was *"evidently due to the decomposition, by the percolation of atmospheric waters to them, of bed of pyrites, which, reacting on the combustible materials such as lignites and other substances of vegetable matter in their vicinity, give rise to spontaneous combustion"* (Allen 1874).

Allen (1874) has provided a good description of the effects of the burning of horizontal to gently dipping lignitic coal seams on overlying sediments (mainly quartzose sandstones and shales) in 'badlands' of the upper Missouri of Dakota and Montana. Where the burnt lignite occurs as a thin layer of "several inches to two feet or more in thickness" it remains as "ashes and cinders and clinkers". Clay below the burnt seam is only slightly discoloured and hardened. The overlying clay-rich sediment is bright brick red, varying in thickness from a "few feet to twenty or more", and above this the strata also show the effects of heating by colour changes. The red layer typically forms the capping of buttes and mesas and can be traced over large distances. In deeply eroded areas several such red layers can be seen, each separated by between 15 to 46 m of unbaked sediment.

Where the burnt lignite is "several feet thick", the overlying sediments have been "more or less fused or at least reduced to a plastic condition" as indicated by their vitreous, porcellanic and vesicular appearance. Highly porous varieties can best be described as pumice. The associated clinker rocks exhibit a variety of colours from white, through yellowish-white, yellow, olive, dark brown, purple, all shades of red, and black and they are often finely banded. In general, red, maroon, pink, orange, yellow colours indicate oxidising conditions and green, grey and black reducing conditions. Rogers (1917) records that red and bright yellow, green or black mottling in common in shales whereas sandstones is generally altered to a uniform pinkish-red. In the generalised sequence described by Allen (1874), the variously coloured clinker rocks are overlain by between 1.2 and 6 m of baked fissile claystone that resemble bright red bricks having a metallic resonance. Overlying sandstones are variably baked.

Allen also describes scattered "jagged, chimney-like mounds" and "narrow walls of ragged lava-like rock" that appear like volcanic breccia, and were produced by "the breaking through to the surface of these subterranean fires". The erosion-resistant chimneys surmount clinker mesas and bluffs and are typically only a meter or so in height and diameter but can form masses 3 to 5 m diameter and up to 6 m in height, the size being proportional to the thickness of the burnt lignite bed below. The rock forming the chimneys is typically highly vesicular and scoria-like (paralava). Adjacent sediments have become fused to a depth of 2.5 to ~8 cm giving the walls of these chimneys and fissures a glazed surface. The melt has clearly been mobile solidifying as pendents, flows and rounded, botryoidal masses, resembling "in structure and general appearance viscous matter that has been pulled, twisted and folded while in a plastic state", and intrudes along cracks in the walls. Many sandstones in contact with the melt rocks have developed a five or six-sided columnar structure in which the columns range from 0.3 to 1 m long, caused by the development of a cleavage oblique to bedding. The areas of fused rock masses present a very broken and chaotic appearance with blocks of "scoriaceous material" that have fallen from the tops of buttes and ridges, lying in jumbled heaps in the adjoining valleys and also extending for some distance out onto the plains.

The age of the burning and pyrometamorphism has been determined in the 37 000 km^2 Powder River Basin of Wyoming and Montana. Paleomganetic reversal data (Jones et al. 1984) indicate that some clinkers in the northern part of the basin formed more than 1.4 Ma ago, the oldest clinker being dated at 2.8 ± 0.6 Ma (Heffern et al. 1993). Zircon fission track ages from the northeastern part of the basin indicate that burn zones become older eastwards from 0.08 to 0.77 ± 0.39 Ma (Coates and Naeser 1984). In Northern Montana Powder River Basin, the clinker (locally up to 60 m in thickness) covers approximately 2700 km^2, has an average thickness of 15–25 cm with a volume estimated to be 40–70 km^3 (Heffern et al. 1993).

Changes recording a mineralogical, textural and chemical continuum over an interval of 2 m from unaltered shale to paralava of a chimney associated with a burnt coal seam in the southern part of the Powder River Basin are described in detail by Clark and Peacor (1992). The horizontal section sampled corresponds to a temperature range of about 1300 °C. The lowest temperature (i.e. that of burial diagenesis) shale consists mainly of silt-sized quartz, K-feldspar, together with detrital muscovite, biotite, chlorite, kaolinite, and pyrophyllite up to 20 microns in length, within an illite-smectite matrix. The following changes are observed with increasing temperature:

1. At low to moderate temperatures, delamination of the phyllosilicate matrix occurs with dehydroxylation. Dehydroxylation progresses layer-by-layer leaving behind a mixture of crinkled/rolled parts of layers in enlarged voids. Detrital minerals remain unchanged.
2. With increasing temperature, silt-sized grains, especially quartz, contain abundant cracks due to thermal expansion. Larger detrital phyllosilicate grains show evidence of delamination. The clay matrix becomes more homogeneous with unresolvable boundaries. Unoriented grains of mullite up to 0.5 microns in length appear.

3. At high temperatures of ~1000 °C within 20–30 cm of the paralava, feldspar and quartz are severely cracked. In the case of quartz, this results from the large volume change caused by inversion to tridymite/cristobalite. Delamination of large detrital phyllosilicates is apparent. Large areas of pore space form within the matrix caused by shrinkage due to dehydroxylation of illite/smectite and contain Fe-oxides. No melt has formed.
4. At the shale clinker/paralava contact the only remaining detrital grains are cristobalite. K-feldspar is absent and has presumably reacted with the clay matrix to form a homogeneous non-crystalline matrix that contains abundant mullite together with grains of magnetite-ulvospinel and spinel-magnetite-hercynite. Although expected, no evidence of melt is found and this maybe due to loss of nearly all H_2O by dehydroxylation of the clay minerals.
5. The change from unmelted clinker to paralava is relatively sharp. The melt appears to have been produced along a well-defined interface by reaction of the altered clay matrix, K-feldspar and silica fluxed with vapour-transported Fe, Mg, Ca and Mn possibly derived from carbonate-rich iron nodules in the shales.

The paralava at this locality and elsewhere is characterised by a large variety of textures and minerals (Cosca and Peacor 1987; Foit et al. 1987; Cosca et al. 1988; Cosca et al.1989) (e.g. Fig. 3.46). Features such as spinifex and hopper crystals, radial clusters of minerals in glass, and alignment of crystals around vesicles that indicate crystallisation from a melt. Phenocrysts and/quench crystals identified include cristobalite, tridymite, anorthite, K-feldspar, barian feldspars, nepheline, fayalite, enstatite, clinopyroxene (diopside–esseneite), wollastonite, dorrite, andradite, mullite, gehlenite-akermanite-sodium melilite, Fe-cordierite, apatite, spinel-magnetite-hercynite-ulvospinel, hematite-ilmenite, hematite-magnesioferrite, and pseudobrookite solid solutions, sahamalite and unidentified Cu- and Fe-sulphides. In addition to single crystal phases, symplectitic intergrowths of spinels and cristobalite are also found. Quartz, zircon and possibly xenotime remain as unmelted relics.

Fig. 3.46.
Drawing from BSE image photograph (Fig. 4E of Cosca et al. 1989) showing textural relations between cristobalite, mullite, Fe-Ti oxides (black) and glass in paralava, Powder River Basin, Wyoming, USA. The texture indicates early crystallisation of Fe-Ti oxides and mullite followed by that of cristobalite. Shrinkage cracks occur in the cristobalite

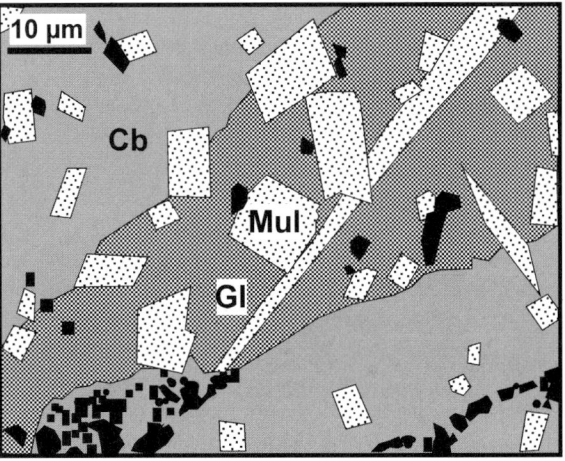

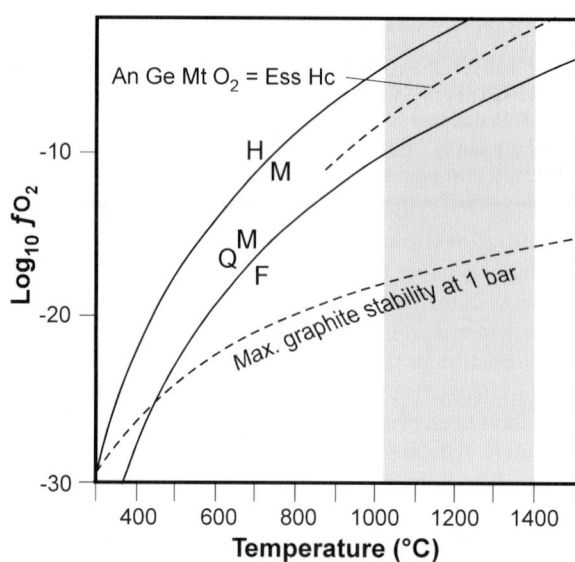

Fig. 3.47.
T-fO_2 plot showing relationship of quartz-fayalite-magnetite (QFM) and hematite-magnetite (HM) buffer curves, maximum graphite stability at 1 bar (after Fig. 8 of Cosca et al. 1989) and the reaction anorthite + gehlenite + magnetite + O_2 = essenite + hercynite determined by Cosca and Peacor (1987). *Vertical grey strip* = estimated temperature range of paralava formation. See text

Mineral assemblages and compositions of the paralavas may be in part controlled by the composition and partial pressure of gas (dominated by C and O) produced during combustion of the coal. Paralavas from a single outcrop have markedly different mineral assemblages that reflect different oxidation states and often steep gradients in fO_2, i.e. within the coal ash layer (see Fig. 2.9), paralava contains an assemblage of Fe^{3+}-rich clinopyroxene (essenite), melilite, anorthite, magnetite-ülvospinel-hercynite solid solutions and glass; paralava occurring in chimneys above the coal ash zone typically contains fayalite, tridymite, cordierite, abundant magnetite-ülvo-spinel-hercynite solid solutions, late hematite and less glass. Figure 3.47 shows log fO_2-T curves for graphite stability in relation to the HM and QFM buffers where Ptotal = PCO + PCO_2 = 1 bar. Oxidation conditions near that of the HM buffer are necessary to stabilize Fe^{3+}-rich clinopyroxene (Cosca and Peacor 1987) in the former assemblage, whereas coexistence of fayalite, tridymite and magnetite in the latter assemblage indicates fO_2 conditions controlled by the QFM buffer. Gradients in fO_2 between the two assemblages thus require variation of several log units at inferred crystallisation temperatures between 1020–1400 °C. Paralava in equilibrium with graphite, e.g. at the sedimentary-coal interface, would have formed under significantly lower fO_2 and with CO as the dominant gas species.

A sample of slag consisting of a fine grained groundmass of feldspar intergrown with an undetermined higher birefringent mineral, dusty magnetite and glass and with laths of anorthite and blebs of ?pyrrhotite, was subjected to incremental heating each of two days duration in evacuated silica glass tubes by Brady and Gregg (1939). A small amount of feldspar persisted at 1212 °C and at 1232 °C complete fusion occurred. These temperatures lie within the 1020–1400 °C range of melt temperatures deduced for the Powder River paralava glass compositions plotted in the system $CaAl_2Si_2O_8$-$KAlSi_3O_8$-SiO_2 (Fig. 3.48), the highest temperatures represented by the cristobalite-mullite-magnetite buchite shown in Fig. 3.46.

Fig. 3.48.
Normative compositions of paralava glass (*filled circles*), Powder River Basin, Wyoming, USA, plotted in the system $CaAl_2Si_2O_8$-$KAlSi_3O_8$-SiO_2 of Schairer and Bowen (1947) (after Fig. 6 of Cosca et al. 1989)

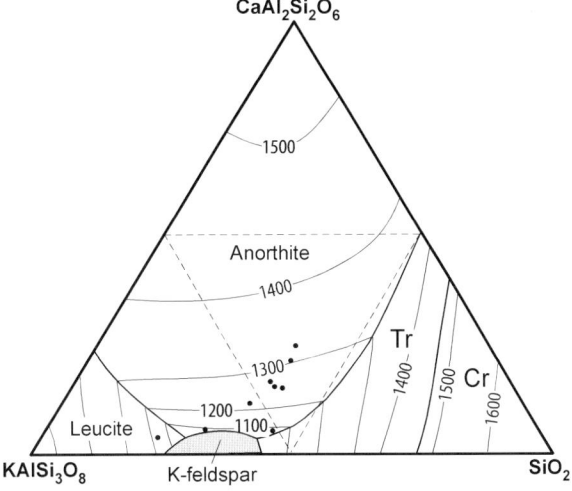

Spectacular evidence of burning bituminous sediments in *southern California* has been noted since the latter part of the eighteenth century and variously described as 'solfataras', 'well-fires' and 'volcanoes' with eruptions of fire, ash and rocks (see references in Bentor et al. 1981). An interesting description of the phenomenon is quoted in Arnold and Anderson (1907) from the vicinity of the so-called "Rincorn Volcano", in the oil fields of Santa Barbara County:

> "I found hot gases burning from numerous apertures in the shales, accompanied in some cases by melted bitumen that hardened into concretionary masses upon cooling…Crystals of sulphur had also formed upon all objects near the issue, and naptha appeared to be present. A few years ago a tunnel was run into the cliff at its base to a depth of 200 feet in search of oil. At this depth the workmen were obliged to cease operations in their endeavor to penetrate further on account of the great heat. Upon entering the tunnel I found the temperature still high but noticed only weak sulphurous gases. Near the entrance for 50 or 60 feet the roof and sides were thickly covered with attenuated colourless crystals of epsomite [$MgSO_4 \cdot 7H_2O$] hanging in tufts and masses."

Most of the combustion localities are associated with bituminous sediments of the Middle Miocene Monterey Formation (e.g. Bentor and Kastner 1976; Bentor et al. 1981) which are dominantly mudstone with less common opaline mudstone, siltstone, diatomite, and rare shale, limestone and dolomite. The organic carbon content of the Monterey sediments is typically > 10 wt.% and a special feature is the abundance of F-apatite (phosphorite) which forms up to 16 vol.% of the mudstones.

The pyrometamorphic products range from baked rocks to glassy rocks (buchites) resembling obsidian, pitchstone or bricks, to recrystallised slag-like silicate and phosphate-rich rocks (Bentor et al.1981). Characteristic features are:

1. Glassy rocks are either; dense, sometimes slightly vesicular, black, dark grey, deep brown, red-brown or orange-red in colour, with a concoidal to hackley fracture and have a vitreous or waxy to resinous luster; or brick-red coloured, hard, dull and porous resembling unglazed pottery. The 'glasses' and 'bricks' form rootless, irregular pods and as larger masses form networks of veins, centimeters to deci-

meters thick and tens to hundreds of meters in length. The glass-rich rocks exhibit both chilled margins and contact-metamorphose the rocks they intrude.

2. Vesicular scoriaceous slag forms the most abundant pyrometamorphic rock type and ranges in colour from black, brown and violet in more siliceous types to shades of olive-green in less siliceous varieties. These rocks are devoid of glass and have a holocrystalline texture made up of tiny crystals ranging in size from < 1 to a few tens of microns. The slag occurs as isolated 'specks' barely 1 cm diameter in unmelted rocks, as banded rocks which mimic the original mudstone laminae in which millimetre to centimeter-thick bands of recrystallised melt alternate with thinner bands of unmelted rocks, as recrystallised *phosphatic* layers, and as 'chimneys' a few meters across to 'stocks' that may have an outcrop area of > 1 km^2. In both these intrusive bodies the slag is frequently brecciated and contains xenoliths of country rock exhibiting all stages of fusion.

Non-hydroxyl-bearing high temperature minerals have crystallised from the various melt compositions derived from fusion of minerals in the mudstones, i.e. dominantly clay minerals (mainly sericite, illite, chlorite) with detrital quartz (opal in diatomites), plagioclase (An_{35-65}), K-feldspar, apatite and authigenic carbonate (mainly calcite) (Bentor et al. 1981). In the glassy rocks, α-cristobalite, tridymite, Na-plagioclase, anorthoclase, sanidine, cordierite, hematite, pyrite, magnetite and rare corundum, garnet and wollastonite, have been identified in the buchites. Additional high temperature phases such as Al-rich augite, pigeonitic pyroxene, plagioclase (An_{58-75}), gehlenite, mullite, spinel, and fluorite occur in the crystalline slag-like rocks. Secondary minerals lining vesicles and filling amygdules include plagioclase (An_{36-49}), gypsum, calcite, aragonite, zeolites (analcite, phillispsite, harmotome, possible clinoptilolite and gmelinite), pyrite, goethite, and rare alunite, chlorite, illite and Na-jarosite.

The glasses are compositionally variable, often flow banded and typically contain lenticular-prismatic, more rarely triangular or forked shards. These may exhibit flow alignment, infrequently in the form of whirlpools, or have blurred boundaries due to welding so that they assume a felt-like appearance. The structureless glass 'matrix' contains abundant microscopic crystals of hematite, pyrite and magnetite that impart a black to dark brown colouration. Glass shards are conspicuously devoid of opaques. Vesicular phosphate-rich segregations thought to result from the melting of F-apatite and now crystallised to essentially isotropic apatite occur as 'droplets' in vesicular recrystallised silica-rich melt rocks. They are questionably inferred to represent immiscible phosphate melts formed at an unsubstantiated temperature of at least 1650 °C. From the evidence presented, however, they could also be recrystallised (annealed) sedimentary phosphorite that has not necessarily undergone melting at such a high temperature.

Canada

Pyrometamorphism related to burnt Early Tertiary coal measures in the *Hat Creek* area, British Columbia, has been described by Church et al. (1979). Evidence of burning is shown by the presence of yellow and red partly fused shales referred to as *boccanebuchites* above the Hat Creek Coal Formation. A trench excavated for bulk sampling of the coal encountered deformed layered clinker-like material that proved to be continuous with bedding planes in adjacent coal. Drilling established that the burnt zone averages 25 m in thickness and extends over an area of some 3.5 km² beneath glacial sediments. From one locality, the fused residue of burnt coal has the appearance of a volcanic agglomerate consisting of welded scoriaceous clasts composed of microlites of anorthite, cordierite, tridymite, cristobalite, hematite and glass. Coal near the combustion zone has taken on a waxy luster, a hard clean surface and a conchoidal fracture not typical of the unaffected low rank coal. It has a reflectance value (Ro_{max}) of 0.42% compared with 0.36% away from the combustion area.

The field relations, and progress, mechanics and products of combustion of an actively burning 35–50°-dipping thick coal seam in the *Aldrich Creek* area, southeastern British Columbia, have been described by Bustin and Mathews (1982). The burnt area is marked by a line of reddened, partly fused sandstones and at one end of which active burning, as evidenced by up to 1 m high flaming gas, from three vents in the sandstone roof of the coal seam was observed between August 1979 and February 1981. Combustion evidently began in 1936 from a forest fire and from reports, air photos and site inspection, the progressive advance of the coal fire is shown to have been at a uniform rate of about 13.5 m yr^{-1}. The advance zone of combustion is marked by the development of sulphur-lined open cracks at the ground surface through which water vapour and other gases are emitted. Within the zone of active combustion, most volatile elements are driven off, mix with air and burn en-route to the surface. These flaming vents may develop locally along earlier cracks or at larger openings caused by collapse of roof rocks. Behind the combustion zone are abandoned vents, some of which serve as air intakes. The vitrified walls of these vents indicate that very high temperatures were attained and in one of the flaming vents, temperatures in excess of 1000 °C were measured.

The stratigraphy in the vicinity of the burning coal seam (Fig. 3.49) shows that it is up to 6 m thick, overlain by thin siltstone followed by alternating fine grained sandstone and siltstone containing detrital quartz and carbonate (calcite, dolomite) in a dominantly kaolinite-illite matrix. Within the burned area the upper 2–3 m of the coal has been consumed leaving a 5–60 cm thick layer of residual ash that is separated from the underlying unburnt coal by 10–50 cm of natural coke. Sediments overlying the coal have been baked, discoloured and locally melted. In response to the combustion, the overlying sediments have become fractured and have locally collapsed to form a welded breccia. Several pyrometamorphic zones can be distinguished; a near surface fused, scoriaceous zone developed within a few centimeters of the active or former vents; a zone marked by the breakdown of carbonate (temperatures possibly > 350 °C) where the rocks have been fissured and blistered but not fused; a zone of discolouration where carbonate is stable. Glass of the fused rocks contains crystals of cristobalite, tridymite, diopside, anorthite, an anorthite-sanidine eutectic mixture, and magnetite. Some of the fused rocks are parabasalts with diopsidic pyroxene, anorthitic plagioclase, magnetite and glass (Fig. 3.50). Fusion experiments of the roof rocks indicate temperatures must have reached values of between 1150 °C (Si-poor sediments) and 1200 °C. These temperatures are almost the same as those of an anorthite-diopside-tridymite assemblage in the system $CaO-Al_2O_3-SiO_2$ with 5% (Fig. 3.6) and 10% MgO, i.e. at 1260 and 1175 °C respectively.

Fig. 3.49.
Stratigraphic section through burnt coal measures at Aldrich Creek, Upper Elk River valley, Canadian Rocky Mountains (after Fig. 4a of Bustin and Mathews 1982)

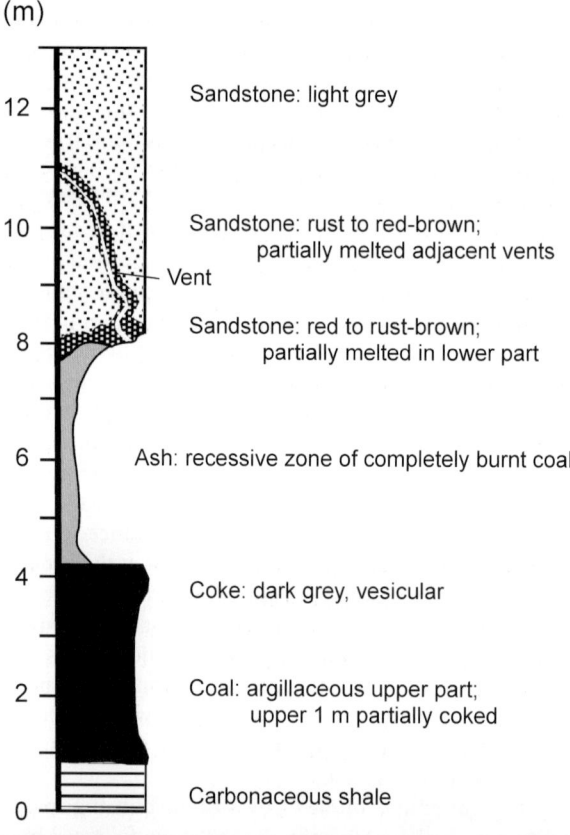

Fig. 3.50.
Drawing from scanning electron photomicrograph (Fig. 6a of Bustin and Mathews 1982) showing texture between diopsidic augite, Ca-plagioclase and magnetite (*black*) in parabasalt, Aldrich Creek, Upper Elk River valley, Canadian Rocky Mountains

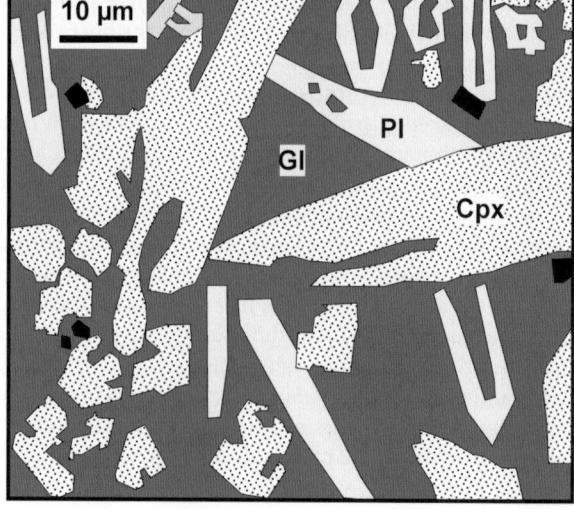

With respect to the coal, the unheated part below the coke consists predominately of vitrinite, with minor semi-fusinite and inertinite and a vitrinite reflectance (Ro_{max}) of ~0.96%. The burned coal immediately below the ash has a Ro_{max} of 7.04%, whereas the coal directly underlying the coke has a Ro_{max} of 1.07% and shows no evidence of devolatilation or heating except for the slightly higher reflectance than that of the unheated coal and the presence of highly reflecting oxidation rims on some grains. These data indicate that very little heat is carried downward by conduction. Rather, heat is carried forward of the combustion zone and upward by convecting gas, thereby coking the coal, baking and locally melting the roof rocks. Bustin and Mathews (1982) calculate that the coal, with a mean heat content of ~32 500 kJ kg^{-1} could yield 32×10^{12} J yr^{-1} at an average rate of 1000 kW, and that with an advance of the burning front by 1 m (a 4 week period) some 2.5×10^{12} J would be released with the baking of 1750 t of rock. Given an average temperature rise of 500 °C in this process, and a heat capacity of ~10^6 J kg^{-1} of the roof rocks, the baking would absorb 1.3×10^{12} J, or about half the energy available to be released later with cooling to ambient temperature.

Australia

At *Ravensworth*, New South Wales, a number of small outcrops of black, vesicular slag-like rock are mapped as volcanic necks or minor intrusions within freshwater shales and sandstone containing thin sideritic bands of the Main Permian Coal Basin. The slag-like rocks do not extend to any great depth and appear to lie along a single horizon (Whitworth 1958). Various sized pieces of hardened shale occur within the slag and give the appearance of a volcanic agglomerate, and the slag also intrudes the surrounding sediments, often forming thin stringers along cracks in the shales. The sandstone shows signs of being heated and ranges from slightly baked, through partially fused to completely melted over a distance as little as a few centimeters. The buchitic sandstones occur as roughly circular grey vesicular masses up to 15 m in diameter and traces of bedding can be seen even where fusion is highly advanced. Shale and sandstone above areas of fused rock are hardened and exhibit a pale pink to terracotta-red colour, such material being much more abundant in areal extent than the fused material. The slag outcrops therefore most probably mark the sites of chimneys/fissures though which "flames and hot gases escaped from an underlying fire and about which intense heating was localized" (Whitworth 1958). Dark coloured basic shale-derived? slag was apparently less viscous than more siliceous fused sandstone as it intrudes the latter as thin (mm to cm wide) veins (Hensen and Gray 1979).

Dark brown glass in the fused sandstones contains quartz (inverted from tridymite), unspecified pyroxene, bytownite, cordierite, possibly mullite, and magnetite. Finely crystalline (glass-poor) Fe-rich slag is largely composed of pale green pyroxene, most probably the clinohypersthene reported by Hensen and Gray (1979), together with magnetite, small amounts of cordierite and andesine, so that it closely resembles a basalt. Whitworth (1958) considers that bands of siderite and pyrite occurring near the tops of coal seams in the area, and which are a constant source of danger because of oxidation causing mine fires, are the source compositions of these Fe-rich slags. Fused arenite contains orthopyroxene together with tridymite, cordierite and rare bytownite with colourless to pink glass. A thin black vein intruding the arenite contains clinohypersthene, bytownite, cordierite, and spinel in brown glass.

Heating samples of sandstone and shale from the vicinity of the slag occurrences in a muffle furnace shows that they remain virtually unchanged at 1100 °C. The first signs of softening occur at 1250 °C, marked softening takes place at temperatures higher than 1350 °C, and the rocks become completely fused at ~1370 °C. The heated samples all assume a pale pink or red colouration as a result of oxidation of Fe-oxides present.

At *Burning Mountain, Wingen,* New South Wales, described by Rattigan (1967), the surface above the burnt zone is characterised by subsidence features such as normal faulting resulting in graben formation, open gash-like fissures, and small areas of breccia (Fig. 3.51). The area of burning is a highly fissured zone heated to red and white heat over an area of about 100 m² from which aqueous fumes have deposited sinter consisting of mainly β-quartz and α-Fe_2O_3 encrusted with sulphur. In burnt-out areas highly refractory kaolitinitc claystones underlying the burnt coal seam have been little affected except for a narrow selvage of mullite rock. Those overlying the coal are extensively altered to an assemblage of mullite-sillimanite-tridymite-cristobalite forming a dense, cream or mauve coloured porcellanite. These rocks suggest formation temperatures of > 1050 °C in comparison with the thermal breakdown of kaolinite. Studies by Brindley and Nakahira (1959a,b,c), Comer (1960, 1961) Segnit and Anderson (1971) show that kaolinite undergoes a series of transformations with increasing temperature following the loss of structural H_2O to form metakaolin at ~500 °C according to the reaction

$$Al_4Si_4O_{10}(OH)_8 = Al_4Si_4O_{14} + 4H_2O$$
kaolinite   metakaolin

Condensation of the metakaolin layers occurs to form a well-ordered Al-Si spinel-type phase between 925–950 °C according to the reaction

$$Al_4Si_4O_{14} = Al_4Si_3O_{12} + SiO_2$$
metakaolin spinel   silica

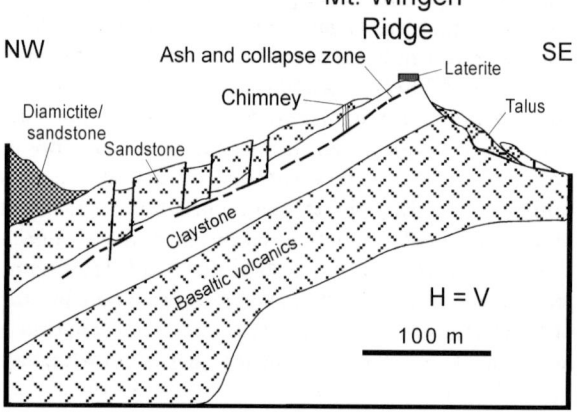

Fig. 3.51.
Cross section 0.6 km north of Burning Mountain, Wingen, New South Wales, Australia (after Fig. 1 of Rattigan (1967))

that transforms through a probable 1:1-type mullite phase and cristobalite at between ~1050–1100 °C, and to 3:2 mullite and cristobalite at higher temperatures according to the reaction

$$1.5\,Al_4Si_3O_{12} + SiO_2 = 3\,Al_2O_3:2\,SiO_2 + 3.5\,SiO_2$$
spinel   silica   mullite   cristobalite

At Wingen, no lower temperature zones containing metakaolin or the spinel-like phase have been recognised.

The thermal effects are not uniformaly distributed in the section above the burning coal seam. Zones of intense alteration extend outwards from the roof of the seam and probably coincide with zones of permeability through which burning gases escaped to produce fused chimneys. Around these, less refractory sandstones have been altered and sintered and completely fused to mauve-coloured tridymite-cristobalite-orthopyroxene buchites suggesting a maximum possible crystallisation temperature of 1470 °C according to the system MgO-Al_2O_3-SiO_2 (Fig. 3.4; Table 3.1). Claystones have been partly converted to a vitrified rock with no mineral structures recognisable on X-ray diffractograms and deformed pellets in these rocks appear to have been heated to softening point.

Pyrometamorphism of the ferriginous zone of the laterite cap that forms Wingen Ridge (Fig. 3.51) has produced a sintered, vesicular black rock composed of hematite-tridymite-andalusite with microlites of sillimanite and mullite occurring within the andalusite. The coexistence of andalusite with hematite in the ferriginous zone suggests that it may be saturated with Fe^{3+}. Transition elements such as Fe^{3+} in andalusite are known to have the potential to shift the And = Sil boundary to higher temperatures, i.e. by 50–100 °C (e.g. Kerrick and Spear 1988; Pattison 1992, 2001) so that andalusite could form in the tridymite field at temperatures > 864 °C. It is clear from the Wingen example that the andalusite is metastable and is decomposing to mullite and sillimanite.

New Zealand

An unusual occurrence of pyrometamorphism thought to have been caused by ignition of a hydrocarbon gas seepage is described by Tulloch and Campbell (1993) from Tertiary sandstones overthrust by a nappe adjacent to the Alpine Fault in the Glenroy Valley area, SE Nelson, South Island of New Zealand. Isolated occurrences of vesicular fused rock occur within calcareous silt and muddy sandstone close to the top of a complex shear zone (Fig. 3.52). The paralava constitutes up to 80% of some outcrops, coats exposed edges of sandstone inclusions and in places forms vertical stalactites. It consists of acicular clinoenstatite and labradorite, radiating sheaths of cristobalite and tridymite, minor magnetite and hematite, within a semi-opaque brownish glass containing abundant fine grained disseminated rutile. Rounded grains of relic quartz are mantled and veined by cristobalite and tridymite. Buchitic sandstone fragments within the paralava up to 10 mm across consist of subequal amounts of corroded quartz, tridymite, cristobalite and patches of pale brown glass together with minor mullite, rutile, hematite and possible Al-ilmenite. As the stable coexistence of tridymite and

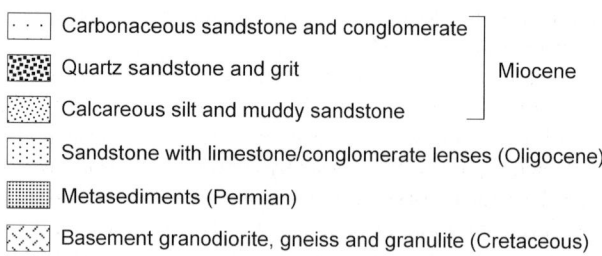

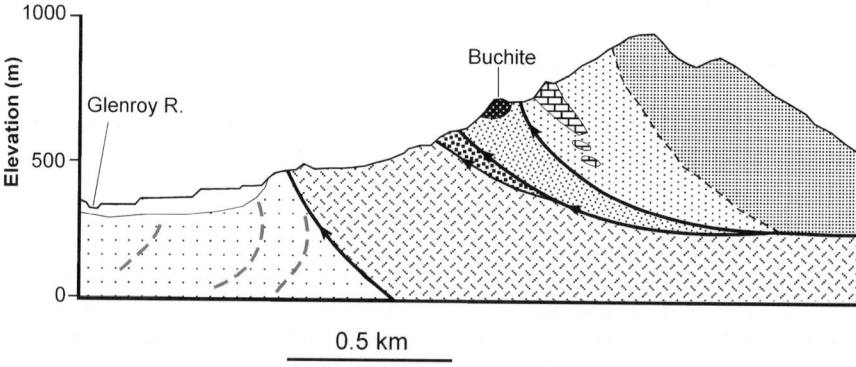

Fig. 3.52. Cross section of the buchite locality at Glenroy Valley, SE Nelson, South Island of New Zealand (redrawn and simplified after Fig. 2 of Tulloch and Campbell 1993)

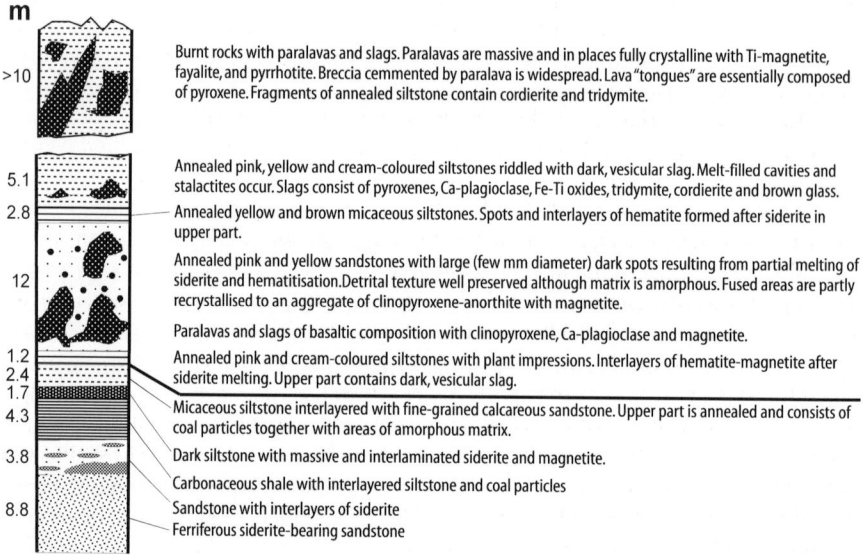

Fig. 3.53. Stratigraphic section through part of the burnt coal-bearing sediments of the Kenderlyk Depression, Eastern Kazakhstan, Russia (redrawn from Fig. 2 of Kalugin et al. 1991)

cristobalite implies a temperature of 1470 °C at atmospheric pressure (Fig. 3.4), it is likely that the cristobalite is metastable.

The close association of the paralava/buchite with a major thrust zone suggests a possible causal relationship. Overpressuring and shear heating could have generated hydrocarbons within the thrust zone that migrated up dip to accumulate in pockets and fractures. Breaching of the shear zone by erosion may have created gas seepages that were ignited by natural forest fire or by lightning. The steeply dipping margins between the paralava containing inclusions of partly fused glass-coated sediment and adjacent largely unaltered sandstone, suggests that the occurrence represents a relatively short-lived explosive vent-like feature.

Russia and Central Asia Republics

One of the largest industrial coal basins of Russia, the Carbonifierous-Jurassic *Kuznetskiy coal basin in southern Siberia*, is characterised by the presence of annealed and melted shales and sandstones that occur as erosion-resistant hills and ridges between 50–100 m high extending over an area of about 45 km² in the Kuzbass district (Yavorsky and Radugina 1932; Belikov 1933). In some places the coal seams have been completely burnt out to the extent of 1–3 km. In other places, combustion has produced caves in otherwise unburnt coal. Up to 100 m thickness of pyrometamorphic rocks are exposed and consist of baked and reddened oil shales, clinker, vesicular slag and melt rocks. Melting of siderite-rich or possibly brown-iron ore material and ferriferous sandstone, has resulted in the formation of magnetite together with smaller amounts of cordierite, fayalite, diopside-hedenbergite, mullite, bytownite and tridymite with rare spinel in a brown glass, and of hematite that is sometimes graphically intergrown with fayalite. Some ferriferous rocks are reported to consist of 40% magnetite and 50% hematite. Argillaceous buchites consist of mullite, tridymite, cordierite, calcic-plagioclase, spinel and glass.

In the Permian-Triassic coal-bearing sediments of the *Kenderlyk Depression, Eastern Kazakhstan*, two burnt horizons, 1.5 and 2.8 km length and up to 70 m in thickness have been traced to a depth of 300 m (Kalugin et al. 1991). The stratigraphic section is shown in Fig. 3.53 together with details of the burnt rocks. Pyrometamorphic products consist of unusual vesicular "iron ore" paralavas, slags and breccias formed from the mutual fusion of sandstone-siltstone and siderite. The paralava contains an assemblage of fayalite, Al-enstatite-hypersthene, augite-fassaite, Al-clinoenstatite-clinoferrosilite, salite-augite, ferrohedenbergite, anorthite-bytownite, K-feldspar, cordierite, tridymite, apatite, spinels (Ti-magnetite, hercynite, magnetite, magnesioferrite), hematite, ilmenite, Fe-hydroxides and pyrrhotite. They are inferred to be the result of selective melting of carbonate-bearing argillaceous sediments and siderite under reduced gas conditions at between 1000–1200 °C. The melt rocks contain clinkers consisting of amorphous, glazed sandstones and siltstones. Lower temperature rocks consist of annealed pink, yellow, brown and cream-coloured siltstones and sandstones which contain dark concentrations up to a few meters in diameter consisting of hematite formed by partial melting of siderite, although the primary detrital texture of the rocks is still preserved.

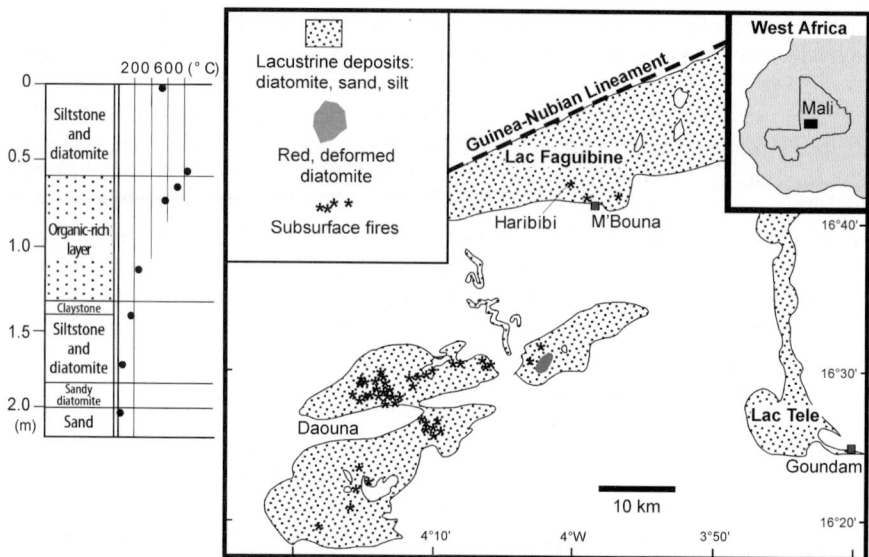

Fig. 3.54. Location map of the Lac Faguibine and Daouna areas, west of Timbuktu, Mali, showing areas of reddened diatomite and subsurface fires (after Fig. 1 of Svensen et al. 2003). Also shown is a section from a trench dug through lacustrine sediments into the heat front at Haribibi. The section contains a combusting organic-rich layer (with 8 wt.% organic C). The high temperatures reflect an increased combustion rate due to direct supply of oxygen to the peat on exposure causing the fire to evolve from smoldering to open flames (after Fig. 3 of Svensen et al. 2003). See text

One of the most interesting and historically significant examples of coal/lignite combustion pyrometamorphism occurs in the mountainous Yagnob River valley in *central Tadjiikistan*, and is known as "Ravat's fire" (e.g. Ermakov 1935; Novilov 1989, 1993; Novikov and Suprychev 1986; Belakvski 1990). Burning coking coal and lignite seams, 0.5–3.5 m in thickness, occur within a steeply dipping, folded and faulted sequence of Jurassic oil shales, multicoloured claystones, arkosic sandstone, siltstone, grit, oolitic limestone and brown iron ore. The area of burning is situated along a thrust and the present day fire covers some 25 m² about 1100 m above the Yagnob River as indicated by the presence of some 200 hot gas jets (40–590 °C). Over the last 300 years the burning front has progressed along strike for between 3.5–5 km to produce ash and coke associated with rose and red coloured clinkers annealed up to "brick-making" conditions.

The pyrometamorphic rocks are yet to be fully investigated by modern petrological methods and published descriptions relate to sulphate, chloride and nitrate mineralisation associated with the fumerolic activity. Of interest are the low temperature minerals that consist of sulphur (up to 590 °C), chalcanthite ($CuSO_4 \cdot 5H_2O$; 170–320 °C), melanterite ($FeSO_4 \cdot 7H_2O$; 200–320 °C), alunogen ($Al_2(SO_4)_3 \cdot 17H_2O$) associated with sulphuric acid, millosevichite ($(Fe,Al)_2[SO_4]_3$; < 200 °C), mascagnite ($(NH_4)_2SO_4$) associated with salammoniac, sulphur and alunogen, halotrichite-pickeringite ($(Fe,Mg)Al_2[SO_4]_4 \cdot 22H_2O$), tschermigite ($NH_4Al[SO_4]_2 \cdot 12H_2O$; 40–60 °C), kremersite ($(NH_4,K)_2Fe^{3+}Cl_5 \cdot H_2O$), and nitrammite ($NH_4NO_3$) as fine grained impurities in salammoniac. Located near to the Great Silk Road, Ravat's fire has been mined for salammoniac (NH_4Cl; < 300 °C) and alum ($Al_2(SO_4)(OH)_4 \cdot 7H_2O$) for at least 2000 years.

Unpublished results on two samples of paralava containing fragments of clinker obtained from the Fersman Mineralogical Museum, Moscow, have been supplied by Dr. E. Sokol (United Institute of Geology, Geophysics and Mineralogy, Novosibirsk). The paralava is unusual in that it is plagioclase-free (see also Cosca et al. 1989). It is characterised by quench textures between fayalitic olivine, Fe-cordierite, Ti-magnetite and tridymite in a matrix of K-Al high silica glass. Clinker fragments are intensively fused and melted to K-Al high silica glass that contains cordierite, mullite and tridymite. With reference to the system $FeO-Al_2O_3-SiO_2$ (Fig. 3.4; Table 3.1), the two mineral assemblages imply temperatures of 1210 °C (Cd Mul Td) and 1083 °C (Fa Cd Td).

Several examples of coal-combustion related pyrometamorphism within a vast area of Jurassic age coal seams associated with sandstone, conglomerate, claystones (some dominated by kaolin) and siltstone sequences are known from *Kirghiza* (Dahergaly, Kok-Moinok, Kok-Yangak, Kyzyl-Kiy and Sulyuktin) and *Uzbekistan* (Angren) (Zbarskiy 1963). Pyrometamorphic products of extensive burning, in some cases down to depths exceeding 500 m, include annealed but unmelted siltstones and grits, sandstones and pelitic rocks transformed to clinkers, and partially melted rocks that indicate temperatures of ~1000–1200 °C.

England

Cretaceous bituminous (oil) shale at *Ringstead Bay* on the Dorset coast spontaneously ignited after landsliding occurred in 1826 and continued to burn for several years, producing *"... volumes of dense, suffocating smoke [H_2S] which, from its specific gravity, seldom rose high into the air. This was followed by bluish flames, rising at times so far above the cliff as to be visible from Weymouth [~6.5 km away]. Through the cracks spread over the surface by the ascending heat the burning substratum beneath was seen. The fissures and other openings were covered with deposits of sulphur."* (Damon 1884). Another landslip locality which was burning between November 1973 and February 1974, is described by Cole (1974). Rocks showed evidence of baking and fumeroles were active from joints and cracks. The absence of kaolinite in the baked shales implies temperatures in excess of 550 °C. In both cases, combustion is attributed to pyrite oxidation after landsliding occurred.

Mali

In the Timbuktu region of northern Mali, Svensen et al. (2003) describe subsurface temperatures as high as 765 °C associated with smoke emanating from holes and fractures in an area of several square kilometers and caused by the combustion of organic material in lacustrine sediments (Fig. 3.54). Numerous thin (2–5 cm) "dikes" (called *daounites*) fill fracture networks within diatomatious sediments and have vitrified walls with the glass containing cristobalite, clinopyroxene and magnetite (Sauvage and Sauvage 1992). A 2.5 m deep trench excavated into the heat front revealed a burning organic-rich (8 wt.% C) layer at 60 cm depth where temperatures reached 830 °C in sediments immediately overlying the burning layer (Fig. 3.54). The residue of combustion consists of elemental carbon and traces of Fe-oxide and mullite. Below the combusting layer the temperature is only 40 °C. H_2O, CO_2 and trace CH_4 are released from the combustion and precipitates of salammoniac (NH_4Cl), ammonium hydrogen sulphate ($NaAl(SO_4)_2(H_2O)_{12}$), sulfur, amorphous silica and sodium alum ($NaAl(SO_4)_2(H_2O)_{12}$) occur at the surface around holes and fractures (fumeroles).

Svensen et al. (2003) propose that initiation and evolution of the subsurface combustion is the result of lowering of water level in the lake, followed by lowering of the water table, drying and microbial decomposition of organic material resulting in heat accumulation, self ignition and slow combustion resulting in baking, reddening and localised melting of overlying diatomite concomitant with formation of surface collapse features. Occurrences of red diatomite of shallow lake deposits throughout the Trans-Saharan region suggests that the current burning phenomena is the latest manifestation of a history of subsurface combustion extending back until at least the Pleistocene.

3.4
Lightning Strike Metamorphism

Winans Lake

One of the most detailed studies of lightning-induced terrestrial pyrometamorphism is the study made by Essene and Fisher (1986) on a glassy fulgurite that extends for 30 m along a morainal ridge in southeastern Michigan (Fig. 3.55). The fulgurite consists of several masses, the largest of which has a maximum diameter of 0.3 m and a length of 5 m, and the fulgurite bodies extend laterally rather than vertically comprising a branching system of subcylindrical structures.

The occurrence is characterised by containing micrometer- to centimetre-sized metallic globules showing diverse intergrowths of metals (Fe, Si), silicides ($FeSi$, Fe_3Si_7, $FeTiSi_2$), and phosphides (TiP, Fe_3P) which have unmixed from a vesicular silica-rich melt. The glass also contains relic crystals of resorbed quartz, zircon that is partly decomposed to ZrO_2 and SiO_2, and graphite presumably originating from reduction of organic matter in the soil. Thermodynamic calculations indicate that temperatures of melting were in excess of 2000 K with reducing conditions approaching those of the SiO_2-Si buffer. The decomposition of zircon indicates temperatures in excess of 1950 K.

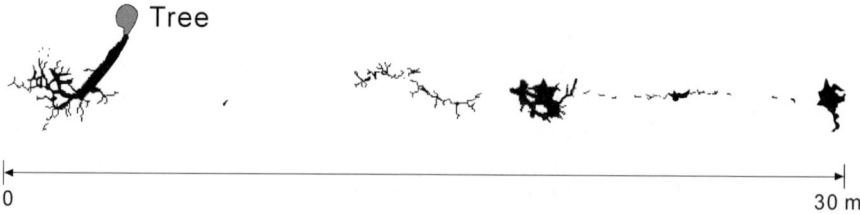

Fig. 3.55. The Winans Lake fulgurite formed along a morainal ridge in southeastern Michigan, USA. Fulgurite development is concentrated in three separate loci with continuity between the loci indicated by charred soil and vegetation (redrawn from Fig. 1 of Essene and Fisher 1986)

Several reduction mechanisms are proposed to explain the formation of the metal globules in the Winans Lake fulgurite:

1. Oxidation of carbon caused smelting of fine grained moraine material to form metallic liquids by reduction of silicates, a process similar to that of iron-refining.
2. Degassing of oxygen or formation of nitrous oxide gases could have enhanced the reduction process. Nitrous oxides which form in milliseconds from air at high temperatures (> 5000 K) during the passage of lightning may scavenge oxygen from the fulgurite.
3. Vapourisation of oxygen during boiling (as evidenced by the vesicular nature of the glass) may have involved the flow of electrons through the melt causing local reduction and nearby loss of oxygen as in electrolysis.

3.5
Vapour Phase Crystallisation

Eifel

Hexagonal crystals of cordierite (indialite) with between 0.55–0.96% K_2O occur in vugs within metapelitic xenoliths in volcanic rocks of the Eifel area, Germany (Hentschal 1977; Schreyer et al. 1990). In one case (Bellerberg), cordierite is associated with tridymite, sanidine, osumilite, biotite, and topaz; in another (Herchenberg) with cristobalite, mullite, pseudobrookite and hematite. The additional osumilite group phases, roedderite ($(Na,K)_2Mg_5Si_{12}O_{30}$) and eifelite ($KNa_3Mg_4Si_{12}O_{30}$) have also been found. Hentschel et al. (1980) describe the first terrestrial occurrence of Na/Mg and Fe/Mg varieties of roedderite that forms euhedral yellow and reddish-brown crystals in melt-coated gas-expansion cavities within pyrometamorphosed sillimanite-quartz gneiss xenoliths (containing tridymite, sanidine, spinel, ortho- and clinopyroxene, hematite) in tephritic lava of the Bellerberg volcano. Colourless, to faint yellow and green idiomorphic eifelite, also occurring in vesicles in gneiss xenoliths from the same locality, is associated with tridymite, pyroxene, amphibole, pseudobrookite and hematite (Abraham et al. 1983). The occurrence of both minerals implies precipitation from highly alkaline, Mg-Si-rich but Al-poor gas phases that are probably related to the well-known alkali metasomatic sanidinite xenoliths, often containing aegirine, described from the Eifel area (see above section on sanidinite).

Vico Volcanic Complex

Sanidine-rich xenoliths (+ minor phlogopite, oxides) possibly derived from the pyrometamorphosed envelope of a shallow magma body of the Vico Volcanic Complex (Latium, Italy) contain vugs with intimately intergrown light blue flattened hexagonal prisms of osumilite and semiradial aggregates of acicular pseudobrookite (with 3.9–4.4% V_2O_5) (Parodi et al. 1989). The vug walls are lined with sanidine, titanite and hematite. Fluids causing high temperature oxidation of titanomagnetite to hematite-pseudobrookite in the host xenoliths were also responsible for the formation of osumilite-pseudobrookite in the vugs under fO_2 conditions > MH buffer at temperatures between 720–850 °C and pressures < 800 bar.

Chapter 4

Calc-Silicates and Evaporates

4.1
Calc-Silicates

Siliceous carbonate rocks can be divided into protoliths that contain variable proportions of mainly dolomite, calcite and quartz that produce high temperature metamorphic assemblages of Ca-silicates, Ca- + CaMg-silicates and CaMg- + Mg-silicates with increasing amounts of primary dolomite. The relationship between Ca-, CaMg-silicates, carbonates, lime and bulk siliceous carbonate compositions is shown in terms

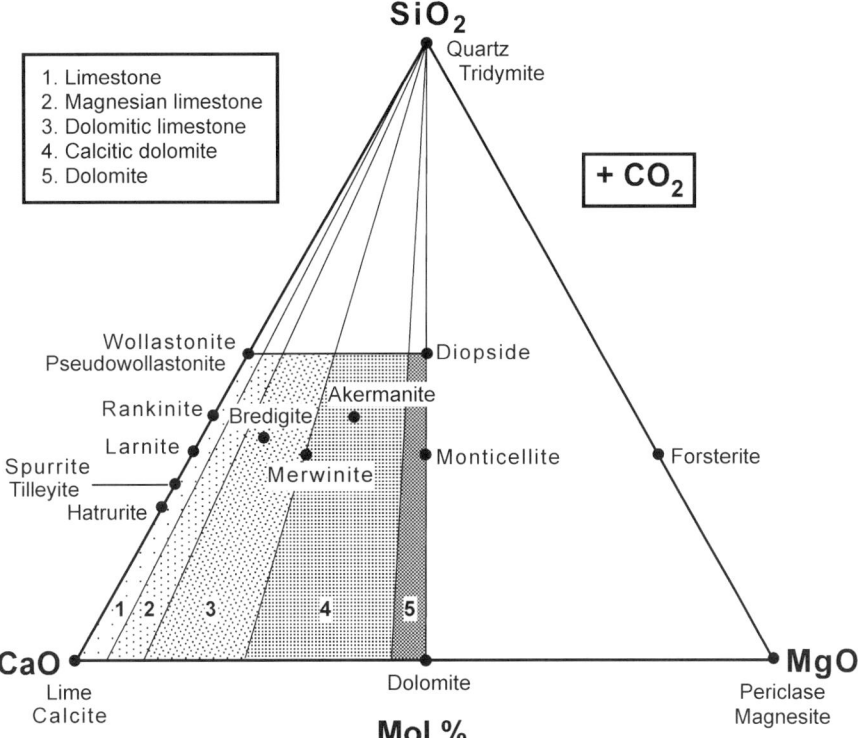

Fig. 4.1. Siliceous carbonate bulk rock nomenclature and sanidinite facies mineral compositions in terms of mol% CaO-MgO-SiO_2-CO_2. Carbonate nomenclature is adapted from Pettijohn (1949)

of mol% CaO – MgO – SiO$_2$ – CO$_2$ in Fig. 4.1. Most primary bulk compositions plot in the area below ~20 mol% SiO$_2$ and those that plot on the Mg-rich side of the diopside-dolomite tie line may indicate contamination from mafic magma during metamorphism. Addition of Al$_2$O$_3$, iron and alkalis, either from accessory (detrital and authigenic) K-feldspar, albite, micas/chlorite, clays and Fe-oxides/siderite or by diffusive interaction with mafic magma, gives rise to marly carbonate, marl, clayey marl compositions and with high temperature metamorphism, Ca-Al-Fe-Mg silicates, oxides (Fig. 4.2) and K-Na-bearing phases such as kalsilite, leucite and possibly phlogopite and nepheline. The common occurrence of accessory perovskite and sulphides in high temperature siliceous marly carbonate assemblages implies a rutile/ilmenite and sulphate source respectively and/or contamination from mafic magma. In many of the examples given below heterogeneous protoliths with variable amounts of argillaceous and carbonate components (often present as fine interlayering) are typical and give rise to the juxtiposition of different sanidinite facies mineral assemblages formed at the same T-P but different XCO$_2$.

In the system SiO$_2$-CaO-MgO-CO$_2$, pyrometamorphism of siliceous carbonates containing quartz, dolomite and calcite involves reactions (7) to (13) of Bowen's (1940) H$_2$O-absent decarbonation series with addition of rankinite and tilleyite (Tilley 1951; Table 4.1) resulting in the formation of sanidinite facies *dolomite + quartz-absent* assemblages (12) to (26) as depicted in Fig. 4.3a and b (Turner 1948; Fyfe et al. 1958; Turner and Verhoogen 1960). Additional decarbonation steps producing larnite and lime (Weeks 1956; Table 4.1) are not represented.

Table 4.1. Decarbonation steps in sanidinite facies (dolomite, quartz and H$_2$O-absent) metamorphism of siliceous carbonate (after Turner 1967)

Bowen (1940)	Tilley (1951)	Weeks (1956)
Step		Step
(7) 2Cc + Di + Fo = 3Mc + 2CO$_2$		(6) 2Cc + Di + Fo = 3Mc + 2CO$_2$
		(7) Cc + Fo = Mc + Pe + CO$_2$
(8) Cc + Di = Åk + CO$_2$		(8) Cc + Di = Åk + CO$_2$
(9) Cc + Fo = Mc + Pe + CO$_2$		
(10) 3Cc + 2Wo = Sp + 2CO$_2$		
	2Cc + Wo = Ty + CO$_2$	
	3Ty + Wo = 2Sp + CO$_2$	
	Sp + 4Wo = 3Rn + 2CO$_2$	
(11) Cc + Åk = Mw + CO$_2$		
(12) Wo + Sp = 3Ln + CO$_2$		(9) Cc + Rn = 2Ln + CO$_2$
(13) Sp + Åk = 2Ln + Mw + CO$_2$		
		(10) Cc = Lm + CO$_2$

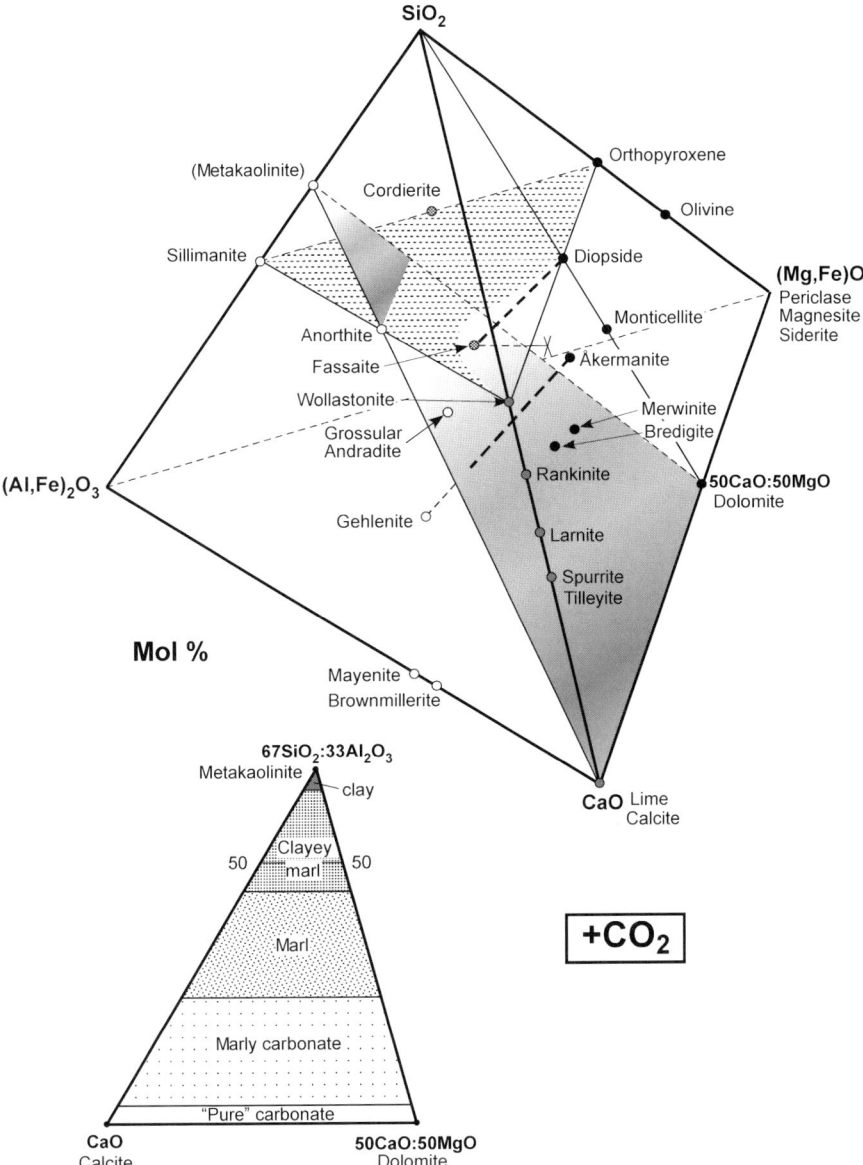

Fig. 4.2. Mineral compositions in siliceous carbonate and marly carbonate rocks plotted in terms of mol% CaO – (Mg,Fe)O – (Al,Fe)$_2$O$_3$ – SiO$_2$ – CO$_2$. *Dashed lines* represent solid solution between åkermanite-gehlenite and diopside-fassaite. Spinel and magnetite (not shown) plot at 50(Mg,Fe)O : 50(Al,Fe)$_2$O$_3$. Nomenclature of the carbonate-clay (metakaolinite) composition plane (*below* and *grey shaded area in tetrahedron*) is adapted from Pettijohn (1949). The sillimanite-orthopyroxene-wollastonite plane (*hatched*) at 50 mol% SiO$_2$ = ACF projection plane. Siliceous carbonate and marly carbonate bulk compositions plot below this plane, mostly within the anorthite-wollastonite-calcite-diopside volume

Fig. 4.3.
Monticellite-melilite subfacies (**a**) and larnite-merwinite-spurrite subfacies (**b**) mineral assemblage fields (numbered *12* to *26*) in terms of mol% $CaO - MgO - SiO_2 - CO_2$. See text

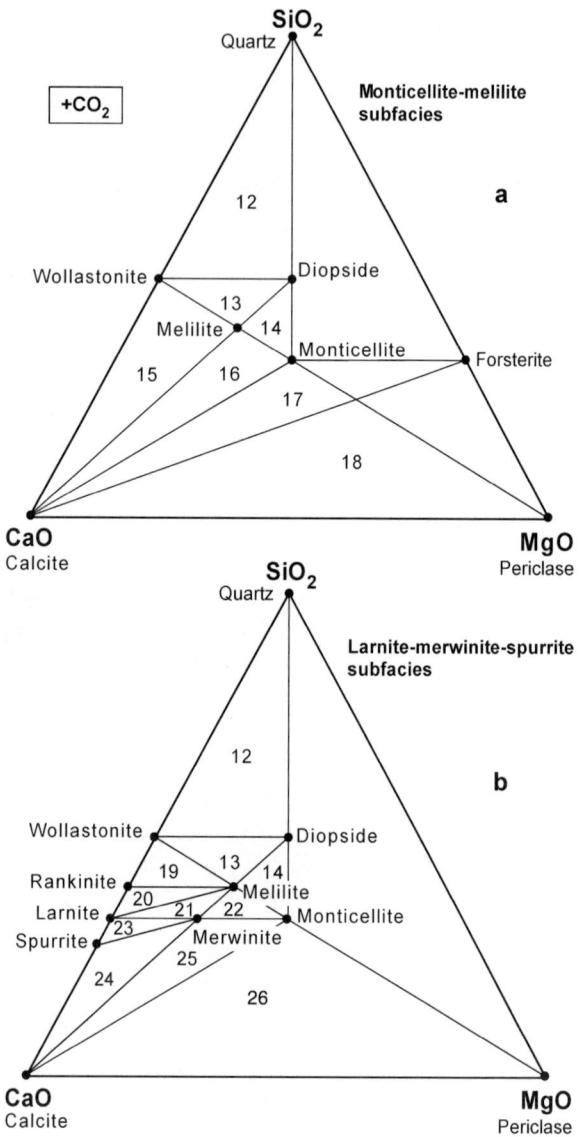

Assemblages

12. Wollastonite-diopside-quartz
13. Diopside-wollastonite-melilite
14. Diopside-monticellite-melilite

are stable at all temperatures (represented in both Fig. 4.3a,b). The remainder have been divided into two subfacies (Fyfe et al. 1958; Reverdatto 1965):

a lower temperature – *monticellite-melilite subfacies*
   15. Wollastonite-melilite-calcite
   16. Melilite-monticellite-calcite
   17. Monticellite-forsterite-calcite
   18. Forsterite-periclase-calcite
b higher temperature – *larnite-merwinite-spurrite or spurrite-merwinite subfacies*
   19. Wollastonite-rankinite-melilite
   20. Rankinite-melilite-larnite
   21. Melilite-larnite-merwinite
   22. Melilite-merwinite-monticellite
   23. Larnite-merwinite-spurrite
   24. Merwinite-spurrite-calcite
   25. Merwinite-monticellite-calcite
   26. Monticellite-periclase-calcite

Simple high temperature marly carbonate compositions can be represented in terms of ACF parameters by the assemblage

anorthite–wollastonite (pseudowollastonite)–diopside–hedenbergite

(Fig. 4.4). Additional silica-undersaturated phases are represented by projection of gehlenite-åkermanite, fassaite, and grossular-andradite as shown in Fig. 4.2. Univariant mineral assemblages relevant to fusion of marl compositions synthesised in experimental oxide and mineral systems at atmospheric pressure are listed in Table 4.2.

4.1.1
CO_2-H_2O in Fluid Phase

The temperature of decarbonation reactions in siliceous carbonates will depend on the ratio of CO_2 and H_2O in the fluid phase. In the presence of an H_2O vapour phase, decarbonation can occur at considerably lower temperatures as illustrated by the reaction

   åkermanite + calcite = merwinite + CO_2

that is important during metamorphism of dolomitic limestone–calcic dolomite compositions. Experimental determination of this reaction has been made by Shmulovich (1969), Bulatov (1974), Zhou and Hsu (1992) and calculated by Walter (1963), Joesten (1976), Zharikov et al. (1977) and Sharp et al. (1986), and some of these data are shown in Fig. 4.5a and b in terms of T-P and T-XCO_2 respectively. In Fig. 4.5a, for example at 500 bar (*P*fluid = PCO_2), there is a maximum temperature difference of ~60 °C between the reaction curves and at 150 bar the difference is reduced to ~30 °C. At 500 bar the reaction temperature of curve 3 is lowered by nearly 200 °C though reduction of XCO_2 in the fluid phase from unity to 0.1. In Fig. 4.5b, the reaction curve is displaced to lower temperatures and higher XCO_2 if *P*fluid is lowered from 1000 to 500 bar. The merwinite-producing reaction thus highlights the difficulty in constraining natural

Fig. 4.4.
ACF plot of minerals typically present in metamorphosed marly carbonates. Compositions plot within the An – (Di,Hd) – (Wo,Pwo) volume and are projected from below the ACF plane (Fig. 4.2). *Thick lines* indicate extent of possible solid solution

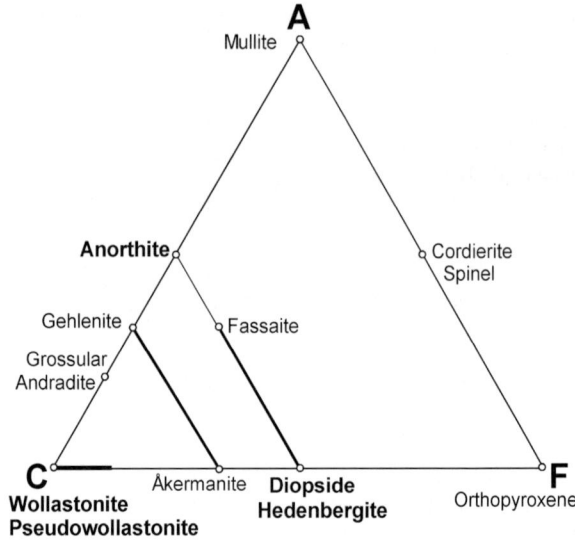

Table 4.2. Invariant assemblages + liquid (L) in ternary oxide and mineral systems relevant to fusion of marly rocks

Invariant assemblage	Ternary system	T (°C)[a]	Reference
Oxide systems			
Wo-Pwo-Mel-L	CMS	1 360	Ricker and Osborn (1954)
Pwo-Di-Tr-L	CMS	1 320	Ricker and Osborn (1954)
Pwo-Di-Mel-L	CMS	1 350	Ricker and Osborn (1954)
Wo-Pwo-An-L	CMS	1 245	Osborn (1942)
Pwol-Mel-An-L	CAS	1 265	Schairer (1942)
An-Wo-Pwo-Mel-L	CMAS	1 205	Chinner and Schairer (1962)
Mineral systems			
Wo-An-Px-L	Wo-Di-An	1 236	Osborn (1942)
Wo-Pwo-An-L	Wo-Di-An	1 245	Osborn (1942)
An-Mel-Px-L	Di-An-Åk	1 226	de Wys and Foster (1958)
An-Mel-Hc-L	An-Wo-Wü	1 130	Schairer (1942)
An-Mel-Wo-L	An-Wo-Wü	1 125	Schairer (1942)

[a] ±5 °C or less unless otherwise stated.
CAS = CaO-Al$_2$O$_3$-SiO$_2$; CMS = CaO-MgO-SiO$_2$; CMAS = CaO-MgO-Al$_2$O$_3$-SiO$_2$.

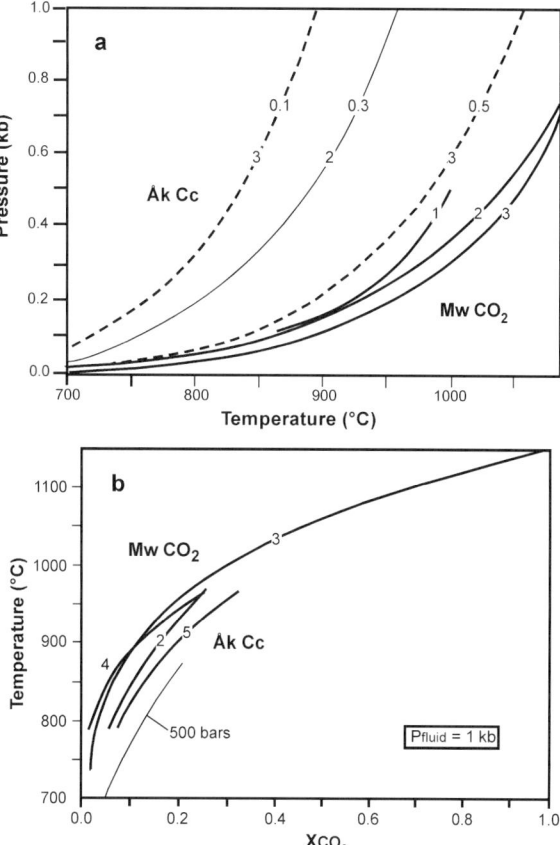

Fig. 4.5.
Experimentally- and thermodynamically-derived stability curves of the akermanite + calcite = merwinite + CO_2 reaction in terms of Ptotal = Pfluid = PCO_2-T (**a**) and XCO_2-T at Pfluid = 1 kb (**b**). In (**a**), *curves 1* (Joesten 1976), *2* (Zharikov et al. 1977), *3* (Sharp et al. 1986) represent conditions where XCO_2 in the fluid phase = 1.0. *Curve 3* is progressively displaced to lower temperatures at 0.5 and 0.1 XCO_2 (labelled *dashed curve 3*) (data from Tracy and Frost 1991), and similarly for *curve 2* at 0.3 XCO_2 (labelled *thin solid curve 2*) (Zharikov et al. 1977). In (**b**), numbered curves are: 2 (Zarikov et al. 1977); 3 (Sharp et al. 1986); 4 (Shmulovich 1969); 5 (Walter 1963). The reaction curve for 500 bar is from Fig. 5 of Tracy and Frost (1991)

divariant assemblages in terms of temperature and fluid composition without an independent pressure estimate or without location of univariant assemblages.

As decomposition and dehydroxylation of H_2O-bearing minerals, e.g. clays, muscovite, chlorite, occur at much lower temperatures that those reached during pyrometamorphism, it would be expected that the CO_2-content of a fluid phase would increase with increasing temperature in carbonate rocks (e.g. Fyfe et al. 1978; Roedder 1984). Although the absence of hydrous phases in textural equilibrium with calc-silicate and CO_2-bearing phases does not necessarily mean that H_2O was absent from the fluid phase during high temperature metamorphism, the almost exclusive presence of CO_2-bearing minerals such as calcite, spurrite and tilleyite in sanidinite facies siliceous carbonates implies that the fluid phase during pyrometamorphism was probably CO_2-rich, in most cases with XCO_2 near unity, and also that temperatures were too high to stabilise any hydrous phase. In the case of igneous pyrometamorphism, as H_2O is significantly more soluble in mafic magma than CO_2 (e.g. Burnham 1979), it will tend to diffuse from country rock and xenoliths undergoing heating into the magma thereby enriching the remaining fluid in CO_2. In the absence of an independent estimate of lithostatic pressure at the time of metamorphism, one is usually forced to assume that Pload = Pfluid = PCO_2. The assumption may be justified by the fact that during peak temperatures of metamorphism, influx of H_2O into xenoliths or near contact rocks is unlikely because partial fluid pressures created by decarbonation reactions equalise the total pressure of the system and the partial pressure of the fluid phase within the magma is usually much lower. Furthermore, rapid heating and concomitant degassing during decarbonation leads to a volume loss. Provided that volume loss occurs simultaneously with compaction, magmatic H_2O-rich fluids will be inhibited from invading the rocks. Nevertheless, in several cases, e.g. Crestmore, California (Burnham 1959), hydrous silicates such as talc, brucite, tremolite, serpentinite, xanthophyllite, and F-bearing hydrosilicates such as clinohumite, cuspidine, vesuvianite occur in pyrometamorphosed calc-silicate rocks. This suggests that wallrock/xenolith compaction probably lagged behind volatile loss allowing an influx of H_2O, F and Cl-rich fluids leading to rehydration and metasomatism during cooling or perhaps reheating to lower temperatures by subsequent intrusion. Similarly, cooling of carbonate rocks that have undergone combustion metamorphism induces cracking allowing the influx of H_2O (containing some CO_2 and SO_3) resulting in the formation of retrograde assemblages of hydrous silicates, hydroxides, carbonates and sulphates. The end result is that the many metamorphosed calc-silicate rocks are complex high-variance disequilibrium assemblages comprising relic (high temperature, CO_2-rich) and later (lower temperature H_2O-rich) phases.

4.1.2
T-P-XCO$_2$ Relations

Siliceous Limestone

Silicate minerals within the system CaO-SiO_2-CO_2, wollastonite, pseudowollastonite, larnite, tilleyite, spurrite, more rarely rankinite and very rare hatrurite, are diagnostic of sanidinite facies metamorphism of *siliceous limestone* lithologies. Larnite is actually the β-polymorph of the high temperature form α'-Ca_2SiO_4 which inverts to larnite, the other being calcio-olivine (γ-Ca_2SiO_4). Phase relations of these Ca-silicates are well known from many studies (e.g. Rankin and Wright 1915; Tuttle and Harker 1957; Harker 1959; Zharikov and Shmulovich 1969; Joesten 1974; Treiman and Essene 1983) and P-T-XCO_2 stability relations pertinent to sanidinite facies reactions are shown in Fig. 4.6a and b. The vapour-absent reactions

spurrite + wollastonite = rankinite + tilleyite (> 920 °C)

spurrite + rankinite = larnite + tilleyite (> 1020 °C)

wollastonite = pseudowollastonite (> 1125 °C)

larnite + lime = hatrurite (> 1250 °C)

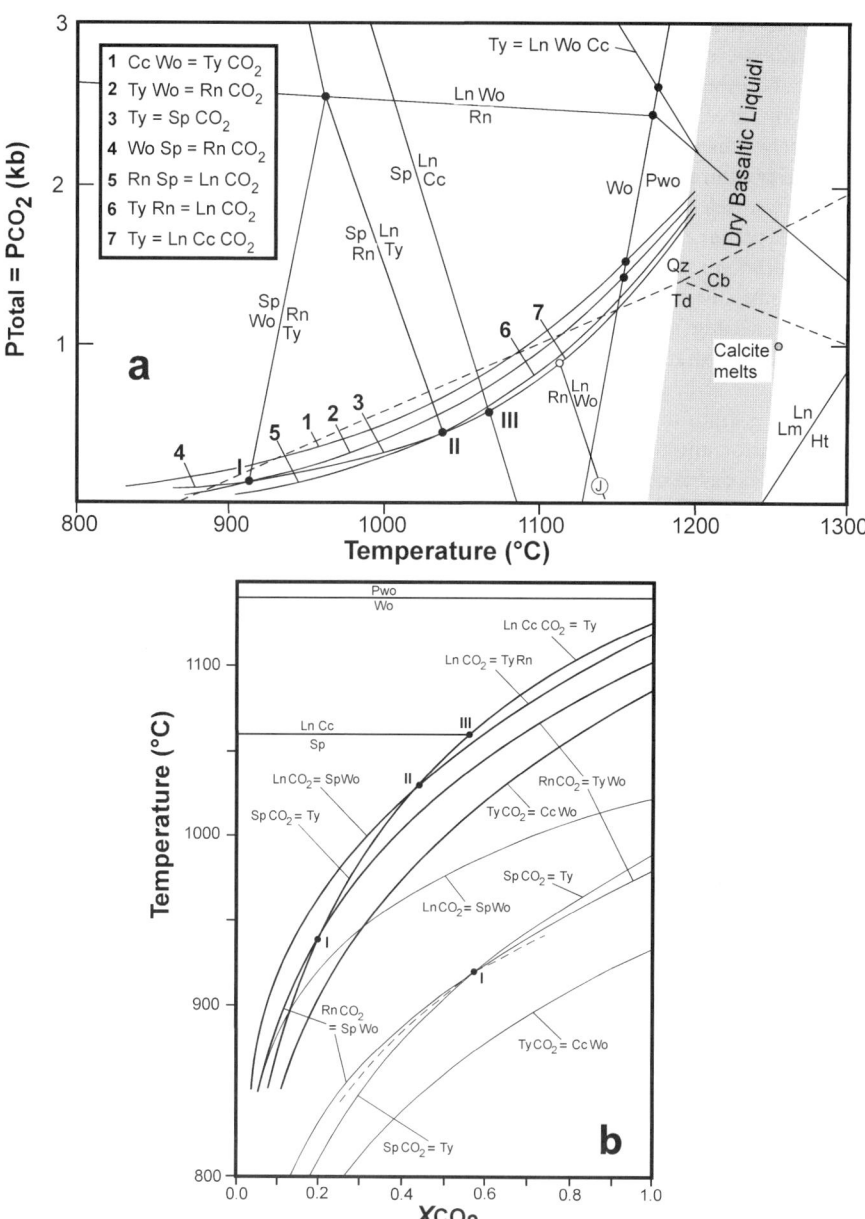

Fig. 4.6. Calculated equilibria in the system CaO-SiO$_2$-CO$_2$. **a** T-Ptotal = PCO$_2$ (after Figs. 8, 9 of Treiman and Essene 1983). The Rn = Ln Wo reaction curve labelled "J" is from Joesten (1974). Quartz-tridymite-cristobalite inversion curves and field of dry basalt liquidi (*shaded strip*) are included for reference. **b** Isobaric T-XCO$_2$ equilibria at Ptotal = 1000 bar (*heavy lines*; after Fig. 10 of Treiman and Essene 1983) and at Ptotal = 300 bar (*thin lines*; after Joesten 1974).

(Fig. 4.6a) are not affected by the ratio of $CO_2:H_2O$ in the fluid phase and provide the most reliable temperature indicators of high temperature metamorphic conditions.

Under isothermal-isobaric conditions stability fields of the calc-silicates are controlled by the chemical potentials of SiO_2 and CO_2 as shown in Fig. 4.7 where each phase represents a variance of four and each reaction line is trivariant. Equilibria involving larnite-rankinite and rankinite-wollastonite fix the value of μSiO_2 while μCO_2 may vary independently. Similarly, μCO_2 is fixed by tilleyite-spurrite over a range of μSiO_2. All other two phase assemblages are stable over a range of μSiO_2 and μCO_2 values, although they both must lie on a trivariant surface.

At the highest temperatures of pyrometamorphism, liquidus relationships in the systems $CaO-MgO-CO_2-H_2O$ (Wyllie 1965) and $CaO-SiO_2-CO_2-H_2O$ (Wyllie and Haas 1966), indicate that melting of calcite according to the reaction

calcite = liquid + CO_2

should be possible although indisputable textural evidence from nature has not been reported (for a possible exception see Schulling 1961). Where the vapour phase is almost

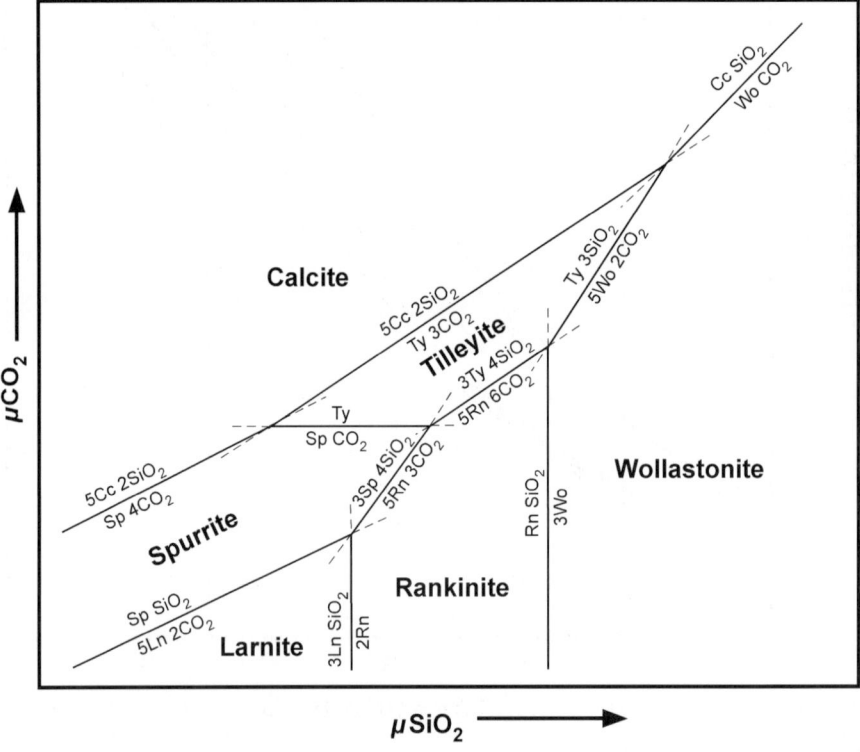

Fig. 4.7. Schematic isothermal-isobaric chemical potential diagram (μSiO_2-μCO_2) for the system $CaO-SiO_2-CO_2$ showing calc-silicate stability fields (after Fig. 7 of Joesten 1974)

pure H_2O, at 1 kb the melting temperature of calcite is 650 °C, whereas if the vapour phase is pure CO_2 calcite melts at just over 1300 °C (Wyllie and Tuttle 1960).

The highest temperature decarbonation reaction without melting that could result from pyrometamorphism is the reaction

calcite = lime + CO_2

that occurs between 900 and 1250 °C at 1 and 40 bar PCO_2 respectively (Treiman and Essene 1983). Natural examples of this reaction are rare but have been recorded by in limestone xenoliths in tephritic lava from the Eifel (Ettringer Bellerberg) (Brauns 1922) and Vesuvius (Zambonini 1935, p. 66).

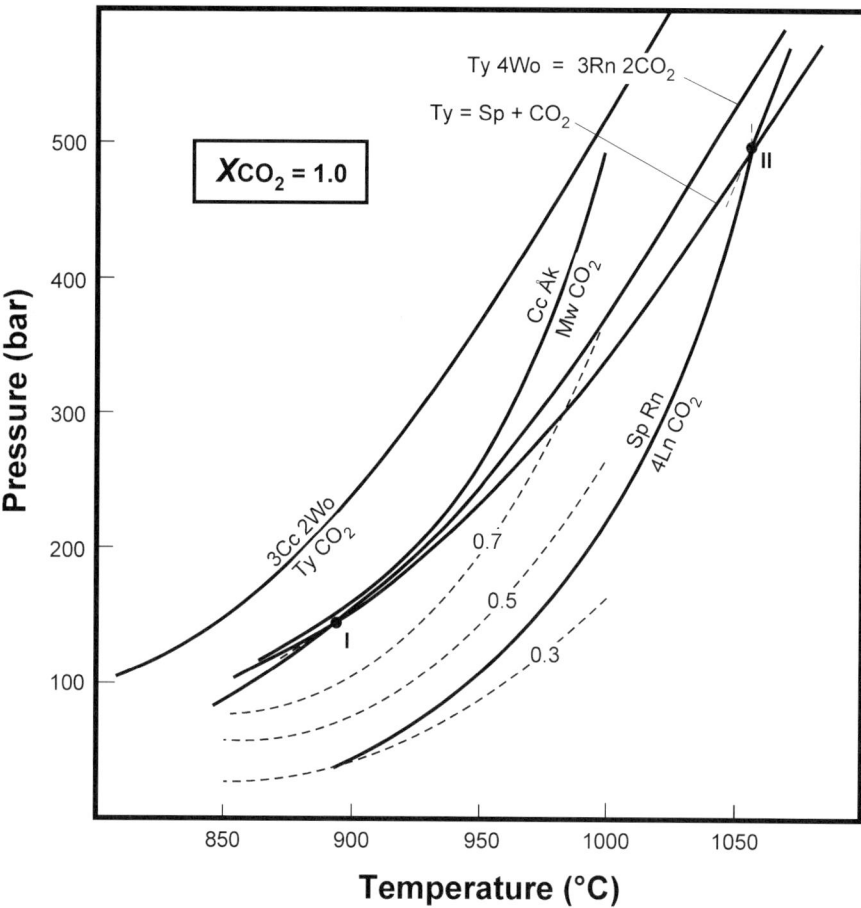

Fig. 4.8. *P-T* diagram at $XCO_2 = 1$ showing the Cc Åk = Mw CO_2 reaction in comparison to calc-silicate decarbonation reactions (after Fig. 2 of Joesten 1976). *Dashed lines* numbered with Åk mol fraction in melilite indicate shift of the Cc Åk = Mw CO_2 reaction to lower P/higher T with *decreasing* Åk component

Siliceous Magnesian Limestone – Dolomite

These compositions (Fig. 4.1), are characterised by:

1. Occurrence of the Mg-Ca silicates, melilite, merwinite, rarely bredigite *with* calc-silicates in *magnesian limestone-dolomitic limestone* protoliths. Where there is sufficient alumina and iron, e.g. at carbonate-calcite-free shale contacts of banded sequences, gehlenitic melilite, fassaitic pyroxene, Ca-rich plagioclase, granditic garnet and spinel occur. Fe^{3+}-bearing peraluminous examples are characterised by the oxides mayenite and brownmillerite.
2. Åkermanite, merwinite, diopside, monticellite, forsterite, ± periclase typically *without* calc-silicates are diagnostic of *siliceous calcitic dolomite and dolomite* protoliths. Addition of alumina and in some cases alkalis results in the formation of spinel, phlogopite, kalsilite, and K-feldspar.

Of critical importance in less magnesian rocks is the decarbonation reaction, Mel Cc = Mw CO_2. In addition to the effect of fluid composition mentioned above, the position of this reaction curve is also affected by the Åk-content of melilite as shown in Fig. 4.8 in relation to stabilities of Ca-silicates. For more magnesian rocks, high temperature reaction curves involving calcite, åkermanite, merwinite, monticellite, diopside, forsterite and periclase are shown in Fig. 4.9a and b in terms of T-P (Pfluid = PCO_2) and XCO_2-T respectively.

Marl

T-P conditions of metamorphism of marls are represented by the mineral stability curves in Fig. 4.10a with upper limits of sanidinite facies metamorphism defined by the melting reactions

   An Wo/Pwo Qtz/Td V = L

   An Wo/Pwo Ge V = L

that at low pressures intersect the field of dry basalt liquidi between ~1175–1250 °C. Reactions involving An, Ge, Wo, Gr and Cc in terms of XCO_2-T at 500 and 1000 bar are given in Fig. 4.10b and show that decarbonation reaction curves shift to higher T and lower XCO_2 at lower pressure with an upper limit of grossular stability given by the invariant point at ~900 °C/0.65 XCO_2 at 500 bar. Where appropriate, these data can be used in combination with Ca-silicate, CaMg- and Mg-silicate stability curves shown in Figs. 4.6, 4.8, 4.9 to better constrain conditions of metamorphism. Temperatures high enough to partially melt marly carbonate rocks are attained during combustion metamorphism and some invariant assemblages produced in ternary oxide and mineral systems at atmospheric pressure are listed in Table 4.2. As bulk marl (CO_2-, SO_3- and H_2O-free) compositions are typically > 96 wt.% SiO_2, Al_2O_3, Fe_2O_3 (as total iron), MgO, CaO, they can be approximated by grossular-rich portions of the grossularite-pyrope and grossularite-andradite joins in the systems CaO-MgO-Al_2O_3-SiO_2 (Chinner and

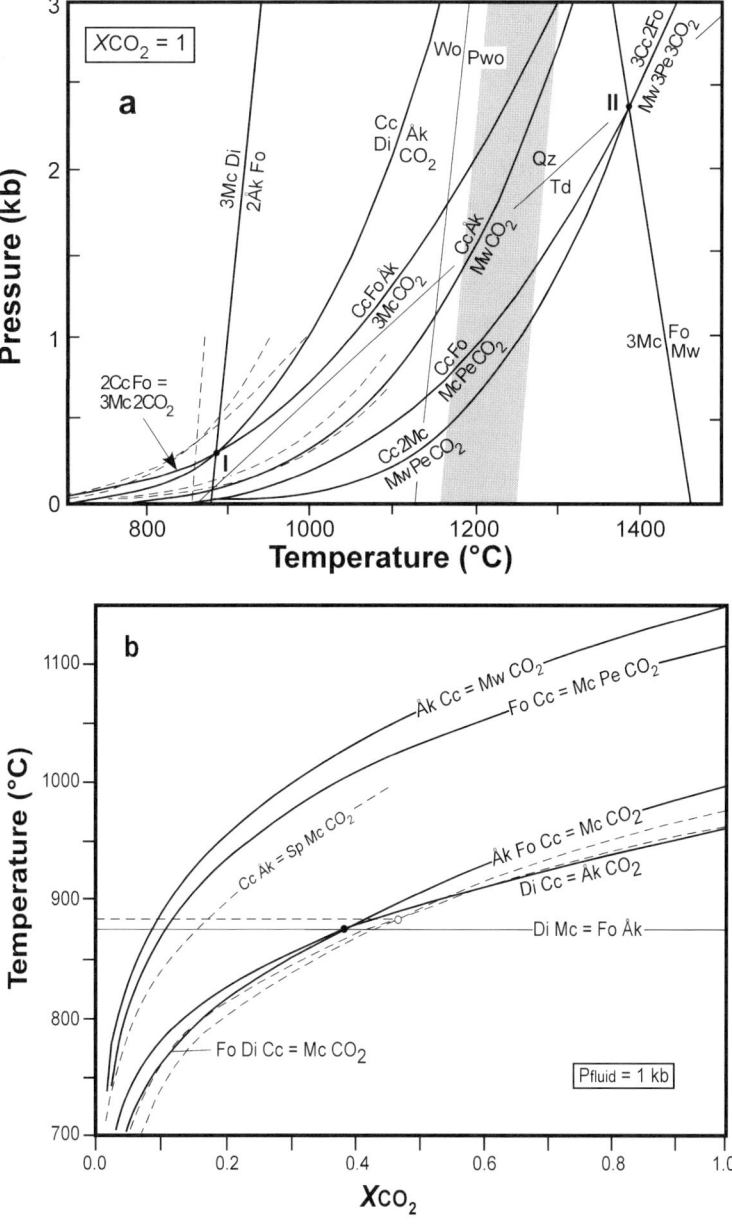

Fig. 4.9. Calculated equilibria in the system CaO-MgO-SiO$_2$-CO$_2$. **a** T-Ptotal = PCO$_2$ (after Fig. 9 of Wallmach et al. 1989). *Dashed reaction curves* from Fig. 4 of Sharp et al. (1986) are included for comparison and are displaced to lower T and higher P with respect to those calculated by Wallmach et al. (1989). *Shaded strip* = dry basalt liquidi. **b** Isobaric T-XCO$_2$ equilibria at Pfluid = 1000 bar. *Solid curves* after Fig. 3 of Sharp et al. (1986). *Dashed curves* after Pertsev (1977) and reproduced in Fig. 6 of Tracy and Frost (1991)

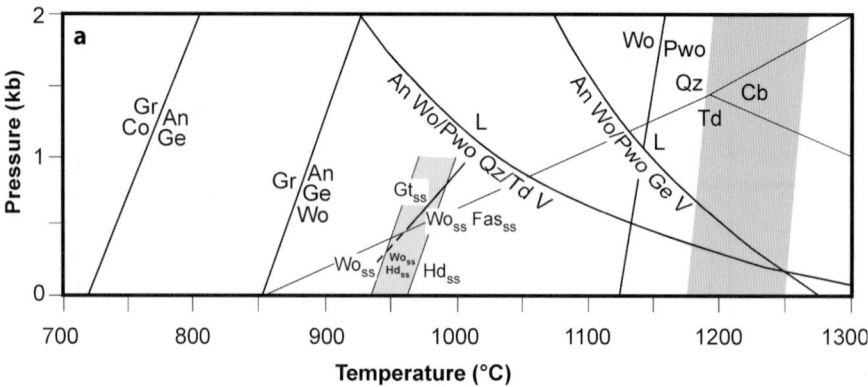

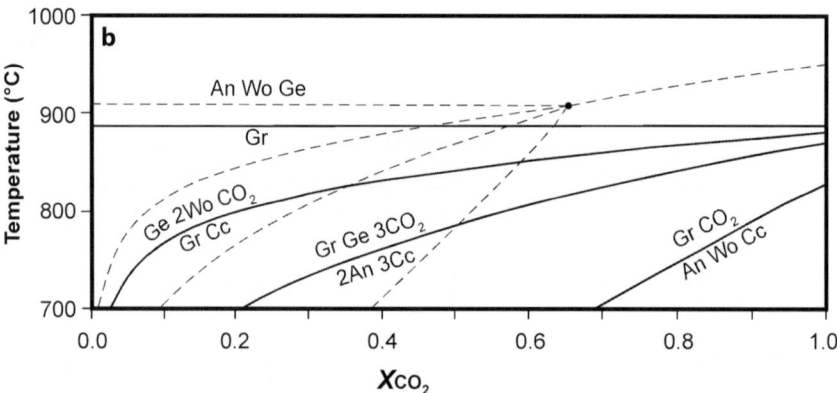

Fig. 4.10. Phase equilibria relevant to mineral stabilities in marly carbonates. **a** Data for reactions in the system CaO-Al$_2$O$_3$-SiO$_2$-H$_2$O (after Fig. 16 of Huckenholz et al. 1975). Wo$_{ss}$ = Hd$_{ss}$ reaction from Lindsley et al. (1969); Gt$_{ss}$ = Wo$_{ss}$ + Fas$_{ss}$ reaction from Huckenholz et al. (1974). *Shaded strip* = dry basalt liquidi. **b** T-XCO$_2$ diagram at 500 bar (*dashed curves*) and 1000 bar (*solid curves*) for the system CaO-Al$_2$O$_3$-SiO$_2$-H$_2$O-CO$_2$ after Figs. 10 and 11 respectively of Tracy and Frost (1991)

Schairer 1962) and CaO-Al$_2$O$_3$-Fe$_2$O$_3$-SiO$_2$ (Huckenholz et al. 1974) shown in Fig. 4.11. In the MgO-bearing system, the garnet join cuts two relevant subsolidus tetrahedra, An-Di-Åk-Wo and An-Åk-Ge-Wo that include pseudowollastonite at temperatures above the wollastonite-pseudowollastonite inversion at 1125 °C (Fig. 4.11). Within the compositional range Gr$_{100}$ and Gr$_{82}$Py$_{18}$, the subsolidus assemblage Wo-Mel-An occurs with an invariant point assemblage Pwo-Wo-Mel-An-L located at ~1205 °C. The composition of melilite ranges from Åk$_{75}$Ge$_{25}$ at Gr$_{82}$Py$_{18}$ to Ge$_{100}$ at Gr$_{100}$. Investigation of this system by Yang et al. (1972), indicates addition of Al-diopside to the assemblage near 1200 °C on the join åkermanite-anorthite-forsterite. In the system CaO-Al$_2$O$_3$-Fe$_2$O$_3$-SiO$_2$, compositions between Gr$_{87-54}$ produce the assemblage Pwo/Wo-An-Mel(Ge-Fe-Ge$_{ss}$)-Fas$_{ss}$ at temperatures > 1135 ± 10 °C and with liquid over a narrow temperature interval between 1203–1208 ± 3 °C. In compositions > Gr$_{87}$ over this temperature range fassaitic pyroxene is not stable (Fig. 4.11).

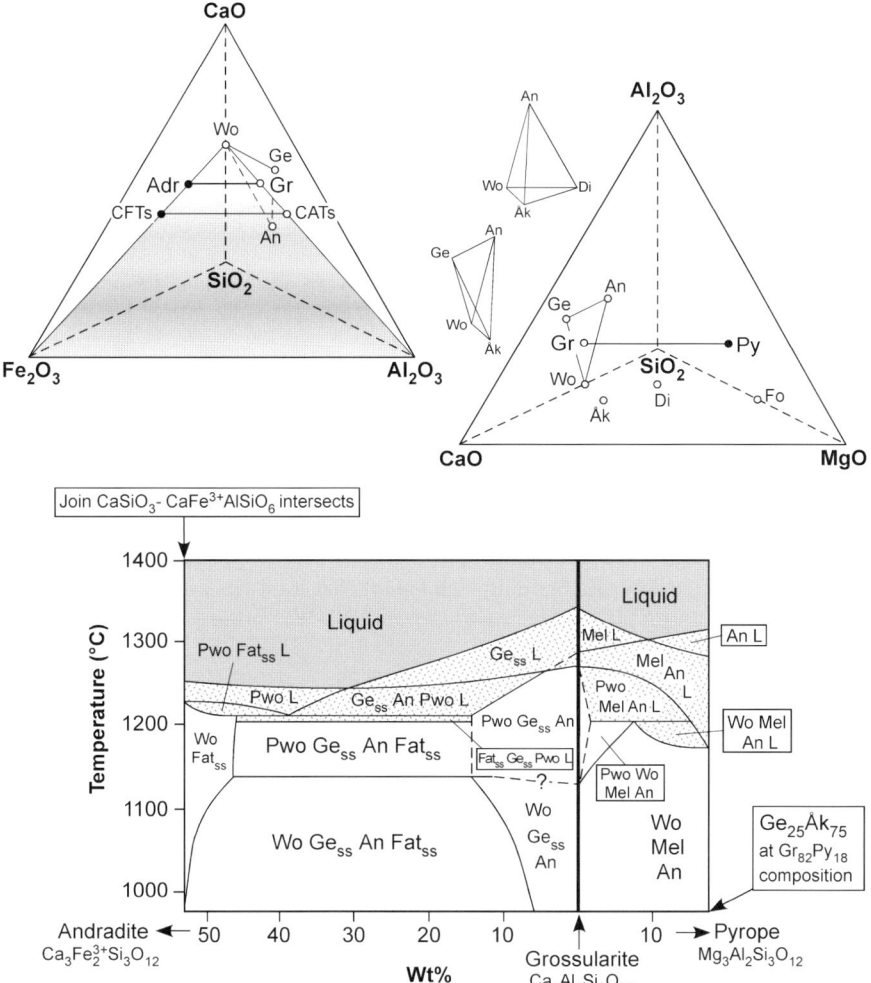

Fig. 4.11. Temperature versus composition plot showing mineral-liquid relationships at atmospheric pressure along part joins grossular-andradite (Huckenholz et al. 1974) and grossular-pyrope (Chinner and Schairer 1969) (*below*) in the systems $CaO-Al_2O_3-Fe_2O_3-SiO_2$ and $CaO-MgO-Al_2O_3-SiO_2$ (*above*)

4.1.3
Contact Aureoles and Xenoliths

4.1.3.1
Siliceous Limestone

Scawt Hill
Flint nodules in chalk pyrometamorphosed to larnite-spurrite-bearing rocks by dolerite at Scawt Hill, Northern Ireland, are surrounded by thin reaction zones of wol-

lastonite separated from larnite by a film of rankinite (Tilley and Harwood 1931; Tilley 1942) that reflects a decrease in μSiO_2 potential between wollastonite-larnite by way of the CO_2-independent reactions

3 wollastonite = rankinite + SiO_2

3 larnite + SiO_2 = 2 rankinite

(Fig. 4.7). Comparison between Scawt Hill and a similar dolerite at Carneal that intrudes Tertiary basaltic lavas overlying the same Cretaceous chalk and flint indicates a lithostatic pressure of ~200 bar (Sabine 1975). Provided Pload = Pfluid during metamorphism, the reactions

rankinite + spurrite = larnite + CO_2

rankinite + CO_2 = spurrite + wollastonite

indicate an intersection temperature of 980 °C from metastable extension of the latter reaction through invariant point (I) in Fig. 4.6a. This is consistent with experimental data of Zharikov and Shmulovich (1969) that indicates *minimum* temperatures of 950°C and 1025 °C at which [Rn Ln] and [Rn Wo] occur at 100 and 300 bar PCO_2, respectively (Fig. 4.12).

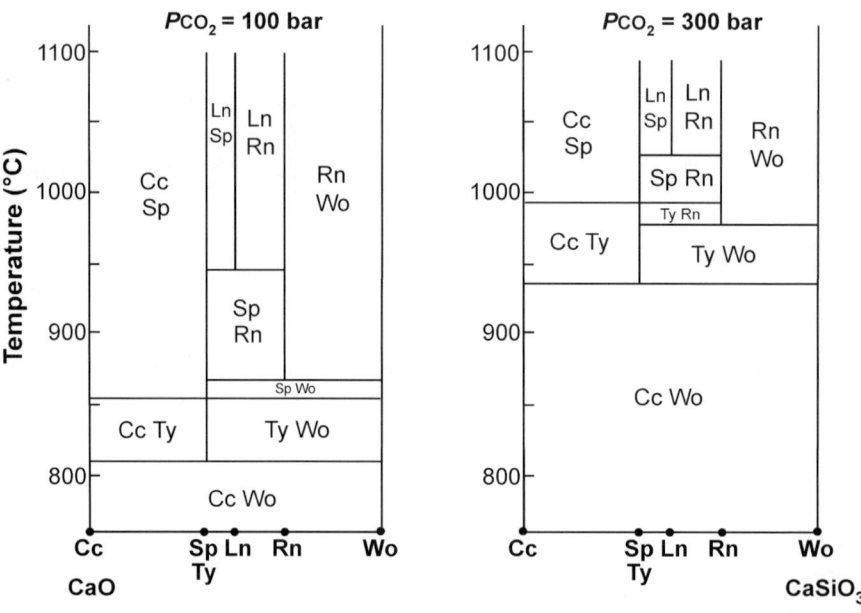

Fig. 4.12. *T* versus CaO-CaSiO$_3$ mineral stability diagrams at 100 and 300 bar PCO_2 in the high temperature part of the system CaO-SiO$_2$-CO$_2$ (after Fig. 9 of Zharikov and Shmulovich 1969)

Christmas Mountains

Metamorphism of nodular chert within limestone to produce sharply-bounded concentrically zoned wollastonite, rankinite, spurrite, tilleyite and calcite bodies within 60 m of a gabbro, Christmas Mountains, Texas, is described in detail by Joesten (1974). The minerals form a series of sharply-bounded, concentric monomineralic and two-phase shells which record a step-wise decrease in SiO_2 from the nodule core to its rim. Minerals and decarbonation reactions within the nodule rims vary with distance and decreasing temperature from the gabbro contact (Fig. 4.13):

- 0–5 m   calcite – spurrite – rankinite – wollastonite
   (Sp 4 Wo = 3 Rn CO_2; Ty = Sp CO_2)
- 5–16 m   calcite – tilleyite – spurrite – rankinite – wollastonite
   (Ty = Sp CO_2; Ty 4 Wo = 3 Rn 2 CO_2)
- 16–31 m   calcite – tilleyite – wollastonite
   (3 Cc 2 Wo = Ty CO_2)
- 31–60 m   calcite – wollastonite
   (Cc Qz = Wo CO_2)
- > 60 m   calcite – quartz

The absence of larnite in the assemblages indicates that the temperature of the reaction

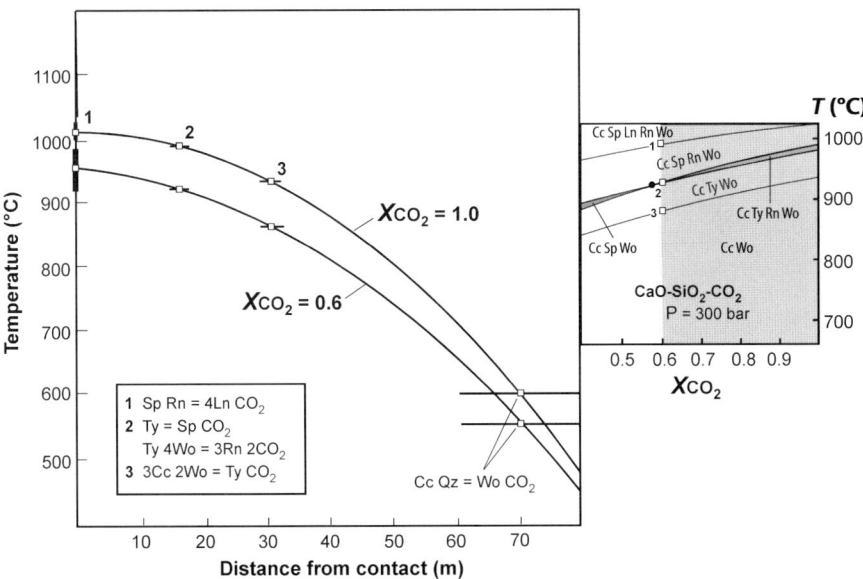

Fig. 4.13. Gradients of maximum temperatures attained in the Christmas Mountains contact aureole determined from positions of reactions divariant in P-T-XCO_2. *Horizontal bars* = uncertainly in location of reaction from sample spacing. *Inset diagram* shows position of reactions in terms of T-XCO_2 at 300 bar within the *shaded area* between XCO_2 = 0.6 to 1.0 (after Figs. 4 and 5 of Joesten 1974)

spurrite + rankinite = 4 larnite + CO_2

was not exceeded.

Idealised composition-distance profiles across chert and calc-silicate nodules with increasing temperature in the contact aureole are shown in Fig. 4.14. Heating of a chert nodule (quartz-calcite in a) to a temperature within the range bounded by reactions

Fig. 4.14.
Composition (mol% SiO_2) versus distance profiles across monominerallic zones for nodules within the contact aureole of the Christmas Mountains gabbro, Texas. The schematic zoned nodule shown is located within 6–16 m of the igneous contact. Closer to the contact, tilleyite is replaced by spurrite. Arrows in composition-distance profiles indicate direction of movement of boundary of each growing zone within the range of its T-XCO_2 stability field, i.e. T increasing from (**a**) to (**d**); XCO_2 = 0.6–1.0; P = 300 bar (Fig. 4.13). *Dashed line* indicates initial position of the calcite-quartz interface of chert nodule

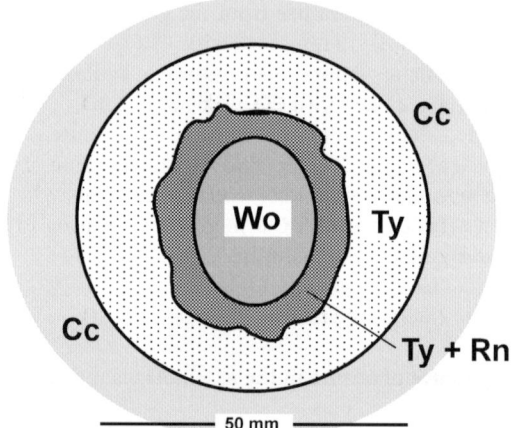

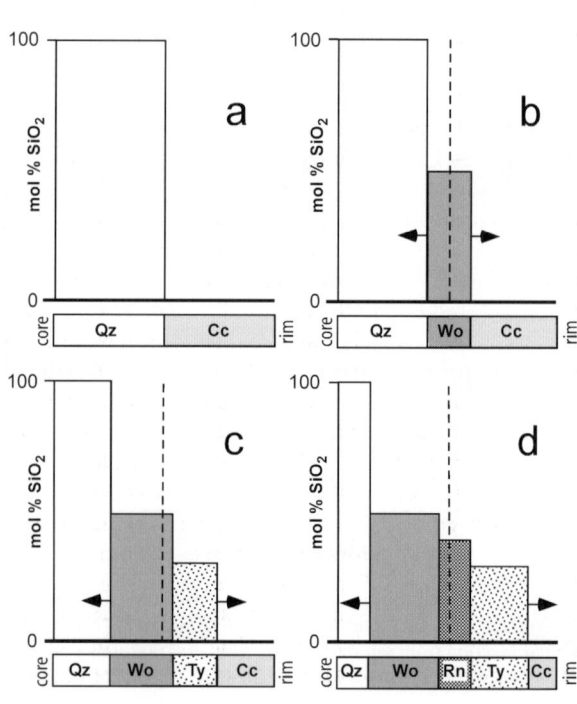

calcite + quartz = wollastonite + CO_2

3 calcite + 2 wollastonite = tilleyite + CO_2

produces a thin zone of wollastonite at the quartz-calcite interface (b) that will grow at the expense of both quartz and calcite until quartz is used up. At a higher temperature within the range bounded by reactions

3 calcite + 2 wollastonite = tilleyite + CO_2

tilleyite + 4 wollastonite = 3 rankinite + 2 CO_2

a tilleyite zone grows at the expense of both calcite and wollastonite (c). The wollastonite zone continues to grow only at the contact with quartz. In d the compositional gradient across the nodule is the result of heating to a temperature within the field bounded by reactions

tilleyite + 4 wollastonite = 3 rankinite + 2 CO_2

tilleyite = spurrite + CO_2

resulting in the growth of rankinite at the expense of tilleyite and wollastonite. Decarbonation of tilleyite to from spurrite does not affect the shape of the compositional gradient for SiO_2 and does not alter the direction of movement of the zone boundaries.

The relative growth rates of each calc-silicate zone depends on the relative diffusion rates of CaO (from calcite) and SiO_2 (from quartz). Thus, widening of the tilleyite zone will occur by reaction of transported SiO_2 with calcite only if the flux of SiO_2 through tilleyite is > 0.6 times that through rankinite. The radial symmetry of the distribution of mineral assemblages in the nodules, transport of CaO and SiO_2 in opposite directions, compatibility of minerals in adjacent zones, and continuity of the chemical potentials across zone boundaries imply that the mineral zoning resulted from diffusion with gradients in the chemical potentials across the monominerallic zones being the driving force for diffusion and element mobility. Prograde mineral reactions that define the zone boundaries involve release of CO_2 which would be expected to diffuse radially outward under the externally generated pressure gradient, mixing with and displacing any water in the vicinity of the reacting nodules to produce a localised CO_2-saturated environment. If the coexistence of tilleyite and rankinite in nodules within 15 m of the igneous contact represents stable equilibrium then the projected position of invariant point (I) limits fluid compositions to those with $XCO_2 > 0.6$ (Fig. 4.13). This precludes the possibility that heat was convectively transferred by a H_2O-rich magmatic fluid outward into the aureole. In the presence of a CO_2-rich fluid and a stratigraphically determined lithostatic pressure of 325 bar, the unusually high temperatures of between 875–1025 °C attained within 32 m of the igneous contact are attributed to conduction as the dominant energy transport mechanism. This could have resulted from the emplacement of a 200 m thick sheet of convecting mafic magma along the contact between marble and a partly crystallised 1400 m diameter cylindrical stock of gabbro/syenite.

Tokatoka

Monominerallic patches and bands of rankinite (as large plates up to 8 mm long and partly altered to kilchoanite), fine grained larnite, spurrite with ubiquitous grossular, accessory gehlenite and Fe-oxide occur in siliceous limestone in contact with dykes of basalt and andesite from Tokatoka, North Island of New Zealand (Mason 1957; Black 1969; Baker and Black 1980). The coexistence of the larnite, spurrite, rankinite, and the absence of tilleyite implies metamorphic conditions equating with the reaction

$$\text{spurrite} + \text{rankinite} = \text{larnite} + CO_2$$

that occurs at temperatures between ~900–1045 °C with $P\text{total} = PCO_2 < 480$ bar (Fig. 4.6a). However, the different grain size and separate occurrence of rankinite with respect to larnite and spurrite may indicate that the three Ca-silicates formed under localised conditions of different XCO_2. For example, at 300 bar (Fig. 4.6b), larnite is stabilised with respect to spurrite and rankinite between XCO_2 0.8 and 0.2 over the temperature range 1000–900 °C respectively. Spurrite is stabilised with respect to tilleyite at $XCO_2 < 0.8 / 960$ °C. In the absence of spurrite, rankinite is stable at temperatures < 920 °C/ $XCO_2 < 0.6$ and on cooling below 725 °C inverts to kilchoanite (Speakman et al. 1967). The presence of grossular indicates a maximum T of ~900 °C and XCO_2 of 0.6 at 500 bar as above this temperature it breaks down to An Wo Ge (Fig. 4.10b).

4.1.3.2
Dolomitic Limestone – Calcitic Dolomite with Al-Fe-K

Kilchoan

Agrell (1965) describes pyrometamorphosed nodular or concretionary dolomitic limestone-calcitic dolomite occurring in a ca. 61 m wide screen between two gabbroic ring dykes at Kilchoan, Scotland. The limestone has undergone two episodes of metamorphism; one characterised by high fluid $CO_2 : H_2O$ and a subsequent event characterised by a high $H_2O : CO_2$ ratio. Bands up to 1.8 m thick show a complex series of mineral assemblages that probably reflect variations in the initial composition of the limestone rather than to compositional modification due to diffusive interaction with mafic magma. The highest temperature decarbonation mineral associations are:

a åkermanite + rankinite + wollastonite
b åkermanite + rankinite + spurrite
c åkermanite + larnite + spurrite
d åkermanite + merwinite + larnite + spurrite
e åkermanite + merwinite + monticellite

as depicted in Fig. 4.15 and reflect variation from Si-rich (Åk Ra Wo) to more Mg-rich (Åk Me Mc) bulk compositions across the bands, and from the coexisting Ca-silicates,

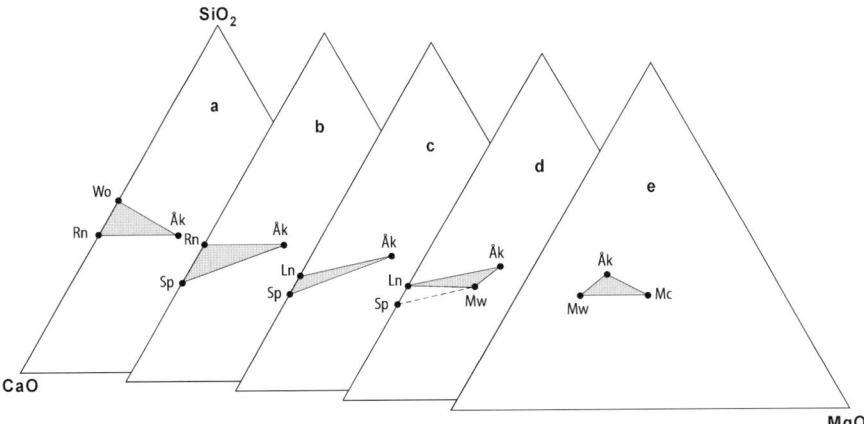

Fig. 4.15. Variation of mineral assemblages (*a–e*) in terms of CaO-MgO-SiO$_2$ (mol%) occurring within individual bands in pyrometamorphosed limestone, Kilchoan, Scotland. Bulk compositions are confined to somewhere within the *shaded areas* defined by mineral *tie lines*. *Dashed line* indicates possibility of spurrite being present in the assemblage Ln Åk Mw

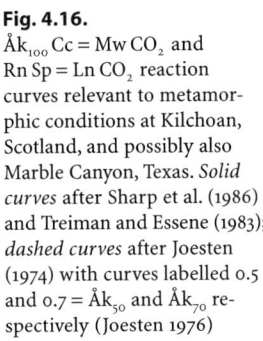

Fig. 4.16.
Åk$_{100}$ Cc = Mw CO$_2$ and Rn Sp = Ln CO$_2$ reaction curves relevant to metamorphic conditions at Kilchoan, Scotland, and possibly also Marble Canyon, Texas. *Solid curves* after Sharp et al. (1986) and Treiman and Essene (1983); *dashed curves* after Joesten (1974) with curves labelled 0.5 and 0.7 = Åk$_{50}$ and Åk$_{70}$ respectively (Joesten 1976)

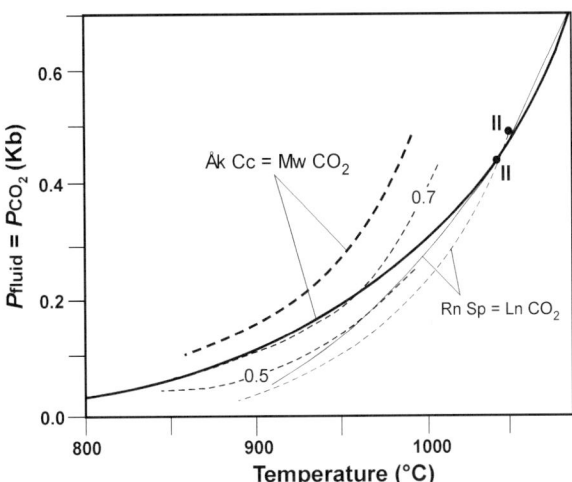

decreasing μSiO$_2$ and μCO$_2$ (Fig. 4.7) from a to e. In Fig. 4.16 at Ptotal = PCO$_2$, mineral assemblages of the bands are consistent with *T-P* conditions below a maximum of ~1050 °C and 48 bar defined by invariant point (II). Optical data by Agrell (1967) indicates a melilite composition of Åk$_{65}$Ge$_{45}$ and he infers a Pfluid = Pload for metamorphism of ~345 bar. This data gives a temperature of ~1020 °C from intersection of the Åk$_{60}$ Cc = Mw CO$_2$ (extrapolated) and Rn Sp = Ln CO$_2$ reaction curves (Fig. 4.16). The high temperature assemblages at Kilchoan were superimposed by OH and F-bearing minerals such as rustumite, kilchoanite, vesuvianite and cuspidine (with additional recrystallisation of wollastonite, grossular and spurrite), when the rocks were re-metamorphosed at lower temperatures (i.e.< ~750 °C) by an intrusion of quartz gabbro.

Marble Canyon

A rare example of three α-Ca_2SiO_5 polymorphs (α'-Ca_2SiO_5 [high temperature unnamed polymorph], larnite [β-Ca_2SiO_5] and calcio-olivine [γ-Ca_2SiO_5]) coexisting with spurrite, rankinite, merwinite and melilite of unspecified composition occurs at Marble Canyon, Texas (Bridge 1966). The Ca- and Ca-Mg silicates occur in localised zones in dolomitic to nearly pure calcium limestone with interlayered cherty horizons near the contact of syenite-monzonite-gabbro. Textural relationships show that the dicalcium silicate may occur in myrmekitic intergrowth with merwinite, rankinite and melilite, and that it cuts across the grain boundaries of these minerals (Fig. 4.17) suggesting that it may have formed by the reactions

spurrite + rankinite = 4 "larnite" + CO_2

merwinite + rankinite = 2 α'-Ca_2SiO_5 + åkermanite

Although the α'-form polymorph is termed "bredigite" by Bridge (1966) (after the high temperature Ca_2SiO_5 polymorph erroneously reported as bredigite from Scawt Hill by Tilley and Vincent 1948), electron microprobe analyses indicates that it does not contain MgO. Therefore, the Marble Canyon locality is one of few natural occurrences of the high temperature α'-Ca_2SiO_5 polymorph despite the fact that transformation of $\alpha' \rightarrow \beta$-$Ca_2SiO_5$ is rapid, with only the β and γ-forms occurring at atmospheric pressure and temperature. In the Marble Canyon rocks the only untransformed crystals of α'-Ca_2SiO_5 and larnite are those enclosed by other minerals. Cleavages are well defined and may be occupied by calci-olivine as the transformation phase (Fig. 4.17). As at Kilchoan, the larnite, merwinite, melilite, rankinite, spurrite assemblage indicates maximum temperatures at $XCO_2 = 1.0$ of less (and probably significantly less) than invariant point II in Fig. 4.16 depending on the Åk component of the melilite.

Re-examination of sanidinite facies assemblages in the Marble Canyon rocks by Anovitz et al. (1991) suggests a contact temperature of 750 ± 50 °C at 370 bar, the rocks equilibrating with H_2O-rich fluids and with most or all of the initial carbonate being

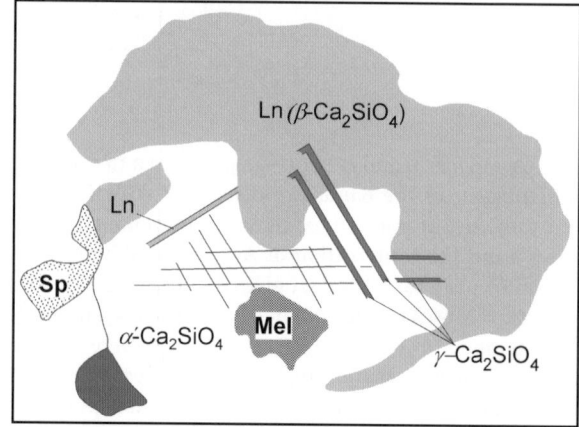

Fig. 4.17.
Textural relationship between α'- (unnamed), β- (larnite), and γ-Ca_2SiO_4 (calcio-olivine), melilite, spurrite in contact metamorphosed marble, Marble Canyon, Texas Cleavage intersects at 60° in α'-Ca_2SiO_4 (after Fig. 1 of Bridge 1966; scale not given in original diagram)

removed. Larnite is not reported and the critical mineral assemblage at the contact is Mel-Mw-Rn-Sp. Assuming an H_2O-rich fluid, this assemblage would lie on the 2Åk Sp = 2Mw Rn CO_2 reaction curve at a temperature above the kilchoanite-rankinite inversion of ~750 °C at XCO_2 between ~0.06/~750 °C and 0.09/~800 °C (Fig. 4.18). Addition of larnite to the above assemblage implies that XCO_2 could have been < 0.02 over this temperature range. Although the possibility of significant interaction between the rocks and aqueous fluids in the aureole is implied by stable isotope data, this could reflect some retrogression at lower temperature because values of < 0.02 XCO_2 suggested by Anovitz et al. (1992) seem unrealistic for the metamorphism of carbonate rocks as pointed out by Tracy and Frost (1991).

Fig. 4.18.
T-XCO_2 plot showing solid-solid and decarbonation reaction curves at low XCO_2 (< 0.1) and 500 bar (Ptotal) for a near contact assemblage of Mel-Mw-Rn-Sp at Marble Canyon, Texas. Kil = kilchoanite (β-$Ca_3Si_2O_7$) (after Pertsev 1977; reproduced as Fig. 5 of Tracy and Frost 1991). The *grey shaded area* = possible T-XCO_2 conditions of metamorphism. See text

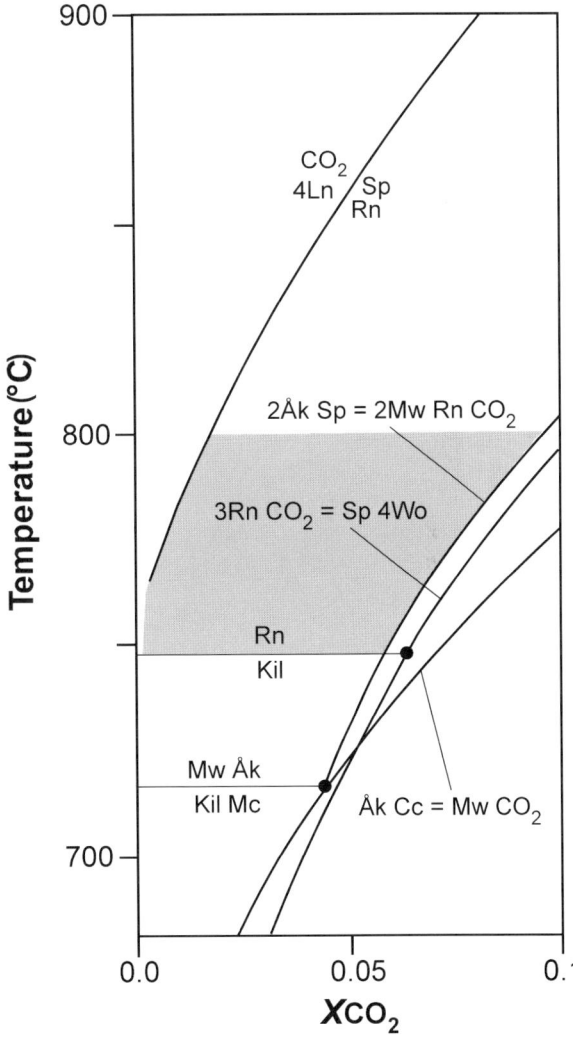

Scawt Hill

At Scawt Hill (see above), chalk within 0.6 m of a dolerite contact is metamorphosed to calcite-bearing assemblages of:

- spurrite
- larnite ± spurrite
- spurrite-larnite-gehlenitic melilite
- spurrite-gehlenitic melilite-merwinite-Fe-Mg spinel ± larnite

with accessory perovskite and magnetite (Fig. 4.19) (Tilley 1929; Tilley and Harwood 1931; Tilley and Alderman 1934). As the chalk is almost pure $CaCO_3$, the mineral assemblages indicate metasomatic addition of Si, Al, Fe, Mg and Ti from the mafic magma. The first two assemblages are the least affected by metasomatism and imply conditions of low μSiO_2 and μCO_2 that precluded the formation of tilleyite (Fig. 4.7). At 200 bar, intersection of the Sp Rn = Ln CO_2 and $Åk_{50} Ge_{50}$ Cc = Mw CO_2 reaction curves

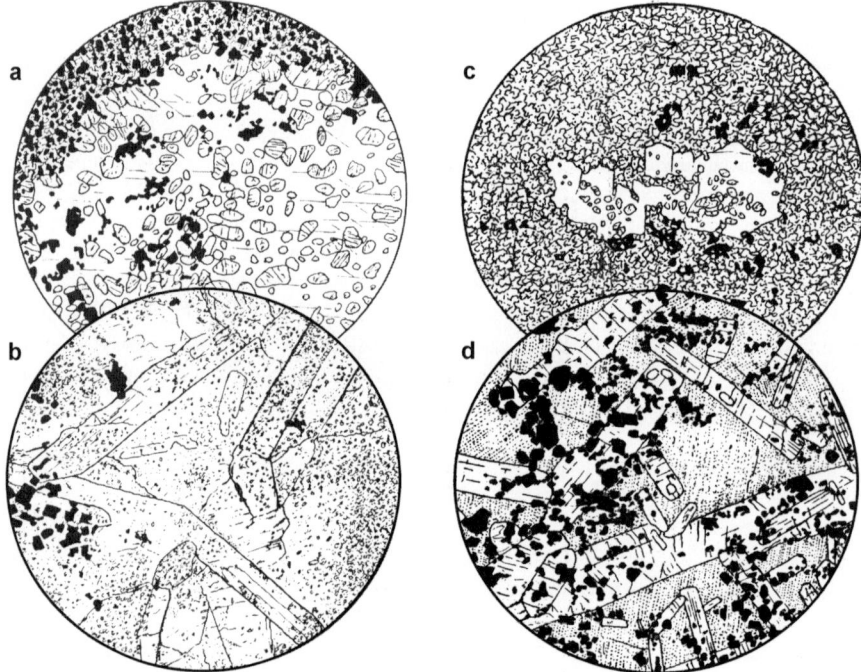

Fig. 4.19. Microtextures of Scawt Hill contact aureole siliceous carbonate rocks. **a** Spurrite-larnite rock. Porphyroblasts of spurrite enclosing grains of larnite and surrounded by fine grained larnite-spinel (×21). **b** Spurrite-merwinite rock. Large tabular crystals of merwinite within a fine matrix of spurrite with both minerals enclosing small grains of larnite (×21). **c** Larnite-spurrite rock. Large porphyroblast of spurrite enclosing grains of larnite and surrounded by fine grained larnite (×21). **d** Gehlenitic melilite-merwinite-spinel-calcite rock. Large tabular crystals of melilite and merwinite within a matrix of calcite. Dark grains are spinel and some perovskite (×21). Reproduced with the kind permission of the Mineralogical Society of Great Britain & Ireland

implies a temperature of ~975 °C (Fig. 4.16) in accordance with the Scawt Hill metamorphic temperature range given above.

Siberian Traps

A number of examples of calc-silicate rocks variously described as marls, marly limestones, limestone with interbedded dolomite, dolomitised limestone and thin discontinuous bands of chert nodules are pyrometamorphosed by dolerite of the Siberian Traps to form wollastonite, spurrite, tilleyite, merwinite, melilite and rare tridymite (Reverdatto 1970).

Chalcedony-carbonate concretions ranging from 1 to 10 cm in length and enclosed in marl within a sequence of banded limestone are largely converted to fine-grained wollastonite as at Scawt Hill, and sometimes to tridymite where wollastonite replacement is incomplete. The concretions are surrounded by an 0.3–0.8 cm thick rim of radially-arranged spurrite that is typically replaced by tilleyite near the wollastonite core. The outer part of the spurrite rim is associated with a fine aggregate of merwinite together with åkermanite and some pyrrhotite that is in contact with the surrounding marble (Fig. 4.20a). Patches thought to represent more argillaceous areas (bands, lenticles, nodules and veinlets) have central parts of merwinite containing numerous grains of åkermanitic melilite. Spurrite (partly replaced by tilleyite) with minor melilite occur in peripheral parts.

At another locality, marly limestone within 0.5 m of a dolerite contact is composed of < 1 cm thick linearly oriented stringers, bands and lenses of spurrite and gehlenitic melilite (up to 35 modal %) with calcite and accessory pyrrhotite. The melilite is commonly concentrated in the central parts of the segregations and spurrite confined to the margins, a distribution that reflects original compositional differences in the limestone.

Stratigraphic relations at the time of magma intrusion suggest a lithostatic pressure of 200–220 bar. At Ptotal = PCO_2, metamorphic temperatures are constrained to within the range 950–970 °C by the coexistence of åkermanite, merwinite (with merwinite replacing åkermanite) and replacement of spurrite by tilleyite according to the reaction

$$\text{spurrite} + CO_2 = \text{tilleyite}$$

with the Åk component of melilite > $Åk_{70}$ (Fig. 4.8). The tilleyite-forming reaction was facilitated by an increase CO_2 in the fluid phase, probably during microfracturing along grain boundaries with cooling (Fig. 4.20b).

Carneal

At Carneal, Co. Antrim, Northern Ireland, a 150 m diameter dolerite plug contains blocks of dense dark grey pyrometamorphosed chalk from Cretaceous rocks underlying basalt (Sabine 1975; Sabine et al. 1985). The xenoliths are both mineralogically variable and texturally complex. "Normal parts" consist of a fine-grained, equigranular aggregate of mainly larnite and spinel. Subsequent reaction has resulted in replacement of spinel by magnetite accompanied by the formation of aggregates of bredigite and gehlenite (94 mol% Ge-Fe-Ge), the bredigite possibly after larnite and/or merwinite according to reactions

Fig. 4.20.
Contact metamorphism of calc-silicates, Siberian Traps.
a Wollastonite (*Wo*) replacement of a chert nodule surrounded by concentric zones of spurrite-tilleyite (*Sp Ty*) and merwinite-melilite (*Mw Mel*). Retrograde cuspidine occurs as veins and replaces Sp and Ty; grossular-andradite locally replaces Sp, Ty, cuspidine etc. (after Fig. 2 of Reverdatto 1970).
b Reaction rims of tilleyite (*Ty*) around spurrite (*Sp*) in a calcite (*Cc*) matrix (× 26) (after Fig. 1 of Reverdatto 1970). See text

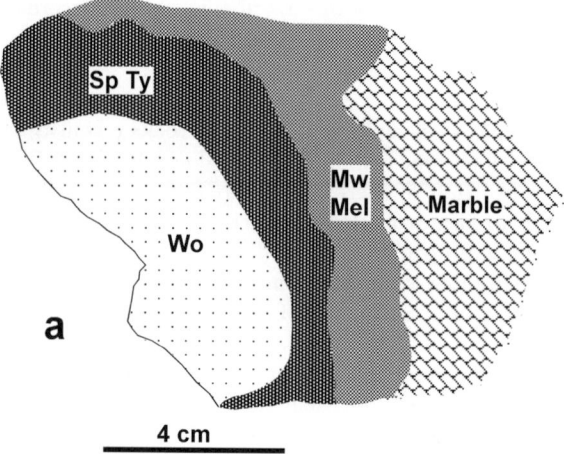

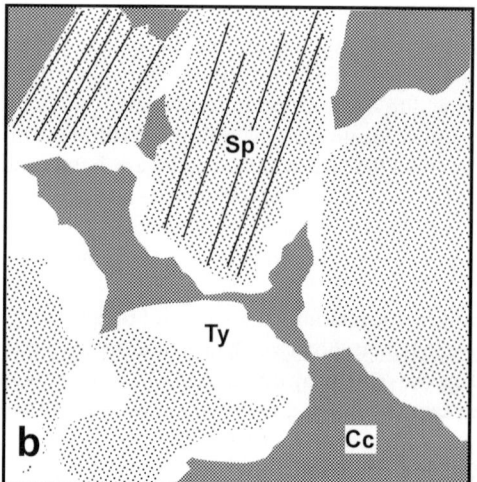

merwinite + 2 larnite = bredigite

4 larnite + åkermanite = bredigite + rankinite

merwinite + rankinite = bredigite + åkermanite

(Essene 1980). Figure 4.21 shows relationships of primary and secondary assemblages at Carneal in terms of $CaO-(Al,Fe)_2O_3-(Mg,Fe)O-SiO_2$. The diagram highlights the wide range of compositions in which bredigite-gehlenite-magnetite is stable in contrast to the restricted stability of larnite in relatively low Al and Mg compositions. More magnesian bulk compositions contain the assemblage merwinite-spinel-bredigite. Spurrite is locally abundant, occurring as porphyroblasts that enclose and replace larnite, spinel,

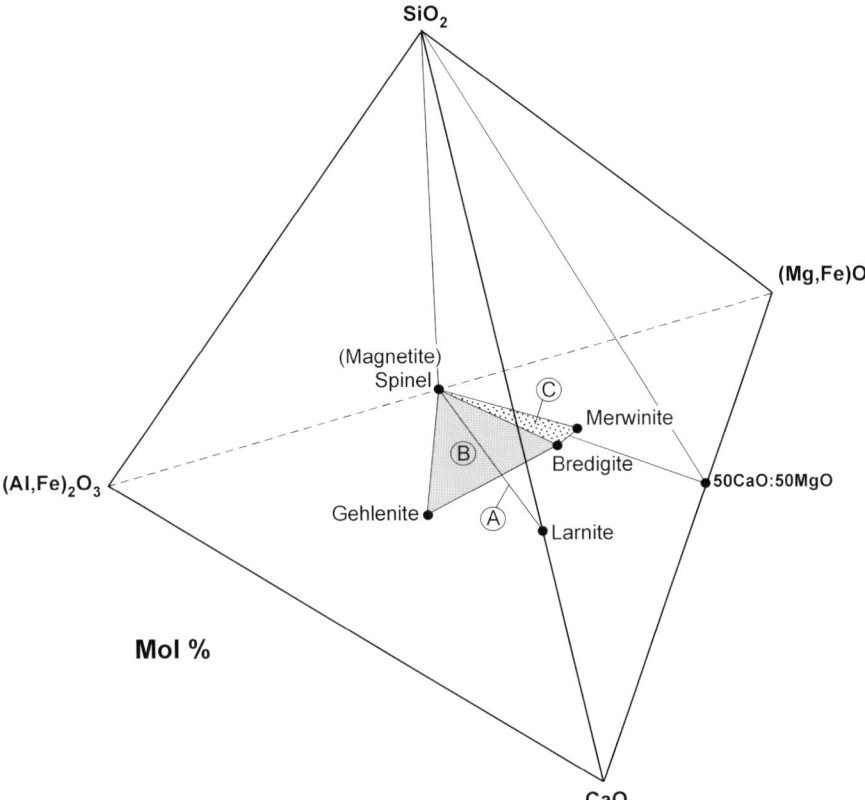

Fig. 4.21. Phase relationships of sanidinite facies mineral assemblages (A, B, C) in terms of mol% CaO-(Mg,Fe)O – (Al,Fe)$_2$O$_3$ – SiO$_2$ at Carneal, Northern Ireland. See text

bredigite and gehlenite, and as veins indicating that it formed during later addition of a CO$_2$-rich fluid. A small amount of calcite is present together with perovskite.

The larnite-spinel association, the absence of rankinite and retrograde formation of spurrite implies a *minimum* temperature of 975 °C where Ptotal = PCO$_2$, i.e. the reaction, Sp Rn = Ln CO$_2$ (Fig. 4.6a) at a lithostatic pressure of 200 bar deduced by Sabine (1975), and a merwinite-producing reaction from melilite with a composition of ~Åk$_{70}$ (Fig. 4.16) in the more Mg-rich protoliths. An upper limit of pyrometamorphism is given by the vapour-absent reaction

spurrite = larnite + calcite

that occurs at 1080 °C (Fig. 4.6b), in which case merwinite would have formed from gehlenite-rich melilite with a mole fraction of Åk < 0.3 (Fig. 4.16). The temperature range is consistent with the stability range of bredigite between 979–1372 °C (Schlaudt and Roy 1966; Lin and Foster 1975; although see Essene 1980).

Muck

The contact zone of a large olivine dolerite dyke intruding limestone on the Island of Muck, Scotland, is documented by Tilley (1947) and characterised by gehlenite-bearing sanidinite facies assemblages of:

a gehlenite-wollastonite-calcite
b gehlenite-monticellite-calcite-spinel
c gehlenite-spurrite-calcite-spinel
d gehlenite-monticellite-merwinite-calcite
e gehlenite-tilleyite-calcite-(spurrite)
f gehlenite-larnite-rankinite (highest temperature assemblage)

(Fig. 4.22) with the assemblage

monticellite-periclase (altered to brucite)-spinel-calcite

forming from interbanded dolomite. Under conditions of $XCO_2 = 1.0$, intersection of the vapour-absent reaction

spurrite + rankinite = larnite + tilleyite

with the decarbonation reactions

rankinite + spurrite = larnite + CO_2

tilleyite + rankinite = larnite + CO_2

tilleyite = larnite + calcite + CO_2

at invariant point II (Fig. 4.6a) gives T-P conditions of 1040 °C and 430 bar. These values are compatible with the formation of merwinite and of monticellite + periclase from forsterite + calcite in the interbanded dolomitic compositions (Fig. 4.9a).

Christmas Mountains

Melilite, merwinite and sometimes bredigite in marble within 1.5 m of the gabbro contact and developed around marble xenoliths in the gabbro at Christmas Mountains, Texas, documented by Joesten (1976) is a further example that probably involved diffusive exchange of Ca, Fe, Mg, Al and Si between magma and carbonate wallrock. A generalised sequence of prograde mineral zones is:

- vesuvianite + wollastonite
- melilite + wollastonite
- melilite + rankinite + spurrite ± wollastonite
- melilite + spurrite
- melilite + spurrite + calcite ± merwinite

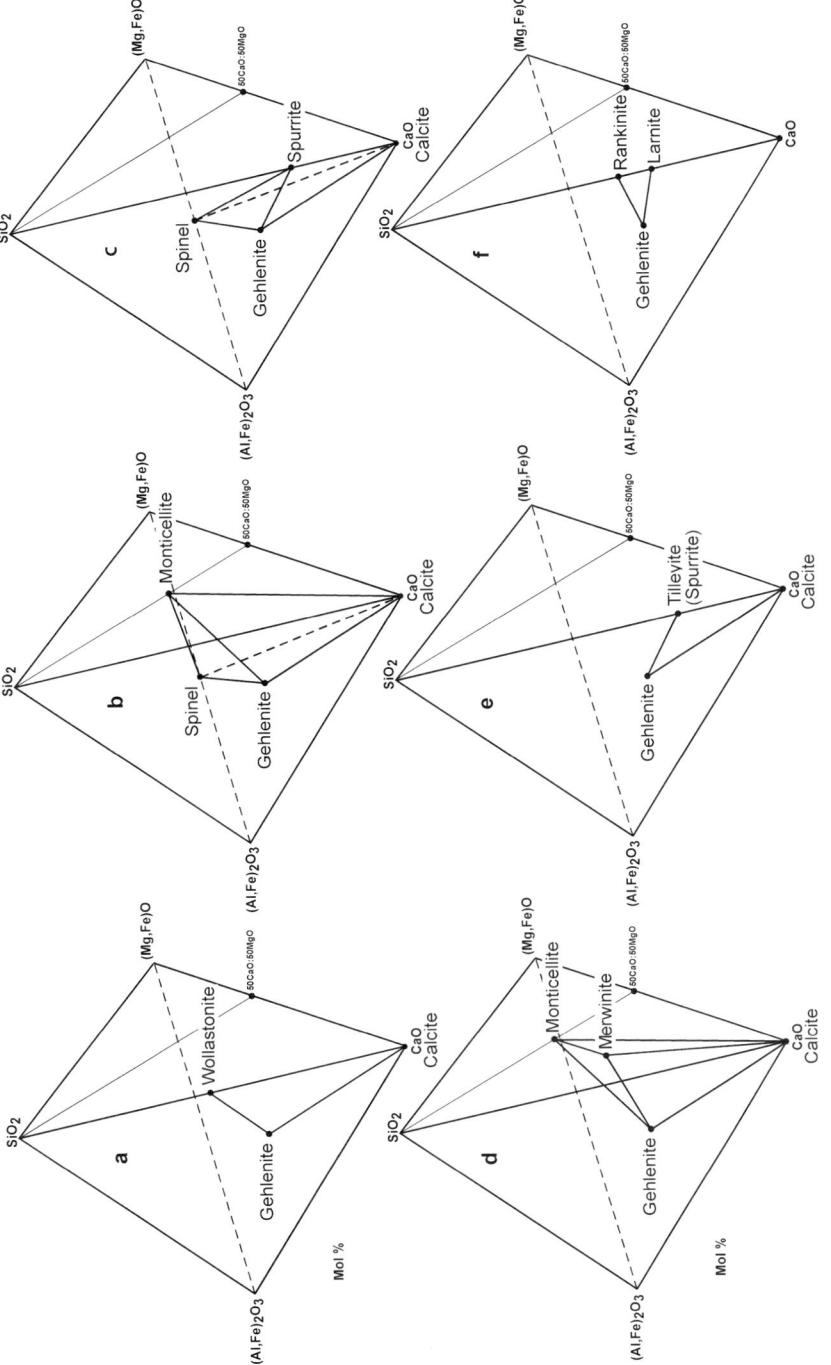

Fig. 4.22. Sanidinite facies gehlenite-bearing assemblages **a–f**, island of Muck, Scotland, represented in terms of mol% CaO–(MgFe)O – (Al,Fe)$_2$O$_3$ – SiO$_2$. See text

Fig. 4.23.
Prograde sanidinite facies mineral assemblages (four-phase volumes) in the Christmas Mountain aureole, Texas, plotted in terms of mol% cation Ca-(Fe,Mg)-Al-Si with melilite composition projected from Al onto the Ca-(Fe,Mg)-Si face (after Fig. 1 of Joesten 1976)

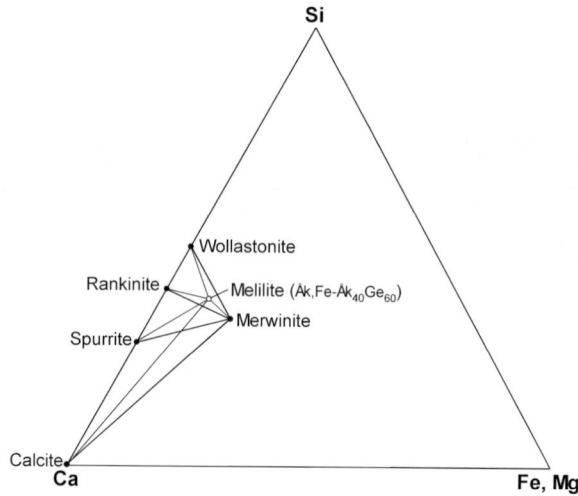

that is graphically depicted in Fig. 4.23. Ti-Zr andradite, perovskite and magnetite are accessory phases. Retrograde monticellite is also sometimes present and larnite coexisting with melilite, bredigite, ± rankinite, ± spurrite occurs in septum between the two gabbros.

At 1.4 m from the contact, the assemblage spurrite-calcite-melilite-merwinite, with merwinite occurring as inclusions in melilite, suggests that the Cc Åk = Mw CO_2 reaction has occurred. Melilite and merwinite are not Mg-end member compositions; melilite has the composition $Åk_{30}$ $FeÅk_8$ Ge_{62} and merwinite has Mg/(Mg + Fe) = 0.9. With reference to Fig. 4.16, at the inferred lithostatic pressure during metamorphism of 350 bar and $XCO_2 = 1.0$ (see above section on chert nodules), merwinite is stable relative to calcite + melilite ($Åk_{30}$) at ~1060 °C but according to Joetsen this decreases by 25 °C from the effect of Fe-substitution in merwinite resulting in a temperature that coincides with the Sp Rn = Ln CO_2 reaction curve at this pressure.

Flekkeren

A ~500 × 100 m finely layered shale-limestone xenolith within a shallow-level (P = 700–1000 bar) larvakite, Flekkeren, southern Oslo Rift, Norway, contains a peak metamorphic assemblage of calcite, wollastonite, melilite ($Åk_{100}$–$Åk_{55}Ge_{45}$), fassaitic pyroxene, Ti-grossular, phlogopite, kalsilite, nepheline, perovskite, cuspidine, baghdadite ($Ca_3ZrSi_2O_8$), Th and LREE-rich silicate apatites, pyrrhotite, ± alabandite (MnS), ± graphite (Jamtveit et al. 1997).

In terms of $KAlO_2$-CaO-MgO-SiO_2-H_2O-CO_2-H-C-O, the observed mineral assemblage Cc-Wo-Ph-Cpx-Åk-Ks corresponds to equilibration near the isobaric invariant points (IA) and (IB) in Fig. 4.24 at ~870 °C / XCO_2 = 0.38 and ~840 °C / XCO_2 = 0.42 at 1000 bar respectively. A pressure reduction to 700 bar results in a lowering of T by ~25 °C and a small increase (0.03) in XCO_2 for the end member system. During cooling to ~700 °C the xenolith was invaded by H_2O-rich fluids resulting in the formation of monticellite, tilleyite, vesuvianite, granditic garnet, diopside, ± hillbrandite ($Ca_2SiO_3(OH)_2$). Further retrogressive phases such as sodalite (veinlets), K-feldspar replacement of

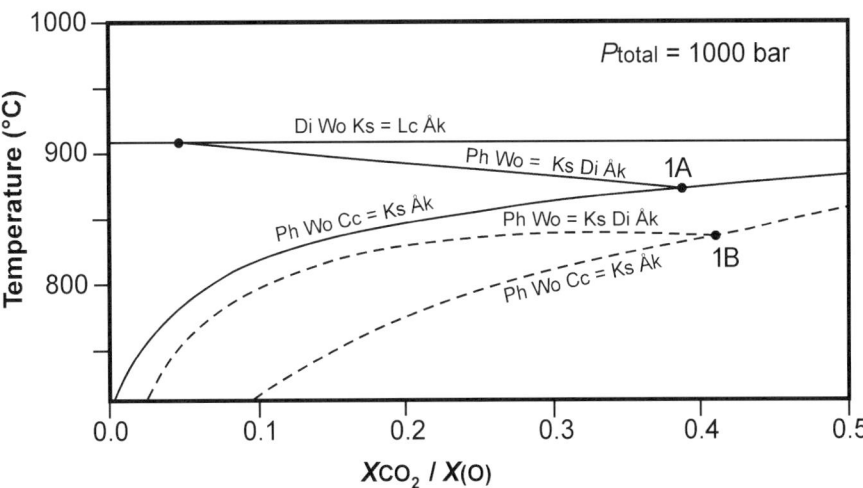

Fig. 4.24. T-XCO_2 diagram calculated for the systems $KAlO_2$-CaO-MgO-SiO_2-H_2O-CO_2 (*solid lines*) and $KAlO_2$-CaO-MgO-SiO_2-C-O-H (graphite saturation) (*dashed lines*) at Ptotal = 1000 bar (after Figs. 7a and 8 of Jamtveit et al. 1997). High temperature reaction products occur to the right of the equality sign. See text

kalsilite and wollastonite resulted from infiltration of Na-rich brines at ~550 °C, with a final assemblage of zeolites, scawite [$(Ca_7Si_6(CO_3)O_{18} \cdot 2H_2O$], hydrogrossular and giuseppetite [$(Na,K,Ca)_{7-8}(Si,Al)_{12}O_{24}(SO_4,Cl)_{1-2}$] forming at lower temperatures.

It is considered that rapid heating caused fluid pressure (CO_2 or CH_4-dominant fluid) in the xenolith to increase through devolatilisation reactions to attain or possibly exceed that in the surrounding partly crystallised intrusion preventing major fluid infiltration into the xenolith. Volatile loss, possibly in excess of 10% of the initial volume depending on the carbonate-silicate ratio, caused a reduction in the solid volume of the xenolith. Volume loss was probably not accompanied by simultaneous compaction, which may have been a relatively slow process at the time scale of heating, so that the increased permeability of the xenolith allowed an influx of C-poor, H_2O-dominated fluid from the magma resulting in the observed initial stage pervasive retrograde mineral replacement.

Santorini

A small xenolith in dacite of Santorini volcano, Cyclades, Greece, consists of a central (primary) assemblage of coarse melilite (cores of Ge_{82} to rims of Ge_{50}), wollastonite and Ti-magnetite, that is replaced near pores and margins by intergrowths of fine fibrous melilite (~Ge_{64})-wollastonite-Ti-andradite (Nicholls 1971). Oxygen fugacity-T °C conditions of metamorphism in relation to the reaction

$$6\,\text{andradite} = 4\,\text{magnetite} + 18\,\text{wollastonite} + O_2$$

(Gustafson 1974) and the dacite magma are shown in Fig. 4.25. The diagram indicates a temperature range of 800–900 °C at 1 bar for the primary and secondary assemblages

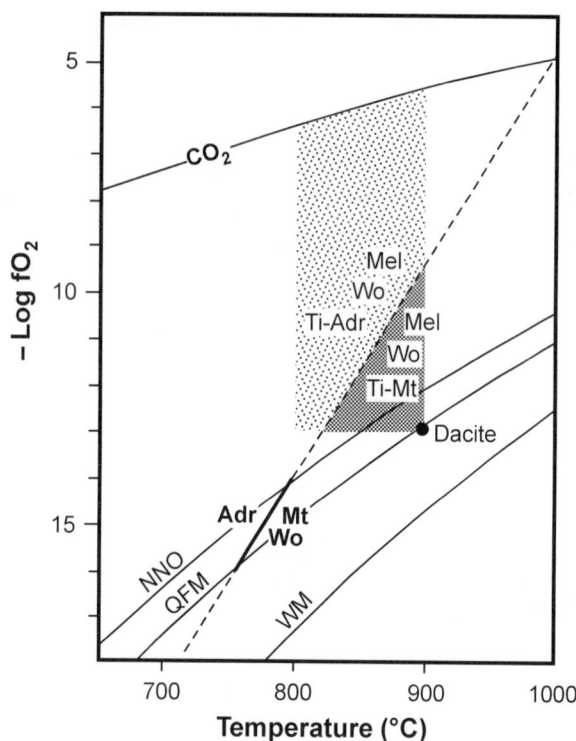

Fig. 4.25.
fO_2-T conditions of high temperature primary (Mel Wo Ti-Mt; *dark shaded area*) and secondary (Mel Wo Ti-Adr; *stippled area*) assemblages in a xenolith within dacite, Santorini, Italy. WM (wustite-magnetite), QFM (quartz-fayalite-magnetite), NNO (nickel-nickel oxide) buffer curves at 1 bar. The 6 Adr = 4 Mt 18 Wo O_2 reaction is extrapolated from Gustafson (1968). Curve $CO_2 = fO_2$-T relationship for pure CO_2 at 1 atm (after Fig. 3 of Nicholls 1971). See text

of the xenolith with a maximum upper limit of fO_2 defined by the breakdown of pure CO_2 to CO and O at Pfluid = PCO_2 at 1 atm. The diagram also demonstrates that it would be possible for the two assemblages to have formed by a relatively small change in T, fO_2 or both, causing the Adr = Wo Mt curve to be crossed. At 500 bar, maximum T for the formation of gehlenite + wollastonite is ~900 °C at XCO_2 = 0.65 (Fig. 4.10a).

La Soufrière

A fist-size calc-silicate nodule described by Devine and Sigurdsson (1980) in a pyroclastic flow of the 1902 eruption from La Soufrière volcano, St. Vincent, lesser Antilles, has an assemblage of fassaitic pyroxene–grandite garnet–wollastonite ± anorthite ± secondary calcite with modal percentages of Fas (73.6%), Gt (23.7%), Wo (2.2%), An (0.5%), Cc (0.1%). The bulk composition of the xenolith (assumed to have been a calcareous sediment) has high Al_2O_3 (18.5 wt.%) and FeO (as total iron = 7.9 wt.%). The modal composition and the fact that wollastonite and anorthite occur as inclusions in garnet and pyroxene indicates that the equilibrium high temperature assemblage in the xenolith is fassaite (Fas_{50}) + grandite (Gr_{78} Adr_{22}). This suggests the possibility of an earlier Fas-rich pyroxene coexisting with wollastonite that reacted to form grandite + less Fas-rich pyroxene as shown by the tie line relationships in Fig. 4.26. Experimental data of Huckenholz et al. (1974) demonstrates that the grossular content of grandite in fassaite$_{ss}$-bearing assemblages increases with decreasing tempera-

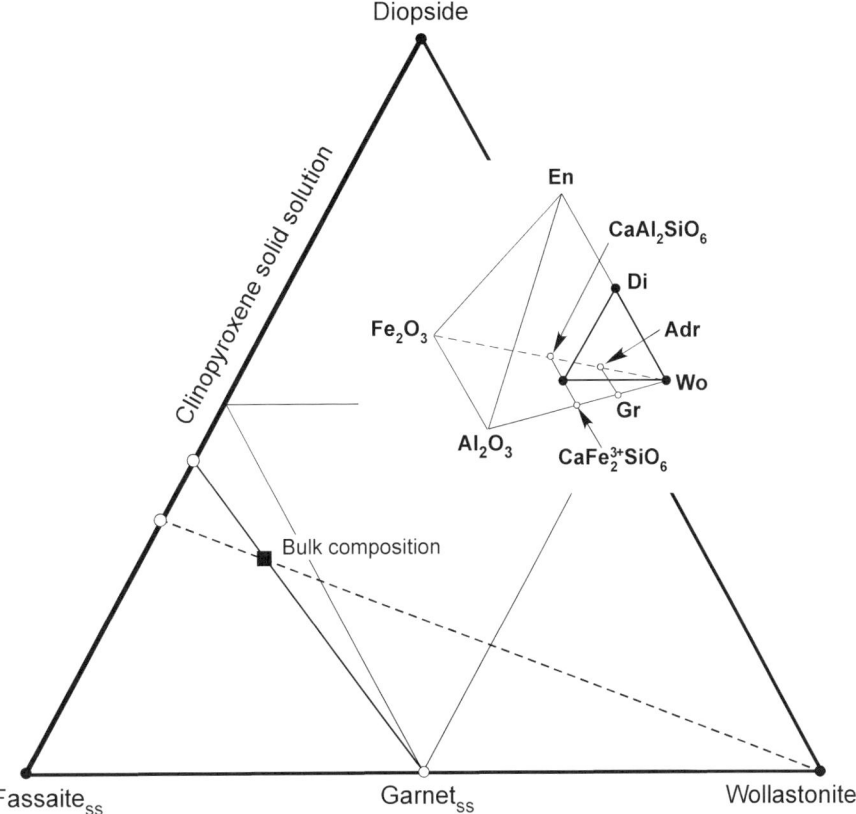

Fig. 4.26. The inferred reaction Wo Fas$_{ss}$ = Gt$_{ss}$ in a calcareous xenolith, La Soufrière volcano, St. Vincent, lesser Antilles, shown in terms of the mol% diopside-fassaite$_{ss}$-wollastonite plane (*inset diagram*) in which Cpx$_{ss}$ + Wo (*dashed line*) reacts to form a more less Fas-rich Cpx$_{ss}$ + grandite in the bulk composition shown (derived from Fig. 20 of Huckenholz et al. 1974). See text

ture at constant pressure and bulk composition. At the same time, the CaFe$_2^{3+}$SiO$_6$ component in fassaite increases as evident in the chemographic relationship depicted in the inset diagram in Fig. 4.26. For the reaction

wollastonite + fassaite$_{ss}$ = grandite$_{ss}$

the formation of grandite on the join fassaite-wollastonite causes the fassaite$_{ss}$–wollastonite tie line to be intersected by that of fassaite$_{ss}$–grandite (Fig. 4.26). At atmospheric pressure, this reaction occurs at 935 ± 20 °C with an increase of ~70–80 °C kb^{-1}. A *T-X* section of the grossular-andradite join in the system CaO-Al$_2$O$_3$-Fe$_2$O$_3$-SiO$_2$ (Huckenholz et al. 1974, Fig. 6) indicates a garnet-fassaite-wollastonite stability field between 935 and 1140 °C. Over a pressure range of 500–1000 bar, the reaction garnet$_{ss}$ = wollastonite$_{ss}$ + fassaite$_{ss}$ occurs between ~950–1000 °C (Fig. 4.10a).

Eifel

"Limestone" blocks in leucite tephrite near Mayen, Eifel volcanic area, Germany, described by Hentschel (1964) and Jasmund and Hentschel (1964), contain assemblages of brownmillerite + mayenite, ± larnite, ± calcite and of wollastonite + gehlenite, together with accessory spinel and pyrrhotite. The largest xenoliths are compositionally heterogeneous with darker brown and green parts rich in brownmillerite and mayenite. Except for the wollastonite-gehlenite xenolith, compositions plot within the larnite and lime fields in the system CaO-Al_2O_3-SiO_2 (mostly on the CaO-rich side of the larnite-mayenite join) and include the composition field of Portland cement (Fig. 4.27). According to Fig. 4.6b, the coexistence of larnite + calcite in one sample, together with the occurrence of wollastonite rather than pseudowollastonite in another, indicates a minimum temperature of 1060 °C, $XCO_2 < 0.55$ at $Ptotal = 1000$ bar.

4.1.3.3
Siliceous Dolomite

Bushveld

Siliceous dolomite xenoliths in the Marginal and Critical zones of the Bushveld Intrusion, South Africa, contain mineral assemblages that indicate extreme temperature conditions of igneous pyrometamorphism (Willemse and Bensch 1964; Wallmach et al. 1989). In the gabbro-norite *Marginal Zone*, high grade paragenesis in the xenoliths are:

1. calcite – åkermanite – monticellite
2. calcite – forsterite – monticellite
3. åkermanite – diopside – monticellite
4. diopside – forsterite – monticellite

as shown in Fig. 4.28a. The first two assemblages have coarse polygonal textures and indicate peak metamorphic conditions. The latter two assemblages are retrograde (connected by dashed lines in Fig. 4.28a) and are characterised by sympletitic texture and are assumed to have derived from an åkermanite + forsterite assemblage that no longer exists. Associated minerals are melilite with accessory spinel, kalsilite, Ba-rich phlogopite, wollastonite and apatite. Mineral assemblages in less siliceous more magnesian xenoliths of the feldspathic pyroxenite *Critical Zone* are:

5. calcite – periclase – monticellite
6. merwinite – åkermanite – monticellite
7. forsterite – periclase – monticellite

as depicted in Fig. 4.28b. Spinel is accessory and periclase is mostly altered to brucite.

With XCO_2 of unity, there is an overlap pressure of ~1.0–1.6 kb between the two highest temperature decarbonation reaction curves pertinent to the Marginal and Critical zone xenoliths (Fig. 4.28c) within the temperature interval 1160–1300 °C (temperature of the Bushveld magma given by Cawthorn and Walraven 1998). In the

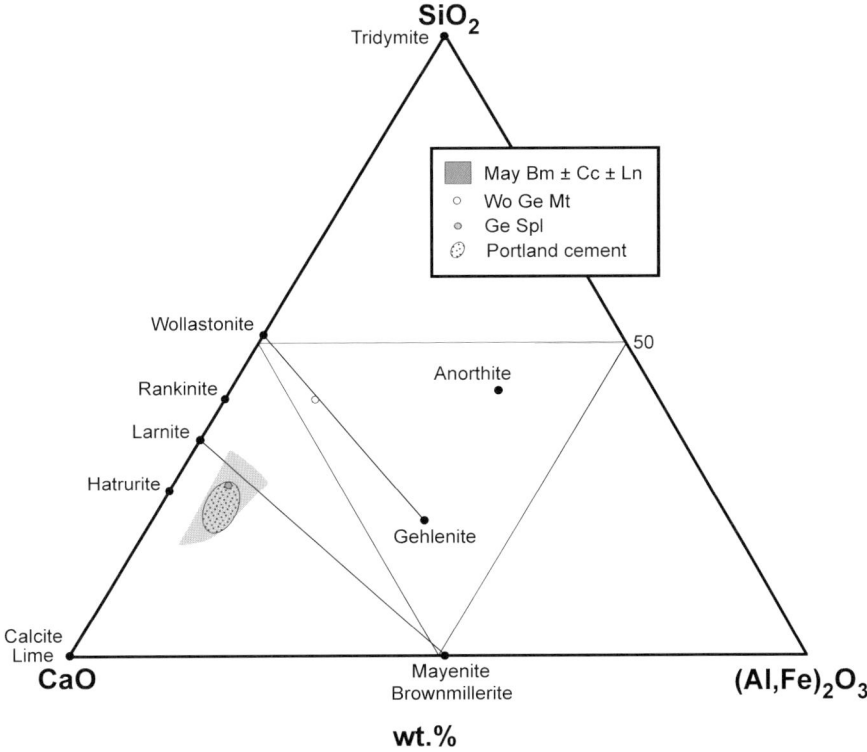

Fig. 4.27. Larnite-mayenite-brownmillerite-gehlenite-wollastonite–bearing xenoliths in tephrite lava, Ettringer Bellerberg, Eifel, Germany plotted in terms of CaO-(Al,Fe)$_2$O$_3$-SiO$_2$ (-CO$_2$)

Marginal Zone xenoliths, textural relations indicate prograde decarbonation mineral reactions textures beginning with

calcite + diopside = åkermanite + CO$_2$

as indicated by åkermanite enclosing calcite in the absence of diopside (consumed). Marginal depletion of Al and enrichment of Mg in the melilite indicates increasing temperature. Assemblages 1 and 2 above indicate the next and highest temperature reaction attained is

calcite + forsterite + åkermanite = 3 monticellite + CO$_2$

with monticellite containing exsolution lamellae of forsterite. Because of the Al-content (gehlenite-component) of the åkermanite reactant (minimum $\alpha_{(Åk)} = 0.75$), the reaction is shifted towards higher temperatures as shown in Fig. 4.28c. The gehlenite-component of åkermanite probably accounts for the formation of additional spinel. The absence of merwinite indicates that the higher temperature reaction

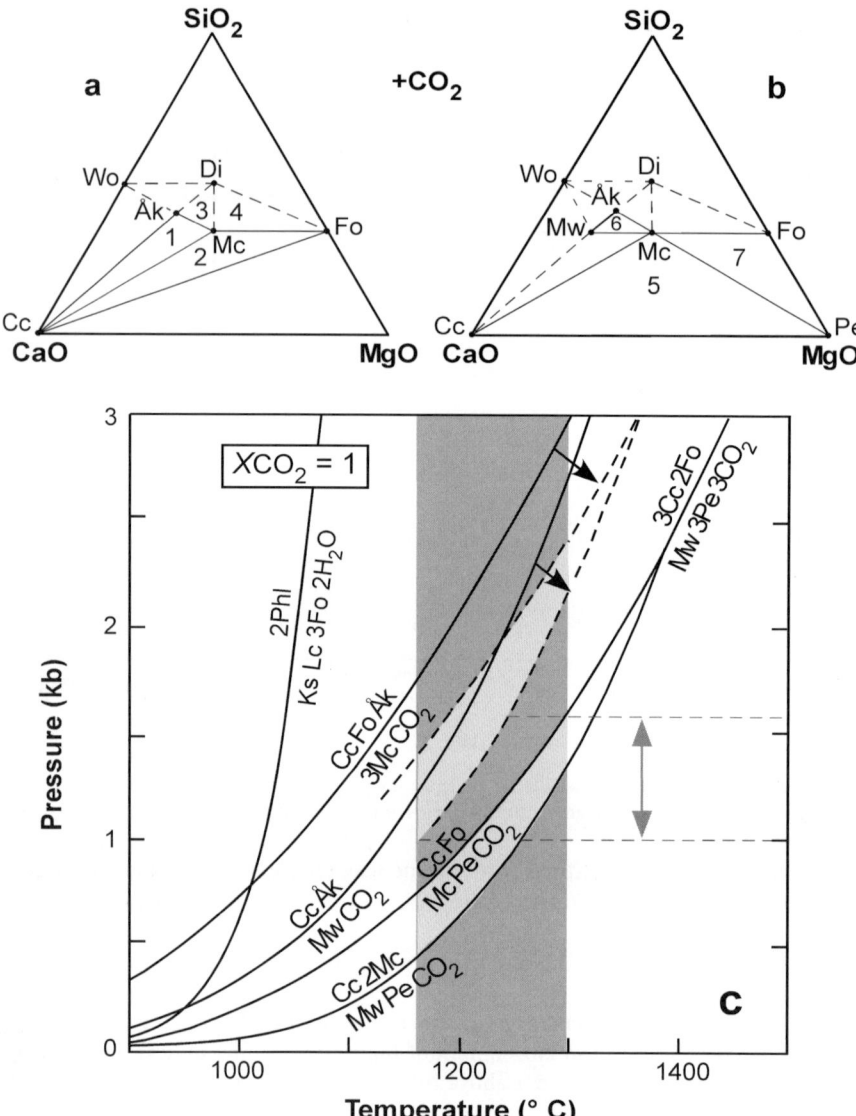

Fig. 4.28. Sanidinite facies siliceous dolomite xenolith assemblages in (**a**) (Marginal Zone) and (**b**) (Critical Zone) of the Bushveld Intrusion, South Africa. *Solid lines* connect minerals formed during prograde and retrograde metamorphism. Minerals connected by *dashed lines* do not coexist in the xenoliths. *Numbers* refer to assemblages listed in text (after Fig. 2 of Wallmach et al. 1989). **c** *P-T* diagram at $XCO_2 = 1$ showing calculated high temperature mineral reactions in siliceous dolomite xenoliths. *Arrows* indicate shift of Cc Fo Åk = 3 Mc CO_2 and Cc Åk = Mw CO_2 reactions curves where the activity of åkermanite = 0.75. *Horizontal dashed lines* separated by *double arrow* = range of P of high temperature metamorphism from overlap of Cc Åk = Mw CO_2 and Cc Fo = Mc Pe CO_2 reactions in relation to range of Bushveld magma temperature (*dark shaded strip*). Upper and lower *light grey shaded areas* = P-T conditions under which the Marginal and Critical Zone magma intruded respectively (after Fig. 9 of Wallmach et al. 1989). See text

calcite + åkermanite = merwinite + CO_2

was not crossed. Again, this reaction is shifted to higher temperatures as a result of Al substitution in åkermanite (Fig. 4.28c).

During cooling of the Marginal Zone xenoliths, retrograde reactions did not reverse the above decarbonation reactions due to insufficient CO_2. The most widespread reaction is

2 åkermanite + forsterite = 3 monticellite + diopside

that produced a second generation of monticellite without forsterite exsolution as symplectitc intergrowths of åkermanite – diopside – monticellite and diopside – forsterite – monticellite depending on the original bulk composition of the xenolith.

In the *Critical Zone* xenoliths with merwinite and periclase, the merwinite-producing reaction above and the reaction

calcite + forsterite = monticellite + periclase + CO_2

were intersected. Because monticellite rather than merwinite + periclase occurs in all the observed stable mineral associations (5, 6, 7) and with calcite in one (5) (Fig. 4.28b), temperatures were not high enough to intersect the reaction

calcite + 2 monticellite = merwinite + periclase + CO_2

(Fig. 4.28c).

The presence of an unusual dehydroxylated Ba-rich phlogopite (up to 15 wt.% BaO) as inclusions within åkermanite and monticellite and in symplectic intergrowths with åkermanite, diopside, monticellite and forsterite, implies very high temperatures of formation that may be due to the stabilising effect of high Ba. At temperatures of ~900–1000 °C at $P < 1$ kb, phlogopite decomposes according to the reaction,

2 phlogopite = kalsilite + leucite + 3 forsterite + H_2O

(Yoder and Eugster 1955) (Fig. 4.28C). In the xenoliths, the presence of diopside and absence of leucite may reflect the relatively low temperature retrograde reaction

2 wollastonite + leucite + forsterite = 2 diopside + kalsilite

(< 600 °C up to 2 kb; Wallmach et al. 1989), evidence of which is provided from a localised association of phlogopite (with < 9 wt.% BaO), kalsilite, forsterite, clinopyroxene and wollastonite in the outermost part of one xenolith.

Brome Mountain

An unusual sanidinite facies kalsilite-bearing impure dolomite xenolith in alkaline gabbro is described from Brome Mountain, Quebec, by Philpotts et al. (1967). The xenolith crops out over an area of ~900 m² and is characterised by 1–5 cm thick alternating buff and dark rusty (weathered) coloured layers. The rock contains melilite ($Åk_{100}$–$Åk_{50}Ge_{50}$) which varies in composition from one bed to another with monticellite concentrated in the rusty-weathered layers and other layers rich in diopsidic pyroxene. Kalsilite is abundant in the melilite and diopside-rich layers where it forms intergrowths with both minerals. It does not occur with monticellite. Accessories include spinel and calcite.

Philpotts et al. (1967) suggest that the kalsilite formed from original feldspar according to the reaction

dolomite + K-feldspar = diopside + kalsilite + $2\,CO_2$

although other probable reactions involve phlogopite together with calcite and wollastonite such as

phlogopite + wollastonite + calcite = kalsilite + åkermanite

phlogopite + wollastonite = kalsilite + diopside + åkermanite

In terms of T-XCO_2 at 1000 bar (Fig. 4.24), maximum temperatures of the above two reactions are ~880 °C/XCO_2 = 0.5 and ~910 °C/XCO_2 = 0.05. Both reactions, together with the reaction Di Cc = Åk, form the invariant point (1A) located at ~870 °C/XCO_2 = 0.38.

Kiglapait

High temperature metamorphism of quartz-bearing dolomite xenoliths within troctolite of the Kiglapait layered mafic intrusion, Labrador, is described by Owens (2000). The xenoliths are massive to foliated and range in size from 30 × 50 cm up to 1 m across. In terms of CaO-(Mg,Fe)O-SiO_2, compositions plot within the diopside-forsterite-åkermanite volume with one on the monticellite-forsterite tie line (Fig. 4.29). The presence

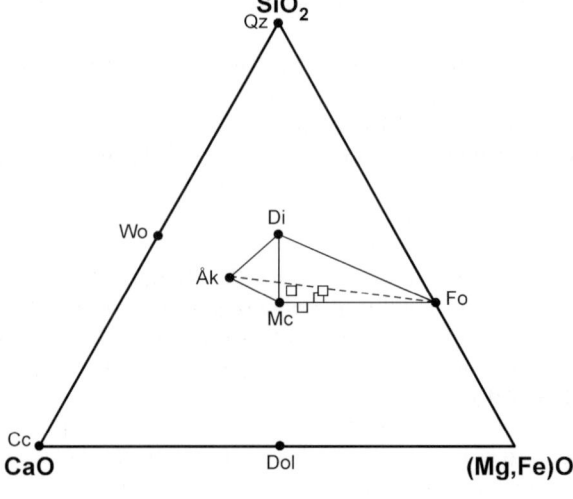

Fig. 4.29.
Bulk compositions of siliceous dolomite xenoliths in the Kiglapait Intrusion, Labrador in terms of mol% CaO – (Mg,Fe)O – SiO_2(-CO_2). *Tie lines* connect coexisting minerals in the xenoliths. *Dashed tie line* indicates the presence of all four phases in one xenolith (after Fig. 7 of Owens 2000)

of significant amounts of Al_2O_3 (6.2–10.2 wt.%) in addition to Fe_2O_3 (0.5–2.6 wt.%) and FeO (3.1–5.2 wt.%) in the bulk compositions and the development of 1–2 cm thick reaction zone of clinopyroxenite around the xenoliths imply diffusive exchange with the mafic magma.

The xenoliths are characterised by complex symplectitic intergrowths of monticellite, diopside (with a significant fassaite component), forsterite (Fo_{82-87}), in one case åkermanite, together with spinel, trace magnetite and rare perovskite, and thus represent the monticellite-melilite subfacies of the sanidinite facies. Several reactions relevant to paragenesis of the xenoliths, namely

diopside + 3 monticellite = 2 åkermanite + forsterite

diopside + calcite = åkermanite + CO_2

diopside + forsterite + 2 calcite = 3 monticellite + 2 CO_2

åkermanite + forsterite + calcite = 3 monticellite + CO_2

are shown in Fig. 4.30. The first of these reactions is independent of fluid composition and its intersection with the other reactions implies a minimum temperature of ~875 °C near 400 bar implying XCO_2 in the fluid phase of ~0.4 (Fig. 4.9b).

Ioko-Dovyren

Abundant pyrometamorphosed dolomitic xenoliths ranging in size from a few centimeters to 20 m in basal dunite of the Ioko-Dovyren Intrusion (a ~26 km × 3.5 km differentiated ultramafic-mafic ?lopolith), north Baikal region, Russia (Wenzel et al. 2001, Wenzel et al. 2002). The unmetamorphosed dolomitic sediments near the margins of the intrusion consist of > 87% dolomite with minor calcite (< 5%), detrital quartz and unspecified sheet silicates. Two types of pyrometamorphically-related mineral associations occur:

1. Concentrically zoned coarse-grained (> 0.8 mm) aggregates of brucite pseudomorphs after periclase, with interstitial forsterite + Cr-poor spinel rimmed by fine grained olivine + Cr-richer spinel (Fig. 4.31).
2. Schlieren of fine-grained (< 0.2 mm) forsterite + Cr-bearing spinel ± monticellite locally intergrown with sub-microscopic diopsidic pyroxene.

No calcite is present except as late stage crosscutting veins. Pentlandite rimmed by chalcopyrite occurs in some of the olivine + spinel associations.

With a magma temperature of at least 1260 °C and an overburden Pfluid = Psolid of < 1 kb, appropriate mineral reactions above the reaction brucite = periclase + H_2O at 600 °C in the system CaO-MgO-SiO_2-CO_2-H_2O are shown in Fig. 4.31. Rapid heating of the xenoliths resulted in the reaction

dolomite = periclase + calcite + CO_2

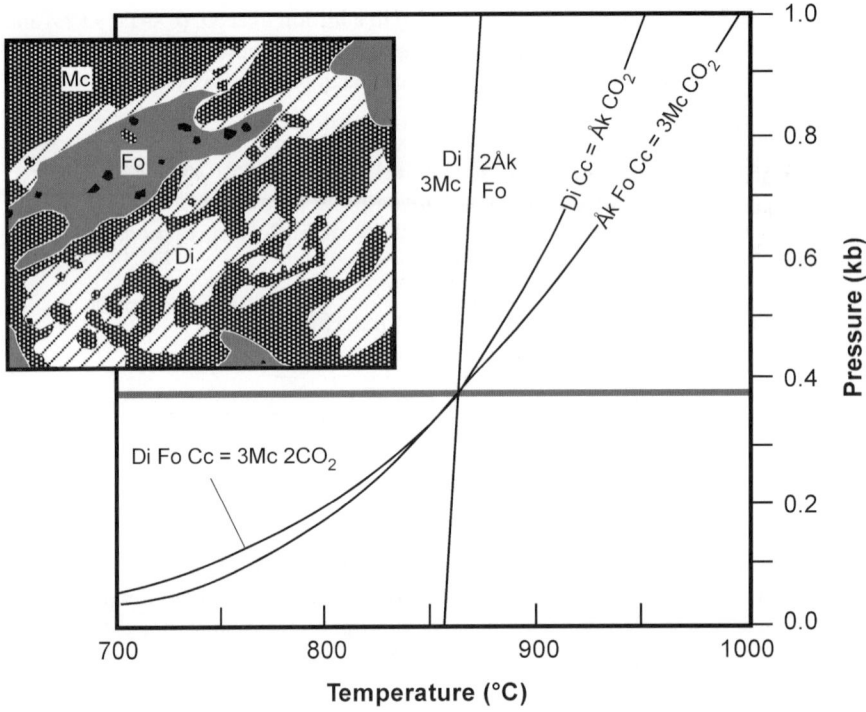

Fig. 4.30. *P-T* diagram of relevant reactions in siliceous dolomite xenoliths, Kiglapait Intrusion, Labrador. High temperature reaction products are to the right of the equalising symbol (after Fig. 8b of Owens 2000). See text. *Inset diagram* shows a symplectic intergrowth of monticellite, forsterite (with black spinel inclusions) and diopside. The texture is inferred to indicate either the retrograde reaction, Åk Fo = Di Mc (Åk consumed) or the reaction Di Fo Cc = Mc CO_2 (calcite consumed). See text. Horizontal dimension = 2.6 mm (after Fig. 2b of Owens 2000)

at temperatures of between 620 °C and 820 °C depending on $X(CO_2/H_2O)$ of the fluid phase. Interstitial olivine + spinel (Fig. 4.32a) probably reflects the final product of up-temperature reactions of dolomite with quartz to form forsterite + calcite at > 570 °C with accessory Mg-Fe-bearing phyllosilicates contributing to the production of pleonaste spinel. As the xenoliths no longer contain calcite, Wenzel et al. (2001, 2002) assume that it has melted, possibly at near magmatic temperatures and relatively high fluid XCO_2, and the melt progressively extracted together with alkalis and CO_2 leading to a ~70% reduction in the volume of the xenoliths (Fig. 4.31).

The olivine + spinel rims around the xenoliths (Fig. 4.32b) are inferred to reflect the redox conditions of a CO_2-rich and consequently highly oxidizing (fO_2 > HM buffer) fluid, developed in the immediate vicinity of the xenoliths that favoured crystallisation of Ca-bearing forsterite and Cr-depleted spinel from the surrounding magma. Patches of monticellite and diopside in forsterite-spinel schlieren without periclase implies Si diffusion (to form diopside) into the margins of what are presumably almost completely assimilated dolomitic xenoliths. At ~1230 °C, periclase formed from dolomite in mafic magma begins to melt (data quoted in Wenzel et al. 2002).

4.1 · Calc-Silicates   155

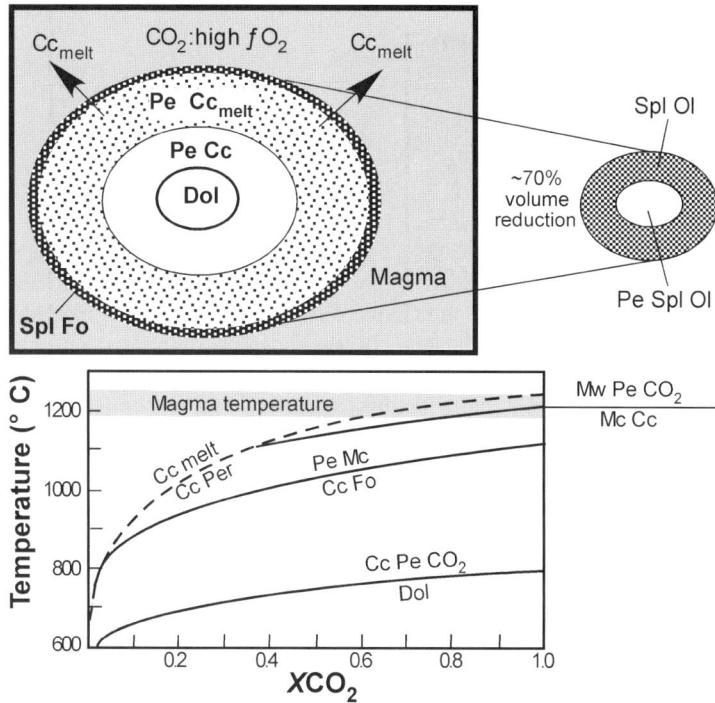

Fig. 4.31. *Above:* Model for metamorphism of dolomitic xenoliths in mafic magma of the Ioko-Dovyren Intrusion, North Baikal region, Russia (after Fig. 5 of Wenzel et al. 2001). *Below:* T-XCO_2 diagram at 1000 bar for high temperature reactions in the system MgO-CaO-SiO_2-H_2O-CO_2 assuming excess calcite. Ioko-Dovyen magma temperature is indicated by horizontal shaded strip (after Fig. 11 of Wenzel et al. 2002). See text

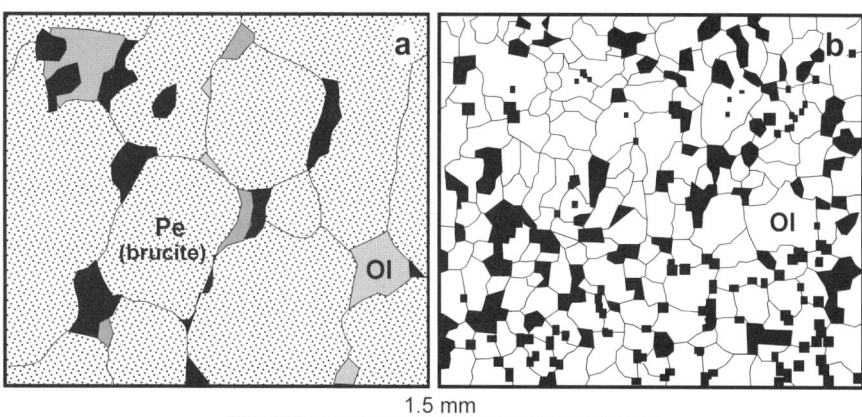

Fig. 4.32. Textures of siliceous dolomite xenoliths in dunite, Ioko-Dovyren Intrusion, North Baikal region, Russia. **a** Periclase (altered to brucite)-rich core with interstitial olivine and spinel (*black*). **b** Olivine-spinel rim on (**a**) and as isolated schlieren (after Figs. 3a,b of Wenzel et al. 2001)

4.1.4
Combustion Pyrometamorphism

4.1.4.1
Marly Compositions

Mottled Zone

The "Mottled Zone" complex resulting from pyrometamorphism caused by the combustion of organic matter (up to 25%) in bituminous calcareous sediments, crops out over a large area in Israel and Jordan (Fig. 4.33). In the Hatrurim Basin adjacent to the Dead Sea, an ~80 m thick Late Cretaceous–Tertiary chalk and marl-dominated sequence, together with sandstone and chert clasts in overlying Neogene conglomerate, has been converted into calc-silicate and calc-aluminate assemblages (e.g. Burg et al. 1991) (Fig. 4.34) according to the overall generalised reaction

$$CaCO_3 + \text{"shale"} = \text{Ca-Al-silicates} + CO_2 + H_2O$$

(Fig. 4.35) at temperatures typically ranging from > 520 to > 800 °C, but locally attaining > 1000 °C (Gross et al. 1967; Kolodny et al. 1971; Kolodny and Gross 1974; Gross 1977; Matthews and Kolodny 1978; Matthews and Gross 1980). High altitude aeromagnetic surveys over the large (~50 km²) burn area in the basin indicate a substantial anomaly that coincides with the thickest sequences of burnt rocks. Fission track dating indicates a 13.6 ± 2.0 Ma for the combustion event (Kolodny et al. 1971) although subsequent $^{40}Ar/^{39}Ar$ dating of high grade K-rich (> 0.8 wt.% K) samples yielded ages in the range of 2.3 to 4 Ma, with one age at ~16 Ma (Gur et al. 1995) suggesting that combustion occurred at different times at different localities following exposure of the bituminous sediments by unroofing.

The metamorphosed rocks are essentially fine grained with mosaic, granoblastic and poikiloblastic textures but have highly variable mineral associations. Some 125 pro-

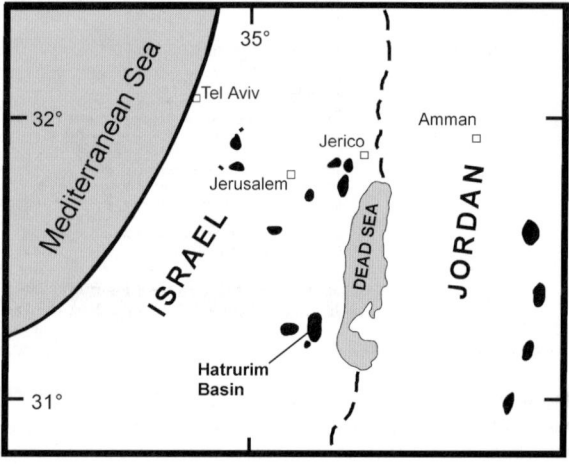

Fig. 4.33.
Map showing distribution of burnt rocks in Israel and Jordan. The Hatrurim Basin is the locality of the well-known "Mottled Zone" of combustion metamorphism

grade and retrograde minerals have been identified, ten of which are new, or known only as synthetic products (Gross 1977). With respect to high temperature pyrometamorphic minerals, spurrite and larnite, less commonly rankinite and wollastonite have formed from siliceous limestone. Melilite (mainly gehlenite–Fe-gehlenite with rare åkermanite), anorthite, grossular-andradite, clinopyroxene (diopside-hedenbergite series, Al-rich fassaite, rare aegirine-augite), accessory merwinite (in melilite-rich rock) and perovskite are developed in marl (Fig. 4.36). Rare monticellite occurs with calcite, spurrite, larnite, melilite, ranknite, brownmillerite in marly dolomitic rocks. Peraluminous silica-deficient compositions contain abundant brownmillerite and mayenite (Fig. 4.37). Other sanidinite facies minerals include calcium-dialuminate, the orthorhombic form of calcium disilicate (α'-Ca$_2$SiO$_4$) in a "pseudoconglomerate" with rankinite, melilite, andradite and perovskite; grossite with larnite, mayenite and brownmillerite; nagelschmidtite associated with gehlenite, rankinite, perovskite, Ti-andradite and magnetite in the lower part of the Hatrurim section and forming up to 30% of the rock. Nagelschmidtite (with 2.67–8.27 wt.% P$_2$O$_5$; Gross 1977), which had previously only been reported from slag (Segnit 1950), together with pseudowollastonite (+ Ti-andradite, gehlenite, ± rankinite, larnite) and hatrurite (+ larnite, mayenite,

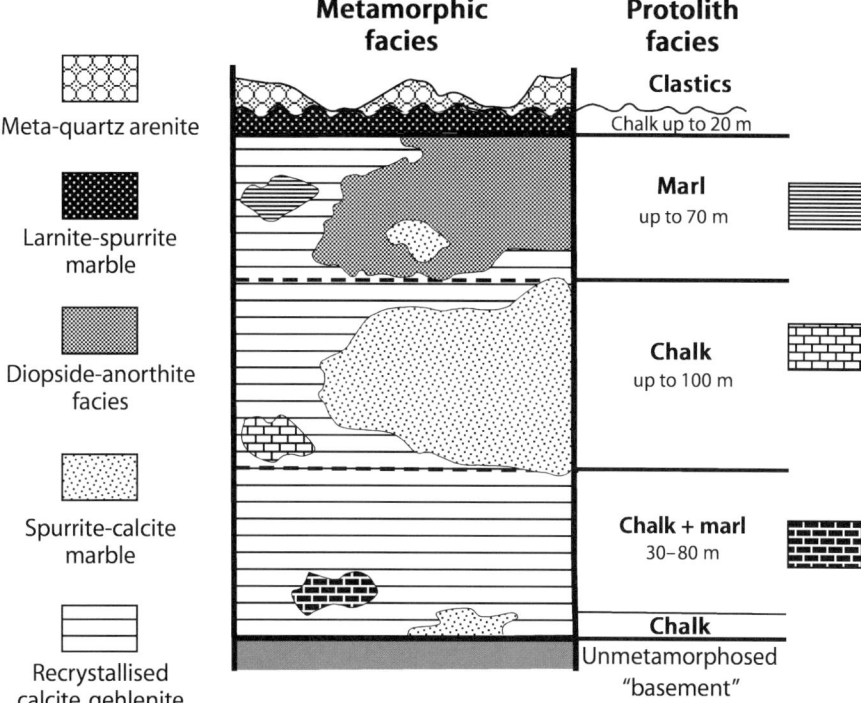

Fig. 4.34. Generalised section through the Mottled Zone of the Hatrurim Basin, Israel, showing the relationship between bituminous protolith rocks and metamorphosed equivalents. Relics of unmetamorphosed protoliths occur throughout the metamorphosed sequence (after Fig. 2 of Gur et al. 1995)

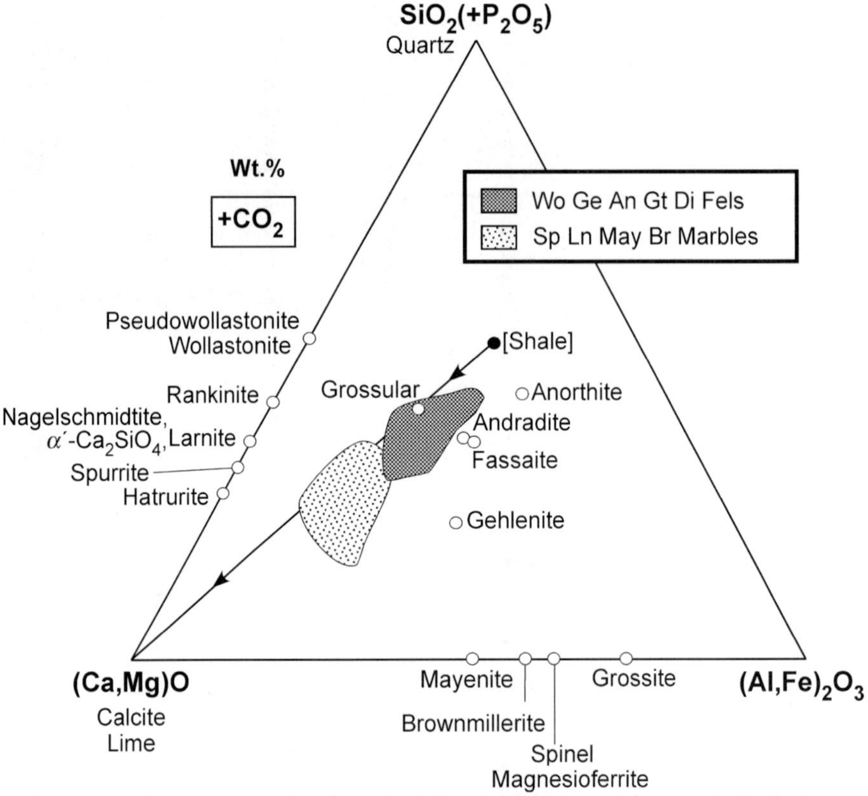

Fig. 4.35. Bulk rock composition fields and mineral compositions from the Mottled Zone, Hatrurim Basin, Israel, plotted in terms of $(Ca,Mg)O$-$(Al,Fe)_2O_3$-$SiO_2(+P_2O_5)$. *Arrowed line* from shale to limestone (calcite) protolith reflects the generalised metamorphic reaction, $CaCO_3$ + "shale" = Ca-Al-silicates + CO_2 + H_2O. Composition fields enclose rock analyses corrected for carbonate content by subtracting and amount of CaO equivalent to CO_2 (after Fig. 3 of Matthews and Gross 1980)

Fig. 4.36. Gehlenite prisms (*stippled*) with anhedral larnite and interstitial andradite, Mottled Zone, Israel (drawn from ordinary light microphotograph Fig. 2, Plate XVII of Gross 1977)

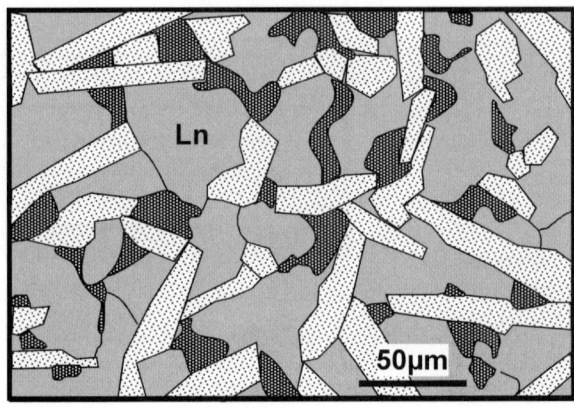

Fig. 4.37.
Larnite-brownmillerite rock, Mottled Zone, Israel (drawn from reflected light section Fig. 4, Plate XIII of Gross 1977)

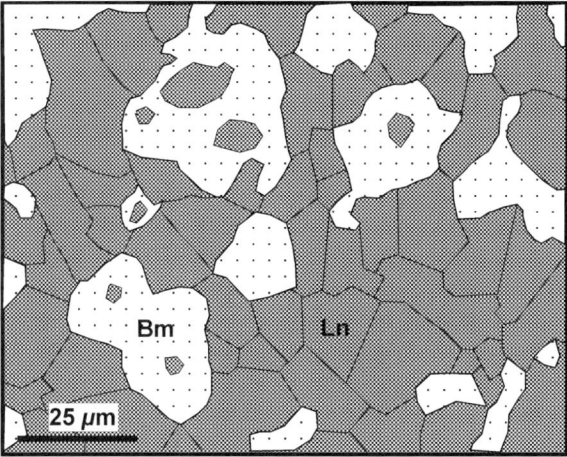

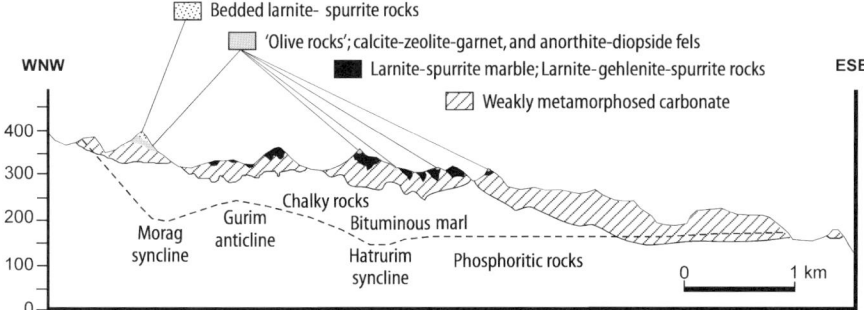

Fig. 4.38. Cross section of the Hatrurim Basin, Israel (after Fig. 7 of Burg et al. 1992)

brownmillerite) indicate extreme temperature conditions of > 1125 °C (Fig. 4.6a). The transition from areas of high temperature recrystallisation to unmetamorphosed rock occurs over a few meters to a few tens of meters and indicates the presence of "hot spots" during the burning process. Highest metamorphic grade occurs nearest the surface (Fig. 4.38) reflecting the abundance of air-supplying joints and the ease at which CO_2 could escape from areas undergoing decarbonation reactions.

Different oxidation regimes dominate the upper and lower parts of the burnt sequence. Hematite-magnetite are widespread in the upper part where there was a adequate supply of oxygen for burning and CO_2 and H_2O were the dominant gas species. At deeper levels (> 30 m) where more reducing conditions (gas species dominantly CH_4, H_2, CO) prevailed, magnetite, ulvospinel, titanomagneite, magnesioferrite, the Fe^{2+}-bearing minerals, hercynite, hedenbergite, vesuvianite, and greigite (Fe_3S_4) (in spurrite marble) were stabilised. The boundary between oxidised and reduced assemblages is characteristically sharp, occurring over an interval of only a few centimeters. Below ~80 m depth the sediments are non-metamorphosed, indicating that there was not enough oxygen for burning to occur.

Temperatures generated in the Hatrurim Basin organic sediments were not high enough to cause melting due to the strongly endothermic nature of decarbonation/dehydration reactions involved. Assuming an approximate ratio of organic carbon (20% of the rocks) to calcite = 1, Matthews and Gross (1977) show that 20% of the heat produced by burning would be used up in driving a decarbonation reaction such as calcite + quartz = wollastonite + CO_2. In rocks with lower organic carbon content, most of the heat budget would be used in decarbonation reactions. Nevertheless, some high grade assemblages are coarse grained and even pegmatitic suggesting recrystallisation in the presence of an H_2O-rich fluid phase. For example, in pseudowollastonite–gehlenite-rich rocks, Ti-rich andradite may form skeletal and dendritic interfingering textures (Fig. 4.39) suggesting a "plasticity" that may indicate partial melting or rapidly cooled sinter.

Low temperature (< 250 °C) hydration and recarbonation has resulted in the development a highly varied retrograde mineral assemblage (see Gross 1977) of carbonates (e.g. calcite, aragonite, siderite, vaterite, dolomite, hydromagnesite, hydrotalcite), hydroxides (e.g. gibbsite, bayerite, boehmite, brucite, portlandite), sulphates (e.g. anhydrite, barite, gypsum, and where S is involved, ettringite), calc-silicate hydrates (e.g. zeolites, hydrogarnet, xonolite, foshagite, hillbrandite, tobermorite-group minerals), phyllosilicates (e.g. serpentinite, apophyllite, pyophyllite, biotite, xanthophyllite, chlorite, illite, smectites) etc. Veins and fissures containing volkonskoite, chromatite and bentorite attest to the mobility of Cr.

Iran

Calcareous (marl) pyrometamorphic rocks associated with the combustion of hydrocarbons are known from several localities in SE Iran and are documented by McLintock (1932). One location is a small conical hill (Tul-I-Marmar) some 76 m high composed of gently-dipping green and red marls. On the western side of the hill is a brecciated mass crosscutting the sediments and consisting of angular blocks in a matrix of porous

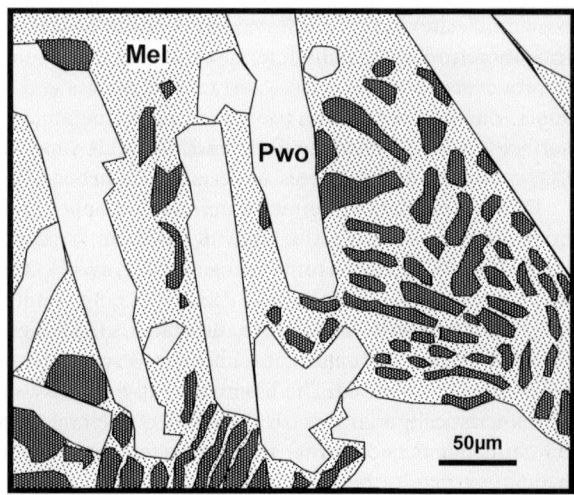

Fig. 4.39.
Andradite inclusions (*dark shading*) in gehlenite (*Mel*) with laths of pseudowollastonite, Mottled Zone, Israel (drawn from ordinary light photomicrograph Fig. 5, Plate XIV of Gross 1977)

reddish-pink gypsum. The blocks consist of country rock marls, "hard, clinkery black limestone, red sandstone and grit" and slag-like rocks that are sometimes coarsely crystalline. Traced eastwards over the crest of the hill the breccia contains greater amounts of shattered red marl in the gypsum matrix and grades into hardened coherent country rock. Around the base of the hill are undisturbed and unaltered mudstones. The field evidence indicates a vent-like structure resulting from explosive escape of oil or gas that became ignited causing fusion of the country rock.

A few kilometers SE of this occurrence there is another small mass of "greenish, vesicular crystalline rock" enclosed within red and yellow, slightly brecciated marls into which it appears to grade. The 'crystalline rock' extends downwards with baked marl underlain by black, bituminous-stained limestone containing bitumen-lined joints. The largest joint extends upwards to an area of intense pyrometamorphism thought to have been caused by the combustion of oil or gas. Other rocks in the vicinity exhibiting jointing, reddening and hardening are ascribed to the same cause.

At two other localities (Zoh-i-Hait and Tang-i-Gogird), spontaneous combustion was in progress (in 1932) with the formation of H_2S, SO_2 and S. The area was covered with partially burnt bituminous matter and a dark-brown powdery ash. Associated limestone and marl are baked sometimes to the point where fusion and recrystallisation has occurred to produce coarsely crystalline rocks. Oxidation of the H_2S has given rise to sulphates that form a grey, powdery gypsum-rich clay containing some $AlSO_4$ known as *sour gypsum*. At another locality (Masjid-I-Sulaiman), fused and fine grained partly recrystallised marls have been produced by jets of burning oil and gas playing on the surface.

Textures of the fused and recrystallised marls vary from very fine-grained, greenish flinty types to coarse grained (2–3 mm length crystals) varieties. The common occurrence of vesicles sometimes lined with calcite or gypsum/anhydrite, give many of the rocks a pumice-like appearance. In weakly altered marls, the only apparent changes are recrystallisation of calcite and the presence of clots and stringers of greenish-yellow pyroxene (possibly a fassaitic variety). A more advanced stage of fusion and recrystallisation results in the formation of a highly vesicular rock composed of clinopyroxene (diopside, aegirine, aegirine-augite and/or possible hedenbergite and fassaitic varieties), wollastonite (more rarely pseudowollastonite), bytownite, melilite (colourless and brown varieties), rare leucite and glass. Relic grains of quartz show no evidence of inversion to tridymite or cristobalite. An ACF diagram (Fig. 4.40) shows that the fused marls plot within the wollastonite-anorthite-diopside subvolume as expected.

Combustion temperatures were clearly high enough locally to cause melting of the marls. Reference to Fig. 4.11 indicates that at ~1208 °C liquids in equilibrium with gehlenite$_{ss}$, anorthite, fassaite$_{ss}$, pseudowollastonite occur in compositions between Gr$_{61-86}$ along the grossularite–andradite join. In examples without clinopyroxene, a similar temperature is indicated for liquids in equilibrium with melilite (up to Ge$_{25}$Åk$_{75}$), pseudowollastonite, wollastonite, anorthite in compositions between Gr$_{98-85}$ along the grossularite–pyrope join. The occurrence of both wollastonite and pseudowollastonite in some samples with textural evidence of the former replacing the latter indicate a crystallisation temperature near the wollastonite-pseudowollastonite transition of 1125 °C at atmospheric pressure.

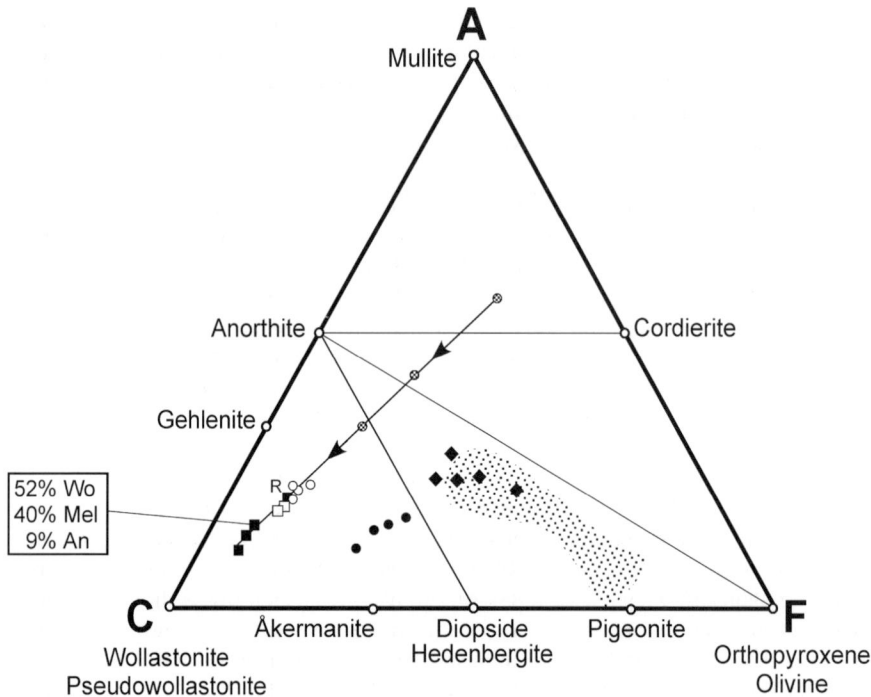

Fig. 4.40. ACF diagram of mostly pyrometamorphosed marl compositions from Iran, Iraq and Central Apennines, Italy. *Filled circles* = Iran; *filled diamonds* = Iraq; *stipped area* = field of basaltic-andesitic rocks from Fig. 4-7 of Winkler (1979). Central Apennines; *filled squares* = core facies + Ricetto slag (labeled *R*); *open squares* = inner facies; *open circles* = spotted facies; *grey-filled circles* = border facies. See text

Central Appennines

Wollastonite and melilitie-bearing paralavas described by Melluso et al. (2003) from two sites in the Central Apennines, Italy, are associated with lignite-rich marls and appear to represent areas of organic combustion at temperatures well above 1000 °C. One body (Colle Fabbri) has an outcrop width of ~10 m and is both vertically and horizontally zoned from:

1. A central (2 m) light grey, medium grained, holocrystalline core facies containing abundant idiomorphic wollastonite and minor plagioclase (An_{98-72}) typically enclosed within melilite, accessory Ti-magnetite, leucite, clinopyroxene, perovskite and rare Ti–Al-rich garnet.
2. An inner fine grained, holocrystalline facies (0.5 m) composed of wollastonite (1.6% FeO), plagioclase (An_{95-83}) and opaques (Ti-magnetite, pyrite, pyrrhotite).
3. A fine grained, hypocrystalline, poikilitic spotted facies (0.5 m) with needles of plagioclase and wollastonite included within fassaitic pyroxene, sometimes arranged in ovoid knots to give a spotted appearance, with interstitial brown and green glass.
4. A marginal facies similar to the spotted facies (0.7 m) but with calcite-filled ocelli and a few samples containing minor hyalophane. Some wollastonite analysed is

Fe-rich (12.5 wt.% FeO) and could be Ca-bustimite, otherwise it contains between 0.9–2.1 wt.% FeO. Near the sedimentary host rocks the facies is strongly heterogeneous and is a mixture of glassy fragments, scoriaceous pockets, both with quartz relics, baked shale and marl country rock. Glass compositions are rhyolitic–dacitic (65–75 wt.% SiO_2; 6–9 wt.% K_2O).

The other outcrop (Ricetto; also described by Capitanio et al. 2004) is ~10 m long and 25 cm thick and is a very hard, black vitreous slag with small spherical vesicles at the top and having a transitional contact with siliciclastic sandstone country rock. The slag contains microphenocrysts (skeletal, dendritic to spinifex crystals) of wollastonite, plagioclase ($An_{55.3} Ab_{11.9} Or_{32.8}$), melilite (av. $Åk_{48} Ge_{37}$ Na-Mel_{15}), fassaite pyroxene, apatite and titanomagnetite with grains of relic quartz, plagioclase, K-feldspar, epidote, zircon and ilmenite-rutile in a SiO_2-poor (43–45 wt.%), CaO-rich (21–22 wt.%), K_2O (3–4 wt.%) glassy groundmass. Partially melted microxenoliths of the country rock are also present.

A plot of the Colle Fabbri and Ricetto compositions projected from plagioclase onto the larnite-forsterite-silica pseudoternary of Pan and Longhi (1990) (Fig. 4.41), indicates that the melilite-bearing samples fall within the anorthite + melilite liquidus field, the inner facies rocks plot close to the ternary eutectics with wollastonite, clinopyroxene, anorthite, and the spotted facies samples plot within the anorthite-diopside liquidus field. The quartz-normative border facies rocks (abundant high Si-glass) plot within the silica-rich part of the diagram. Similar relations are shown in the ACF plot (Fig. 4.40). Although crystallisation temperatures must have been less that that of the wollastonite-pseudowollastonite transition at 1125 °C at atmospheric pressure, liquids in equilibrium with anorthite-gehlenite-pseudowollastonite-fassaite occur on the grossularite-andradite join between Gr_{61-86} at ~1208 °C (Fig. 4.11), similar to those inferred for the Iranian rocks.

Iraq

In the Injana area some 140 km NE of Baghdad, baked and fused calcareous sediments crop out as six prominent hills or knolls representing erosion remnants 6 to 14 m high (Basi and Jassim 1974) (Fig. 2.12). The hills are composed of steeply dipping interbedded grey calcareous sandstone and marl that become yellowish to reddish due to baking and recrystallisation in the middle to upper parts of the hills and are in sharp contact with black and red vesicular rocks sometimes exhibiting a ropy structure and breccia that caps their summits. Baking has caused the normally friable calcareous sandstones to become hard by recrystallisation of calcite grains and cement. Quartz grains are fractured and optically strained and partial fusion is indicated by the presence of colourless and brown glass. Partially melted marl is harder and red to brown in colour. The cap rocks are largely holocrystalline and vesicular with an ophitic to spherulitic texture consisting of colourless diopside, yellowish-green pyroxene (tentatively identified as aegirine-augite but possibly a fassaitic variety), and labradorite (An_{79}). Wollastonite has not been identified. Partially melted quartz and polycrystalline silica remain as relics. Calcite and gypsum occur in vesicles. In terms of ACF parameters, bulk compositions plot within the anorthite–diopside–orthopyroxene subvolume and within the field of basaltic-andesitic rocks (Fig. 4.40).

164   Chapter 4 · Calc-Silicates and Evaporates

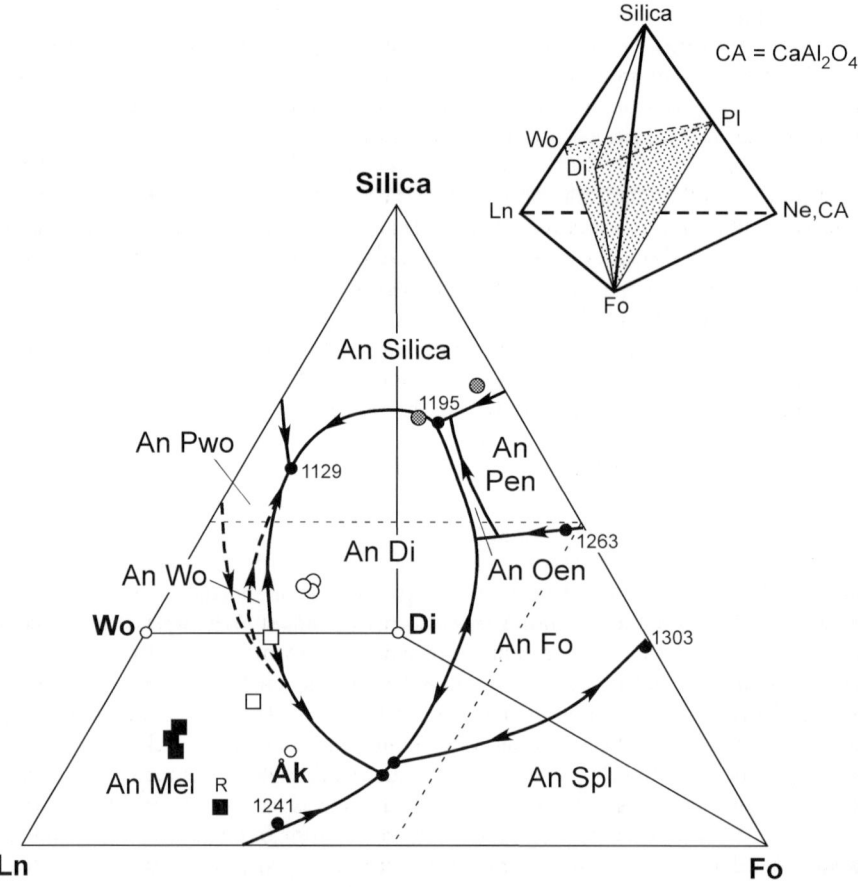

Fig. 4.41. Plot of Central Apennine melilite- and wollastonite-bearing paralava compositions (oxygen units) projected from anorthite in a larnite-forsterite-silica liquidus diagram (Ne/(Ne + CA) = 0) of Pan and Longhi (1990). *Filled squares* = core facies + Ricetto slag (labeled *R*); *open squares* = inner facies; *open circles* = spotted facies; *grey-filled circles* = border facies (after Fig. 6b of Melluso et al. 2003). See text

Continuously burning hydrocarbon seeps in the form of small pits define a lineament along which the six hills are situated and probably defines a major reverse fault up which gases are ascending. It is evident that the gas seeps have been burning for a considerable time and that the hills are the erosion remnants of products from an earlier combustion. Fusion may have been repeated at some localities from evidence of blocks of fused rock becoming imbedded in molten material (as indicated by its ropy structure) produced by a later pyrometamorphic event. The naturally fused rocks are similar to building bricks formed from the same marls, where yellow to brown bricks are produced at a low heating stage, and black vesicular bricks (parabasalt) containing well developed crystals of clinopyroxene and calcic plagioclase are formed at a higher heating stage. In the brick-making process, fusion begins at between 1050–1100 °C.

Czech Republic

Ca-Fe–rich buchites (with up to 32.1 wt.% CaO and 32.4 wt.% Fe_2O_3) are described from the North-Bohemian brown coal field, Czech Republic, by Zacek et al. (2005). They are considered to be pyrometamorphosed carbonate-rich loessic sediment containing siderite nodules and mixed with residual coal ash derived from former chimneys filled with fused sandy gravel that contained the loess, and slag. The buchites are best described as melilities. They are dominated by melilite (Ge_{21-93} Fe-Ge_{0-34} $Åk_{1-49}$ Fe-$Åk_{0-25}$) and Na-melilite up to 15%, fassaitic clinopyroxene (Ti-rich essenite), anorthite, pleonaste spinels, magnesioferrite, hematite, perovskite, calcium ferrite, srebrodolskite and barium hexaferrite ($BaFe^{3+}_{12}O_{19}$). The last three phases are associated with ferrigehlenite-rich melilite but not with essenite.

Mineral assemblages of the buchites imply temperatures within the range 980–1150 °C and high oxidation conditions between those of the QFM and HM buffers as indicated from the stability of essenite according to the reaction

6 An 6 Ge 7 Mt O_2 = 18 Ess 3 Hc

(Cosca and Peacor 1987) (see Fig. 3.47).

Australia

The Leigh Creek Coalfield, South Australia, provides unusual examples of fused coal ash resembling clinker or slag resulting from the burning of sub-bituminous coal (Baker 1953). The clinkers have the composition of marls with wt.% 18.4–32.2% SiO_2, 19.7–25.5% Al_2O_3; 4.3–21.3% Fe_2O_3; 0.8–5.9% FeO; 3.9–5.0% MgO; 21.2–24.7% CaO. They are fine grained, vesicular to scoriaceous ranging in colour from black through dark grey, greenish grey to yellowish-grey. Although largely holocrystalline they can contain small amounts of glass, are variably magnetic and have densities ranging from 3.14 to 2.62. Some of the clinker occurs as surface outcrops, but most was formed under an overburden of Triassic sediments and occurs down to 13.3 m below the surface where it lies beneath baked and brecciated shale.

The mineralogy of the clinker is highly variable. Parts are rich in magnetite together with hematite, pyrrhotite, pyrite, chalcopyrite and native iron which demonstrates that the burning of coal can readily create reducing conditions analogous to the process of iron smelting. In other parts, the opaque minerals are variably associated with gehlenite, titanaugite, fassaite, albite, Ca-rich plagioclase, hercynitic spinel, perovskite, biotite, quartz and apatite. Greenish to yellow coloured clinker is essentially a fassaite-gehlenite-spinel assemblage with only minor iron sulphides and oxides. Gas cavities in the clinkers contain hematite, "limonite", calcite, cryptocrystalline and opaline silica, and occasionally gypsum. All these mineral assemblages reflect significant variations in fO_2 similar to those of the Powder River Basin paralavas described in Chapter 3.

The fusibility of the Leigh Creek sub-bituminous coal ash to form slag occurs between 1250 °C to 1290 °C. Initial deformation in a reducing atmosphere occurs at 1210 °C with "blobbing" at 1295 °C. Under oxidizing conditions, initial deformation and blobbing takes place at 1300 °C.

4.2
Evaporates

Irkutsk

A rare example of the pyrometamorphism of evaporate deposits has been reported by Pavlov (1970) from Irkutsk where a sequence of salt, carbonate and sulphate-carbonate rocks have been intruded by dolerite. The halite is very pure with less than 1% anhydrite, dolomite, quartz and/or opaques. Two drill holes indicate that between 2000 to 2756 m depth one (upper) dolerite has intruded massive halite and another (lower dolerite) intrudes the contact between halite and a parting of carbonate-sulphate.

Within ~10 m of the dolerite, the halite has been recrystallised with some crystals up to 6 cm in size at the contacts suggesting that with the melting point of halite at 800 °C and a minimum crystallisation temperature of 950 °C for the dolerite, melting of the halite should have occurred. At the upper contact of the lower intrusion rounded grains of anhydrite partly rimmed by thin zones of small halite crystals occur within halite which is itself mantled by anhydrite exhibiting a more characteristic rectangular habit. Sporadic interstitial grains of sylvite are also present. The halite/anhydrite textures suggest fusion (rounded grains) of anhydrite in molten halite. With decreasing temperature, crystallisation of small amounts of halite on partly fused anhydrite occurred followed by crystallisation of anhydrite (rectangular habit). Sylvite crystallised at temperatures below its melting point of 778 °C, the potassium derived from a small amount (~0.25 wt.% K_2O) in the original halite. Near the lower contact of the dolerite, the dolomitic rock shows a patchy alteration to calcite associated with spinel and forsterite. Metamorphism of sulphate-carbonate rock has produced an anhydrite, forsterite (partly serpentinised), spinel assemblage with cubes of periclase that are partially replaced by brucite, similar to that described above from the Ioko-Dovyren intrusion.

Martian Meteorite

Evidence of salt-melts from a Martian meteorite (Nakhla) formed by the pyrometamorphism of presumed evaporate-like sediments by basaltic magma is described by Bridges and Grady (1998). The olivine-bearing clinopyroxenite meteorite contains interstitial areas of halite together with minor chlorapatite, anhydrite, and siderite-rich carbonate (23.2–87.0 mol% $FeCO_3$). In some places, the salt melt has intruded cracks extending from the interstitial areas into surrounding olivine and augite. The halite-dominant assemblage is considered to have crystallised from a trapped immiscible ionic melt at temperatures between 800–1000 °C. It is evident that the atmospheric pressure high melting temperature of anhydrite at 1450 °C must have been lowered in the presence of molten halite.

Chapter 5

Mafic Rocks

In metabasaltic rocks, the disappearance of pargasitic-rich amphibole and stabilisation of an anhydrous olivine, diopside-augite, pigeonite, orthopyroxene, plagioclase, Fe-Ti oxide assemblage is taken to indicate sanidinite facies conditions of metamorphism (Fig. 5.1). The distinction from a pyroxene-hornfels facies assemblage is shown in Fig. 5.2 and can sometimes also be made on the basis of whether melting has occurred as evidenced by glass and sanidinite facies minerals in associated rocks such as lithomarge or siliceous carbonate. Maximum P (~1.2 kb) at ~920 °C for metamorphism not involving melting is defined by intersection of the wet basalt solidi and amphibole-out curves within the quartz stability field (Fig. 5.1). Although the presence of olivine in particular, coupled with the absence of amphibole, is indicative of highest temperatures of metamorphism, its occurrence is also controlled by silica content of the mafic rocks. In silica undersaturated rocks, the breakdown of amphibole to form olivine rather than quartz is dependent on the edenite ($NaAlSi_{-1}$) and tschermakite ($Al_2Mg_{-1}Si_{-1}$) content of the amphibole as both substitutions result in Al-enrichment and Si-depletion favouring the production of olivine as discussed by Tracy and Frost (1991).

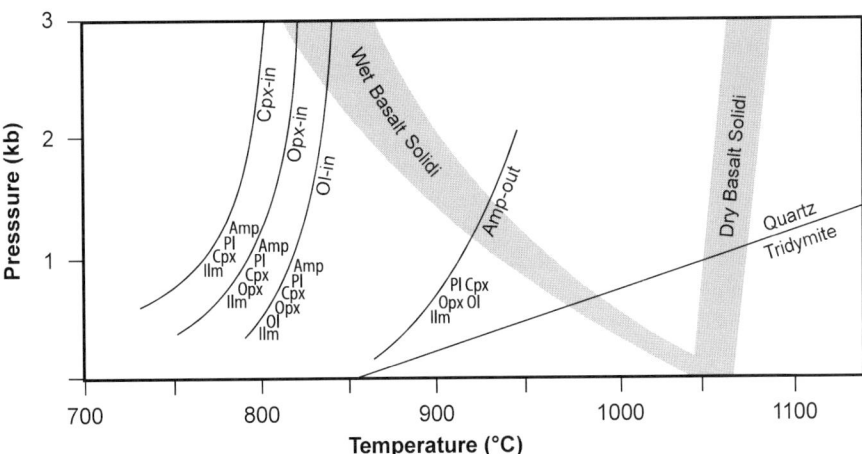

Fig. 5.1. Petrogenetic "grid" for sanidinite facies metamorphism of mafic rocks, i.e. at T > ~900 °C. Cpx-, Opx-, Ol-in curves and Amp-out curve at fO_2 defined by the quartz-fayalite-magnetite (QFM) buffer for a basalt composition (after Spear 1981). Additional symbols are: Amp = calcic (pargasitic) amphibole; Pl = plagioclase

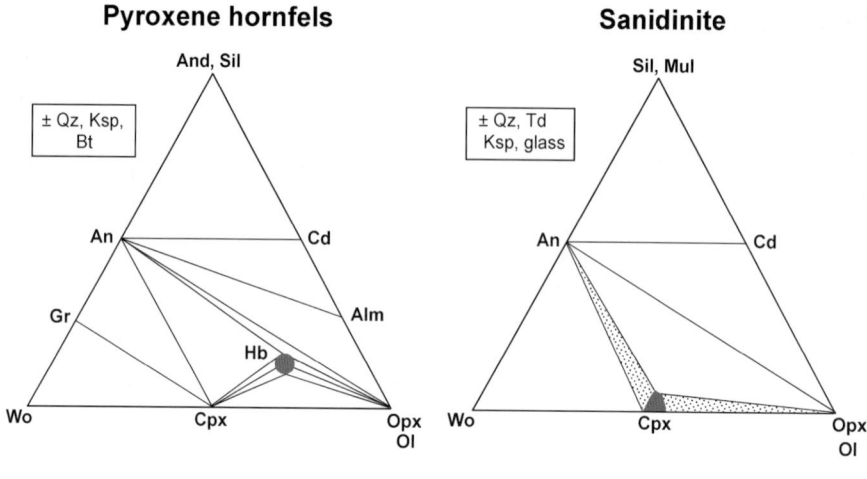

Fig. 5.2. ACF diagrams illustrating pyroxene hornfels and sanidinite facies assemblages (plotting below the An-Alm and An-Opx tie lines, respectively) in mafic rocks

5.1
Basaltic Rocks

5.1.1
Contact Aureoles

Smartville Complex

Greenschist facies metabasaltic rocks at the contacts of gabbro plutons in the Smartville Complex, Sierra Nevada, California, appear to be an example where partial melting has occurred at low pressure (≪3 kb and possibly as low as ~800 bar based on Al-content of amphibole in associated tonalite and granodiorite; Beard 1990). The highest temperature metamorphosed mafic rocks lack newly-formed olivine and are characterised by a well-developed hornfelsic texture between clinopyroxene, orthopyroxene, plagioclase (An_{50-59}), magnetite and ilmenite. Although two pyroxene and oxide thermometry yields temperatures of 705–765 °C and 740–792 °C respectively, the absence of calcic amphibole implies that temperatures along the pluton contacts were at least 900 °C (Fig. 5.1). Partial melting is suggested by the presence of veins and layers of "diorite" within the mafic hornfels that have an igneous (hypidiomorphic granular) texture between orthopyroxene, clinopyroxene, reversed zoned andesine-labradorite, magnetite, ilmenite and accessory quartz, that suggests the melting reaction

hornblende + plagioclase + quartz
= clinopyroxene + orthopyroxene + ilmenite + magnetite + melt

Fig. 5.3. Experimentally determined mineral stability fields of H₂O-saturated greenschist facies mafic rock melting from the Smartville complex, California. *Dashed line* = dehydration melting solidus (modified after Fig. 3 of Beard and Lofgren 1991)

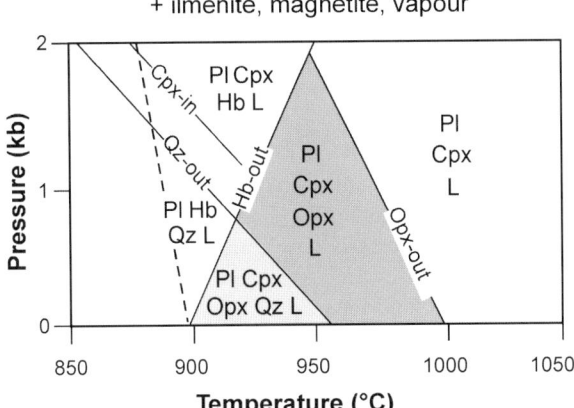

Experimental studies by Beard and Lofgren (1991) on greenschist facies metabasite from the Smartville complex delineate low pressure (< 2 kb) stability fields for quartz-absent and quartz-present assemblages of plagioclase-clinopyroxene-orthopyroxene-Fe-Ti oxides + tonolite melt at between 900 and 1000 °C for H₂O-saturated melting (Fig. 5.3).

Bunowen

Another example that has involved partial melting of basaltic compositions during pyrometamorphism is documented by Nawaz (1977) in a 35 m wide contact aureole associated with a tholeiitic dolerite plug near Bunowen, Co. Galway, Ireland. The dolerite intrudes mafic (amphibole-chlorite) schist and interlayered quartz-rich gneiss that are cut by a thin basaltic dyke within 5 m of the dolerite contact. The mafic schists are composed of variable amounts of quartz, plagioclase, actinolitic hornblende, chlorite, epidote and with a decrease in modal chlorite and amphibole they grade into quartzofeldspathic gneiss. The pyrometamorphosed mafic rocks contain orthopyroxene, clinopyroxene, calcic plagioclase and Fe-oxides. Melting has occurred in quartz-rich areas as indicated by the presence of either a pale brown glass or areas of recrystallised quartz (inverted tridymite), plagioclase, K-feldspar, orthopyroxene, apatite and opaques. In associated quartzofeldspathic rocks, an inferred melt is now present as a graphic intergrowth of inverted quartz, K-feldspar and rare orthopyroxene. Outlines of former amphibole replaced by granular aggregates of clinopyroxene, and an opaque phase (presumably magnetite ± ilmenite) are evident in quartz-absent, plagioclase-rich gneisses. Thermal alteration of the basaltic dyke is indicated by small rounded grains of orthopyroxene and dusty opaques that replace plagioclase, and in some areas by an isotropic phase (possibly glass) containing abundant magnetite.

Textural evidence for the former presence of tridymite and the lack of amphibole implies a minimum temperature of ~980 °C, i.e. at or above the wet basaltic solidus, and ~100 °C above the amphibole-out curve at ~0.6 kb (Fig. 5.1). The absence of olivine and formation of quartz (after tridymite) at this temperature reflects the silica-oversaturated composition of the mafic rocks coupled with a low edenite-tschermakite component of the actinolitic hornblende.

Skye

Shallow (~500 bar) contact metamorphism of alkali olivine basalt by the Cuillin gabbro, Isle of Skye, Scotland, has resulted in the formation of a ~500 m wide (map distance) orthopyroxene-olivine zone of granoblastic olivine (Fo_{75-68}), augite, orthopyroxene, plagioclase (An_{59-71}), ± biotite, magnetite, ilmenite hornfels without evidence of melting (Almond 1964; Spear et al. 1987) (Fig. 5.4). These rocks also occur as inclusions in the gabbro. Localised occurrence of the orthopyroxene-olivine zone rocks along the southern contact of the gabbro may reflect variation in fluid (and therefore heat) flow or indicate an irregular shape of the intrusion at depth. High temperature metamorphism resulted in the formation of two pyroxenes and a significant change in the relative amounts of pyroxene and olivine, i.e. an average ratio of pyroxene to olivine of 2:1 (basalt) to 4:1 (hornfels) with orthopyroxene (derived from both olivine and clinopyroxene) accounting for more than 1/3rd the total pyroxene in the hornfels. Two pyroxene thermometry indicates temperatures of 1000–1030 °C, i.e. just below the dry basalt solidus at 0.5 kb (Fig. 5.1). This temperature range is also compatible with the presence of typical pyrometamorphic minerals such as spurrite and rankinite in associated calcareous rocks and cordierite, mullite, sanidine, plagioclase and Fe-oxides in black, vitreous-like layers up to 0.7 cm thick representing metamorphosed laterite horizons between lava flows.

Skaergaard

In contrast to the ~500 m-wide occurrence of olivine-bearing hornfels associated with the Cuillin gabbro, Skye, the development of a granoblastic-polygonal assemblage of olivine, clinopyroxene, orthopyroxene, labradorite, biotite, ilmenite (olivine zone) occurs within ~10 m or less of the steeply-dipping contact of the Skaergaard gabbro, Greenland (Manning and Bird 1991; Manning et al, 1993). Two pyroxene thermometry indicates an average temperature of 888 ± 67 °C for an olivine-bearing sample 3 m from the contact and temperatures of ~800 °C 240 m from the contact in pyroxene zone rocks without olivine (Fig. 5.5). At an estimated lithostatic pressure of ~2 kb, the olivine hornfels represents a zone of partial melting in the mafic rocks (Fig. 5.1).

Thermal modelling by Manning et al. (1993) indicates that the peak metamorphic profile computed for heterogeneous distribution of a volume-averaged latent heat of crystallisation between 1150°–1050 °C for the Skaergaard intrusion as a whole (at ΔH_{xln} = 4184 J kg^{-1}) agrees well with metamorphic temperatures attained in the inner portion of the pyroxene zone, i.e. within 100 m of the contact during a 10 000 yr period, although the ~900 °C average temperature from the olivine zone indicates a steepening of the thermal gradient within 30 m of the contact (Fig. 5.5). This might be a reflection of a change in permeability, the most important factor controlling heat and fluid transport in the aureole. Numerical simulations of temperature versus distance

Fig. 5.4.
Texture of sanidinite facies olivine hornfels developed near contact of the Cuillin gabbro, Skye (redrawn from Fig. 3 of Almond 1964). *Dark grains* = olivine; *stippled grains* = orthopyroxene with larger grains (probably after original clinopyroxene) containing magnetite inclusions; *white grains* = polygonal mosaic of plagioclase; *diagonal stripped pattern* = ?retrograde biotite; *black grains* = spinel

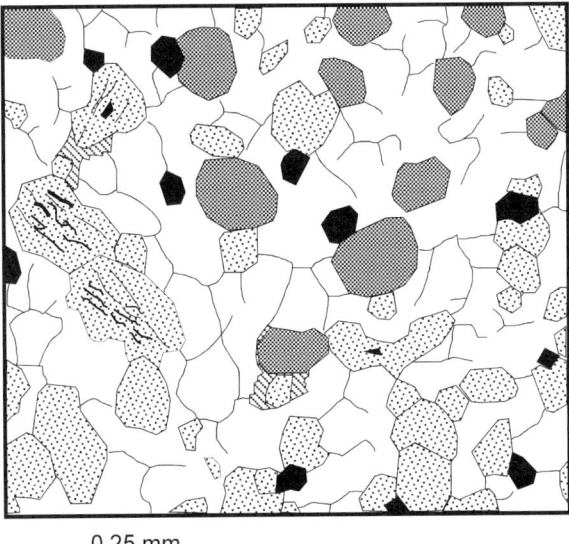

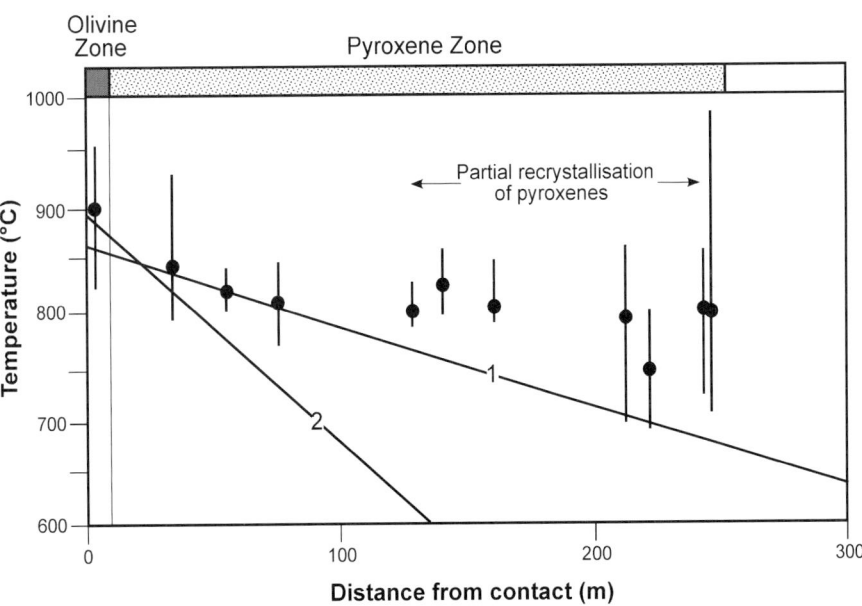

Fig. 5.5. Plot of temperatures derived from coexisting pyroxene thermometry (average and range) versus distance from contact of Skaergaard gabbro, Greenland (data from Appendix 1 of Manning et al. 1993). *Line 1* = simulated 10 000 yr conduction temperature profile for heterogeneous volume-averaged latent heat of crystallisation distribution in Skaergaard magma chamber. *Line 2* = simulated 1000 yr temperature profile in aureole basalt with a homogeneous permeability of 10^{-15} m^2 for the first 1000 yr after magma emplacement (from Figs. 6 and 7B respectively of Manning et al. 1993)

from the contact based on homogeneous permeability by Manning et al. (1993), indicate that peak contact temperatures approaching 900 °C could have been attained in the first 1000 years after emplacement (assumed instantaneous) for a basalt permeability of 10^{-15} m² and with a hydrothermal system dominated by advective heat transport. In pyroxene zone rocks between ~30–100 m from the contact, agreement between predicted and calculated peak temperatures (Fig. 5.5) equates to a lower permeability of 10^{-16} m² and a system dominated by conductive heat transport. This change probably relates to increasing enhancement of vertical conductivity and volume of the pore network (net-like vein swarms) towards the contact resulting in greater fluid flow and heat transport during prograde metamorphism (Manning and Bird 1991).

5.1.2
Amygdules and Mesostasis

Metamorphism of clay mesostasis/replacement and zeolite-bearing amygdules in basaltic rocks begins when complete dehydroxylation of smectites (e.g. montmorillonite, saponite, nontronite) occurs at ~750 °C and zeolites at temperatures of ~450 °C or lower. An abundance of Ca and alkalis in zeolites and Fe^{3+}, Mg, Al in smectites would be expected to produce minerals such as Ca-silicates, Na-pyroxene, melilite in addition to orthopyroxene, clinopyroxene, olivine, plagioclase, Fe-oxides and melts different in composition than that produced by reaction and fusion of the surrounding basalt.

'S Airde Beinn

Metamorphism without melting of zeolite and calcite-filled amygdules in basalt intruded by a dolerite plug at 'S Airde Beinn, northern Mull, Scotland, is described by Cann (1965). With increasing temperature, the sequence of neometamorphic Fe-Mg silicates in basalt and amygdules within about 10 m of the dolerite contact is orthopyroxene followed by clinopyroxene and then olivine.

Unmetamorphosed amygdules consist of:

1. Thompsonite ($Na_4Ca_8[Al_{20}Si_{20}O_{89}] \cdot 24H_2O$), natrolite ($Na_{16}[Al_{16}Si_{24}O_{80}] \cdot 16H_2O$) and analcite ($Na[AlSi_2O_6] \cdot H_2O$) that occur in varying proportions, sometimes with a rim of chlorphaeite (essentially a mixture of chlorite, smectite, goethite and calcite) separating the zeolites from basalt.
2. Dominantly gyrolite ($NaCa_{16}[AlSi_{23}O_{60}](OH)_5 \cdot 15H_2O$) with smaller amounts of thompsonite, analcite and natrolite.
3. Dominantly calcite.

At the lowest temperature, (orthopyroxene-forming stage; maximum oxidation of the basalt), *thompsonite, natrolite and analcime amygdules* have reacted to form *felsic* varieties dominated by plagioclase in which chlorphaeite may be converted to orthopyroxene, and *mafic* varieties with little feldspar containing orthopyroxene (replaces primary olivine), Fe-oxide and brown hornblende that may, in part have resulted in an infilling of the amygdules, i.e. radial arrangement of orthopyroxene crystals growing inwards from the amygdule walls suggests a vapour phase origin, rather than replacement. At higher grades, orthopyroxene is replaced by clinopyroxene and finally by olivine. Similar meta-amygdules consisting principally of plagioclase, pale green clinopyroxene, orthopyroxene, and sometimes with olivine and Fe-oxides are documented by Almond (1964) within olivine-pyroxene hornfels along the contact of the Cuillin Gabbro described above.

In the *gyrolite-bearing amygdules*, the formation of wollastonite at an early stage persists to the highest metamorphic grade accompanied by an increase in grain size, where it is sometimes replaced by melilite and rarely monticellite. The low grade appearance of wollastonite is accompanied by the formation of a thin rim of aegirine-augite around the inside of the amygdule that is in turn, surrounded by chlorophaeite. At higher temperature, reaction of chlorophaeite to orthopyroxene results in zonation of the aegirine-augite to augite towards the orthopyroxene. The characteristic sequence between basalt and amygdule involves Ca $\rightleftharpoons$ Na,Fe exchange and is

basalt → augite-rich basalt
  → augite ± plagioclase grading to aegirine-augite ± plagioclase
  → wollastonite ± aegirine-augite ± plagioclase

At the highest grade of metamorphism, outer parts of the aegirine-augite in contact with the basalt react under reduced conditions to form a vermicular intergrowth of augite, Fe-oxide and plagioclase so that the basalt-amygdule sequence is

basalt → augite + Fe-oxide + plagioclase → vermicular replacement
  → aegirine-augite → wollastonite ± aegirine-augite

In *calcite-dominated amygdules*, lowest grade indicators of metamorphism are rare examples of wollastonite + calcite. Within ~1.8 m of the dolerite plug contact the basalt contains large amygdules up to 76 mm across that contain a outward zoned assemblage of

larnite → rankinite → wollastonite → wollastonite ± melilite
  → melilite ± Fe-oxide rimmed by perovskite
  → rim of aegirine-augite with large plagioclase crystals associated
    with apatite and titanite at corners → augite-rich basalt

(Fig. 5.6). The above sequence represents progressive decarbonation under low PCO_2 (precluding the formation of spurrite and tilleyite), with silica diffusion into the amygdule from the surrounding basalt, outward diffusion of Ca leading to the development of augite-rich basalt around the amygdules, and inward diffusion of Mg and Al to form melilite. The melilite with its Fe-oxide/perovskite inclusions could also represent replaced areas of basalt that were closest to the centre of the amygdules.

The occurrence of larnite, rankinite and wollastonite provides the best indicator of high temperature metamorphism of the amygdules by way of the reactions

3 larnite + SiO_2 = 2 rankinite

rankinite + SiO_2 = 3 wollastonite

Fig. 5.6.
Sanidinite facies mineralogy of an original calcite-filled amygdule in basalt, S'Airde Beinn, Scotland (redrawn from Fig. 4 of Cann 1965)

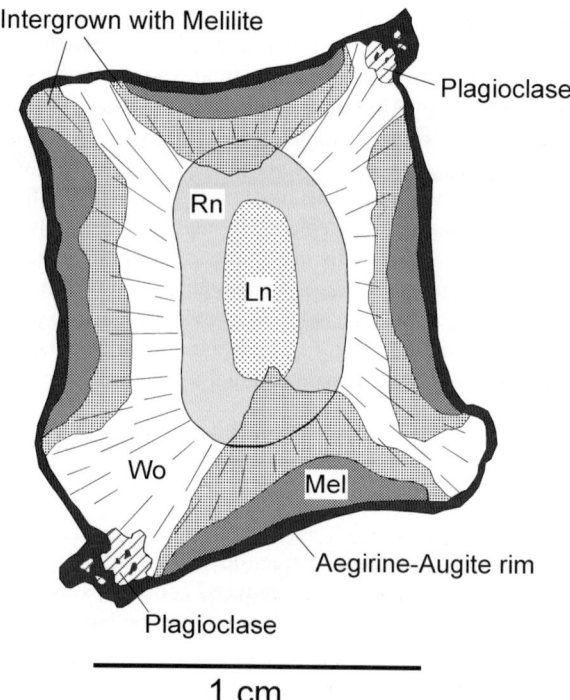

with increasing μSiO_2 (Fig. 4.8) and with the zonal sequence indicating that a wollastonite + larnite assemblage is not stable. The relevant vapour-absent reaction

   rankinite = larnite + wollastonite

calculated by Joesten (1974) has a step negative slope in P-T space and occurs between ~1137–1115 °C at 1 and 780 bar respectively. These temperatures are clearly too high for the amygdule assemblage as they substantially exceed those of dry basalt solidi (Fig. 5.1) where extensive melting would be expected. The stability of rankinite is bracketed by two decarbonation reactions

   spurrite + wollastonite = rankinite + CO_2

   spurrite + rankinite = larnite + CO_2

at between 880–970 °C and 100–200 bar PCO_2 (Fig. 4.6) This temperature range is consistent with the highest grade occurrence of melilite and monticellite in gyrolite-bearing amygdules, and of melilite (åkermanite) with rankinite and wollastonite in calcite-rich amygdules. It is also compatible with the formation of olivine and orthopyroxene, the absence of amphibole and conditions below the wet basalt solidus at pressures < 500 bar (Fig. 5.1).

Scawt Hill

Temperatures high enough to melt amygdule-bearing basalt were attained adjacent a dolerite intrusion at Scawt Hill, Northern Ireland, and the sequence, composition and nature of recrystallisation in basalt, mesostasis and amygdules is described by Kitchen (1985). The unaltered basalt contains partly serpentinised olivine phenocrysts (Fo_{81-65}) enclosing spinel partly mantled by Ti-magnetite within a groundmass of serpentinised olivine, magnetite, augite, plagioclase (An_{76-68}) and ilmenite. Mesostasis consists of saponite-nontronite and unidentified zeolite and 1–5 mm diameter amygdules lined with saponite-nontronite, calcite and zeolite. Compositions of unaltered and metamorphosed basalt and mesostasis are plotted in terms of a $CaO-MgO-Al_2O_3-SiO_2$ (CMAS) diagram at 1 kbar in Fig. 5.7 which shows that the mesostasis represents a highly altered residual glass very different in composition from the unaltered and metamorphosed basalt.

At 2 m from the contact, the basalt is substantially recrystallised and mesostasis and amygdules have melted. Olivine is recrystallised to microcrystalline orthopyroxene and olivine, the later with 1.4 and 1.9 wt.% CaO and Al_2O_3 respectively, as a result of reaction with the mesostasis melt. Augite has reacted with the mesostasis melt to form low-Ca augite and pigeonite. Plagioclase is recrystallised to a microcrystalline mosaic.

Melted amygdules (5 – < 1 mm size) are partly recrystallised to pyroxene, plagioclase and Fe-Ti oxide in a dark brown glass. Pyroxenes show normal zoning and vary from pigeonite through subcalcic augite to augite (Fig. 5.7) similar to that observed in experimental studies of basalts involving rapid growth in structurally and compositionally complex melts (Walker et al. 1976). Such conditions have led to substitution of Ti (0.4–0.9 wt.% TiO_2) and Al (0.4–4.4 wt.% Al_2O_3) that is facilitated by high growth rates and low diffusion rates accompanying rapid cooling of possibly 1–10 °C hr^{-1} for larger pyroxenes to perhaps as much as 84 °C hr^{-1} for Fe-rich skeletal overgrowths (Schiffman and Lofgren 1981). Plagioclase (An_{55-36}) occurs as multiple generations of euhedral crystals with skeletal terminations. Fe-Ti oxide forms skeletal crystals with exsolved lamellae of Ti-magnetite and ilmenite. Glass compositions are rich in SiO_2 (~60.7 wt.%) and K_2O (8.7–9.9 wt.%) with CaO (0.7–0.8 wt.%), Na_2O (~1.9 wt.%) and (Fe,Mg)O (5–7 wt.%) and presumably reflect the initial clay : zeolite ratio of the amygdules.

Basalt *at the dolerite contact* is entirely recrystallised. The groundmass is converted to a fine grained to microcrystalline aggregate of olivine, orthopyroxene, plagioclase, magnetite and ilmenite. Irregular streaks and lenses of devitrified basalt melt contain dendritic olivine, orthopyroxene, skeletal magnetite, ilmenite, pleonaste spinel in a matrix of cryptocrystalline plagioclase of $An_{70.5-60.7} Ab_{17.1-20.9} Or_{12.4-18.4}$ with significant amounts of Fe (1.9–4.0 wt.% Fe_2O_3) and Mg (0.7–2.0 wt.% MgO). High FeO occurs in plagioclase experimentally grown in basaltic melts and is related to Fe-entrapment on crystal interfaces during very fast cooling rates (Schiffman and Lofgren 1981). No amygdules are present. Evidence of plastic deformation suggests the possibility that melted amygdules could have been mobilised and smeared into the surrounding basaltic melt.

Temperatures of pyrometamorphism based on two pyroxene and plagioclase melt thermometry range from 970–1168 °C consistent with an estimated dolerite solidus of

$C = (\text{mol. prop. } CaO - 3^{1/3}P_2O_5 + 2Na_2O + 2K_2O) \times 56.08$
$M = (\text{mol. prop. } FeO + MnO + NiO + MgO - TiO_2) \times 40.32$
$A = (\text{mol. prop. } Al_2O_3 + Cr_2O_3 + Fe_2O_3 + Na_2O + K_2O + TiO_2) \times 101.96$
$S = (\text{mol. prop. } SiO_2 - 2Na_2O - K_2O) \times 60.09$

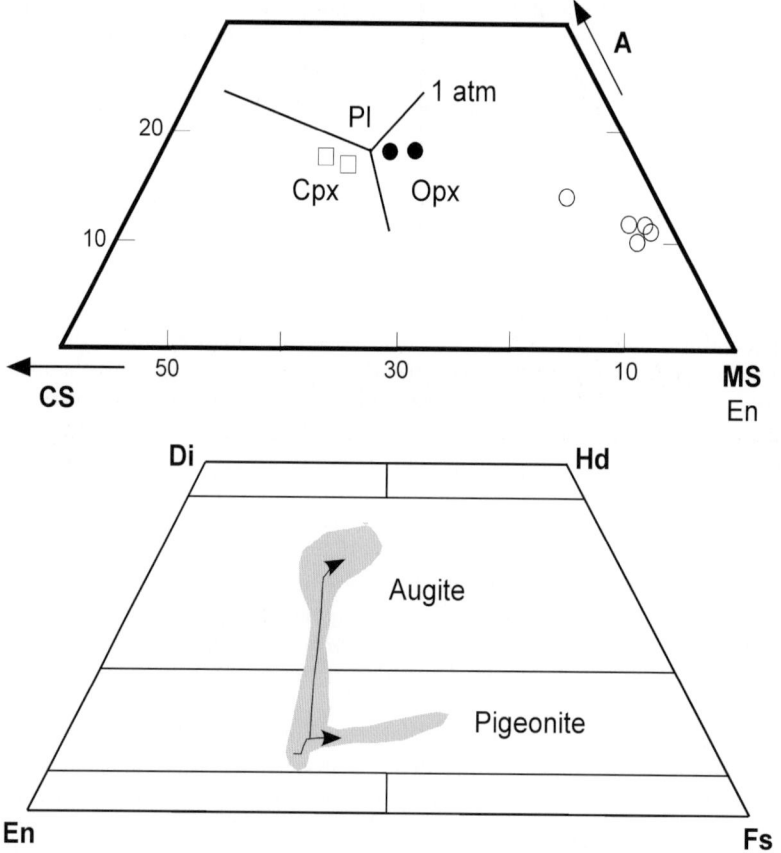

Fig. 5.7. Scawt Hill basalt, Northern Ireland. *Above:* CMAS section at 1 atm projected from olivine (O'Hara 1968) showing compositions of unaltered basalt (*open squares* in Cpx field), metamorphosed basalts (*solid circles* in Opx field), and clay-rich mesostasis (*open circles*). *Below:* Pyroxene quadralateral showing range of pyroxene compositions in melted mineral-filled cavities (*shaded area*) with zonal trends depicted by *arrowed lines* (from Figs. 3 and 5 of Kitchen 1985)

~936 °C. Temperatures > 1050 °C would be required to intersect the dry basalt solidi, with the neometamorphic assemblage olivine-orthopyroxene-clinopyroxene-plagioclase-ilmenite forming at minimum temperatures of 860–900 °C for inferred pressures < 500 bar (Fig. 5.1).

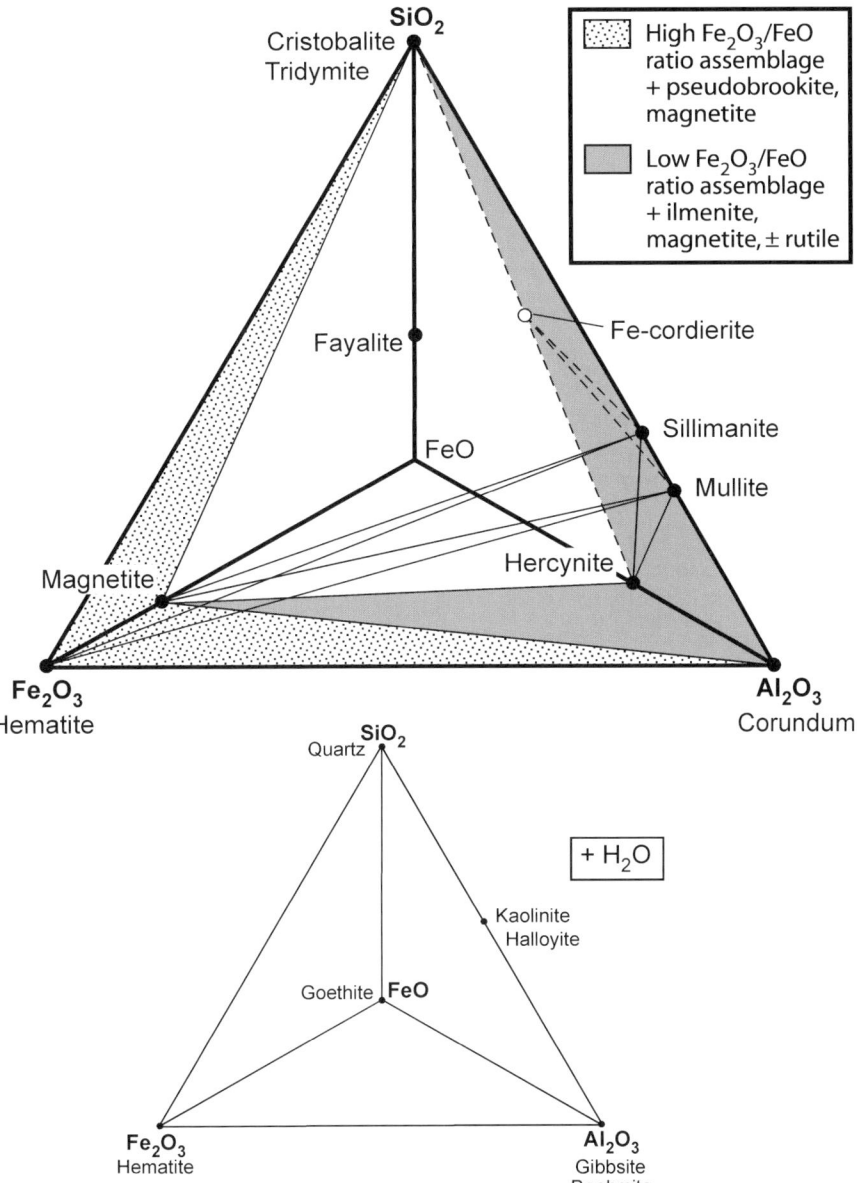

Fig. 5.8. The system $SiO_2 - FeO - Fe_2O_3 - Al_2O_3$ (wt.%) showing composition volumes of high and low Fe_2O_3:FeO assemblages (shown as *tie lines*) related to sanidinite facies metamorphism of weathered mafic rocks characterised below by the presence of quartz, kaolinite/halloysite, goethite, hematite, and gibbsite/boehmite. *Thickened parts* of the mullite-hematite and corundum-hematite tie lines indicate extent of solid solution of Fe^{3+} in mullite and corundum respectively. Phase equilibria (in air) of the SiO_2-magnetite-corundum plane in the tetrahedron after Muan (1956)

5.1.3
Weathered Mafic Rocks

Weathering of mafic rocks produces a bol, lithomarge, lateritic lithomarge or Fe-bauxite characterised by assemblages of quartz, halloysite, kaolinite, montmorillinite, gibbsite, hematite, goethite and anatase (e.g. Eyles 1952; Patterson 1955). During sanidinite facies metamorphism such rocks would be expected to form cristobalite/tridymite, mullite/sillimanite, corundum, hercynitic spinel, magnetite/ilmenite/hematite and pseudobrookite. Small amounts of Mg (in montmorillonite and "chlorite") and Ca (in carbonate) allow the formation of cordierite and anorthite respectively. The high temperature mineral assemblages are similar to those developed in emery rocks, e.g. Sithean Sluaigh described in Chapter 3, that result from removal of silica and alkalis as a granophyric melt phase. In this case, silica and alkalis are variably leached during the pre-metamorphic laterisation process and like emeries, pyrometamorphic mineral assemblages developed from extreme weathering of mafic rocks are closely analogous with those formed in the system $FeO\text{-}Fe_2O_3\text{-}Al_2O_3\text{-}SiO_2$ (Fig. 5.8).

Tievbulliagh

At Tievebulliagh, Northern Ireland, an inclined dolerite plug with a 1.5 m thick basal layer of picrodolerite has intruded basalts of the Antrim plateau that overlie Old Red Sandstone (Agrell and Langley 1958) (Fig. 5.9). Basalt and associated *bols* adjacent to the plug have been metamorphosed to orthopyroxene, plagioclase, magnetite, ilmenite,

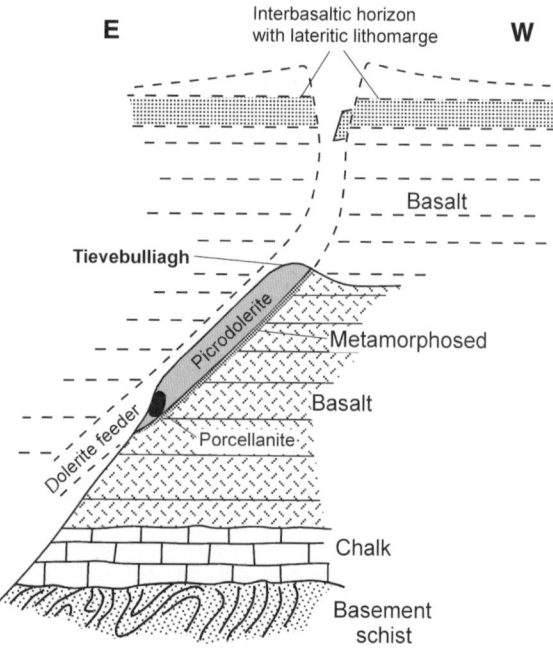

Fig. 5.9.
Diagrammatic section at Tievebulliagh, Northern Ireland (redrawn from Fig. 1 of Agrell and Langley 1958). See text

pseudobrookite and to cordierite, bytownite, magnetite, hematite, pseudobrookite, tridymite/cristobalite ± mullite assemblages (Figs. 5.10 and 5.11). A small outcrop of what remains of the eroded dolerite contains a large block of porcellanite (spotted porcellanite described by Tomkeieff 1940), inferred to represent a metamorphosed fragment of thoroughly weathered basalt or lateritic lithomarge slumped from an interbasalt layer (Main Interbasaltic Horizon) where this alteration is commonly observed (Fig. 5.10b). The porcellanite typically shows nodular and irregular banded structures that consist of various combinations of corundum, hercynite, mullite, tridymite and/or cristobalite, either with hematite, pseudobrookite, magnetite (oxidised), or with a less oxidised assemblage of ilmenite, magnetite, ± pyrite and rutile (Fig. 5.11). Near the dolerite contact, the porcellanite contains cordierite in association with mullite, hercynite and cristobalite of less oxidised assemblages (dashed tie lines in Fig. 5.8). White patchy segregations of almost pure mullite, yellow ones rich in corundum, brown streaks composed mainly of pseudobrookite and black hematite-rich patches, reflect the heterogeneous nature of the original lithomarge. The oxidised assemblages are thought to correspond to ferruginous bauxites, the less oxidised assemblages to a lithomarge in which halloysite, gibbsite and hematite were the original constituents. A

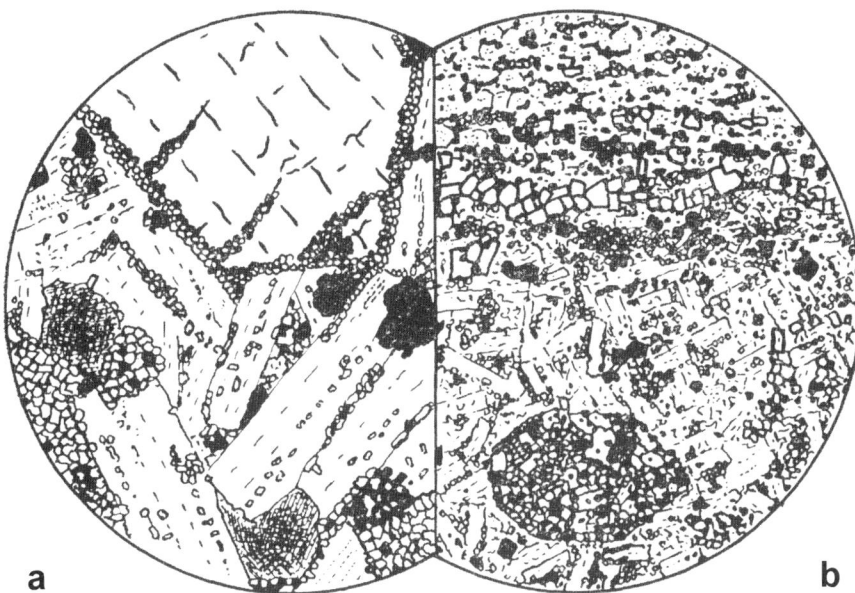

Fig. 5.10. Textures of pyrometamorphosed olivine basalt, Tievebulliagh, northern Ireland (Fig. 2 of Agrell and Langely 1958). **a** Microphenocrysts of olivine with rims and seams of orthopyroxene + magnetite within a recrystallised groundmass of plagioclase (An_{64}), orthopyroxene and clinopyroxene. **b** Contact between metabasalt (*lower part*) and its weathered equivalent (bole) (*upper part*) separated by a thin seam of orthopyroxene. Microphenocrysts of olivine in the basalt are completely replaced by orthopyroxene and magnetite and occur in a groundmass of plagioclase (An_{64}), orthopyroxene and magnetite. The bole consists of cordierite, Fe-oxides, minor plagioclase (An_{62-68}) and orthopyroxene. Reproduced with the kind permission of the Royal Irish Academy

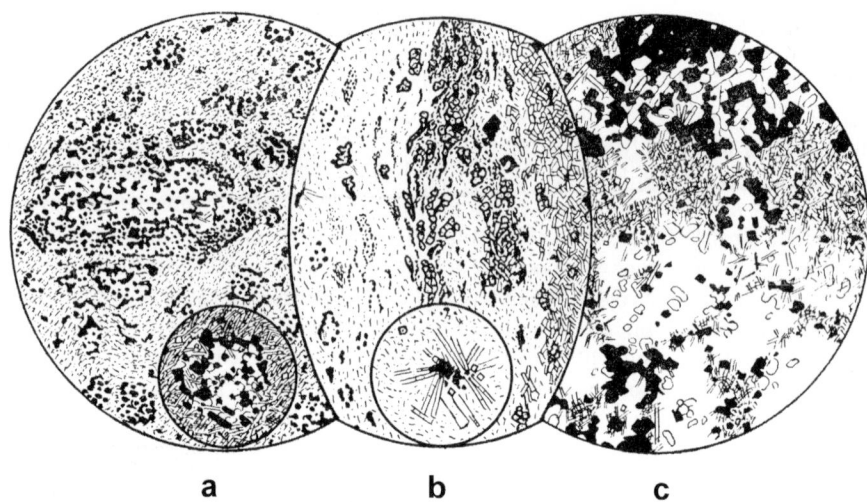

Fig. 5.11. Textures of pyrometamorphosed lithomarge and lateritic lithomarge, Tievebulliagh, Northern Ireland (Fig. 3 of Agrell and Langely 1958). **a** Mullite-cristobalite-hematite porcellanite. Patches of cristobalite-hematite (see *inset* for detail) probably after olivine, within a groundmass of mullite fibres associated with cristobalite, hematite and pseudobrookite. Original rock was a lithomarge. **b** Mullite-corundum-hematite-pseudobrookite porcellanite (*right*) passing though a zone rich in pseudobrookite into mullite-cristobalite-hematite-pseudobrookite porcellanite (*left*). *Inset* shows development of mullite needles around a cluster of hematite crystals. Original rock corresponds to the contact between a lateritic lithomarge and a lithomarge. **c** Mullite-tridymite-cordierite-magnetite porcellanite. Tridymite and magnetite are enclosed in xenoblastic cordierite that replaces mullite. The original rock was a lithomarge at the immediate contact of basalt and has undergone metasomatic exchange involving introduction of MgO either before or during metamorphism resulting in the formation of cordierite. Reproduced with the kind permission of the Royal Irish Academy

plot of Al, Fe, Ti, Mg enrichment versus calculated silica loss from an average original basalt composition (Fig. 5.12) indicates a positive relationship that reflects increasing intensity of weathering (laterisation) of the basalt protolith.

Maximum temperatures of pyrometamorphism are bounded by the dry basaltic liquidi at between ~1175–1250 °C that brackets those of invariant assemblages of mullite-tridymite-hercynite at 1205 °C (Schairer 1942) and mullite-cordierite-tridymite and mullite-cordierite-hercynite at 1210 °C and 1205 °C respectively, in the system FeO-Al$_2$O$_3$-SiO$_2$ (Schairer and Yagi 1952) (Fig. 3.4; Table 3.1). The occurrence of cristobalite in place of tridymite highlights its metastability with respect to the ternary system (Fig. 5.8).

Rathlin Island

At Rathlin Island, Co. Antrim, Northern Ireland, intrusion of an olivine dolerite "plug" has metamorphosed a weathered interbasaltic horizon of probable ferruginous bauxite to porcellanite (Dawson 1951). The rock is mottled in various shades of grey and ranges in colour from black through greyish-purple to lavender. It has a high specific gravity of 3.3–3.6 and consists of variable amounts of tridymite, aluminous sillimanite (64–65 wt.% Al$_2$O$_3$), cordierite (XFe$_{73}$), hercynite and magnetite (Cameron 1976a)

Fig. 5.12.
Plot of SiO_2-loss versus sum of restite elements in metamorphosed basalt and lithomarge, Tievebulliagh and Rathlin Is, Northern Ireland (redrawn from Fig. 4A of Agrell and Langely 1958; calculated silica-loss values from Table 3)

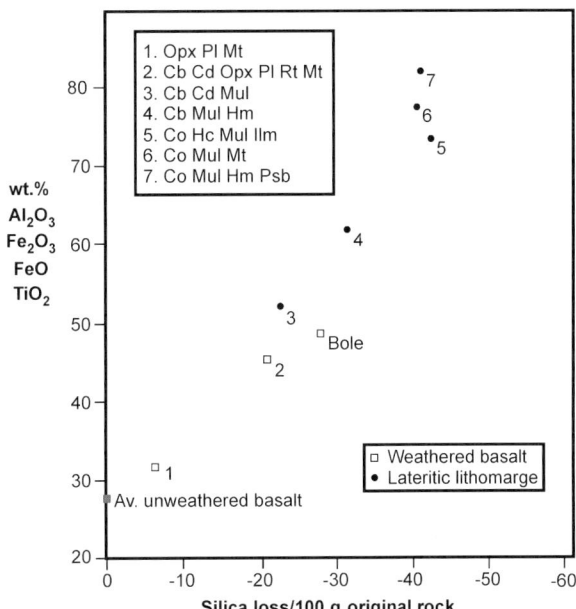

(Fig. 5.12). High TiO_2 of 3.40 wt.% in a porcellanite analysis given by Dawson (1951) implies the presence of pseudobrookite who also records the presence of a small amount of glass (~5%).

Oregon

Very similar rocks to those at Tievebulliagh and Rathlin Island are found as cobble float in the west-central Cascades, Oregon (White et al. 1968). The rocks are essentially black (mullite-sillimanite, cristobalite-rich with hercynitic spinel, magnetite) to light bluish-grey to white (corundum-rich) porcellanites (Fig. 5.13). They also contain varying amounts of tridymite, ± pseudobrookite, perovskite and trace quartz. Flow banding and vermicular features reflect original structures of extensively weathered (laterised) basalt-andesite protoliths; black cobbles representing ferruginous bauxite and grey-white cobbles representing kaolinite and/or halloysite-rich lithomarge. Cameron (1976a) describes one sample with a polygonal structure where polygon cores composed of a fine-grained aggregate of sillimanite, mullite, cristobalite, spinel and pseudobrookite grade abruptly into a tridymite-hercynite margin. Relic igneous texture is discernible in the core with the Al-silicates pseudomorphing plagioclase.

Southeastern China

Tridymite (35–45%), hercynite (40–50%), ilmenite (~2%), colourless to pale yellow glass (10–20%) xenoliths in basalt from the southeast coastal area of China reported by Qi et al. (1994) and Zhou et al. (2004) are an example of pyrometamorphosed lateritic paleosols where complete melting appears to have occurred (Fig. 5.13). The tridymite

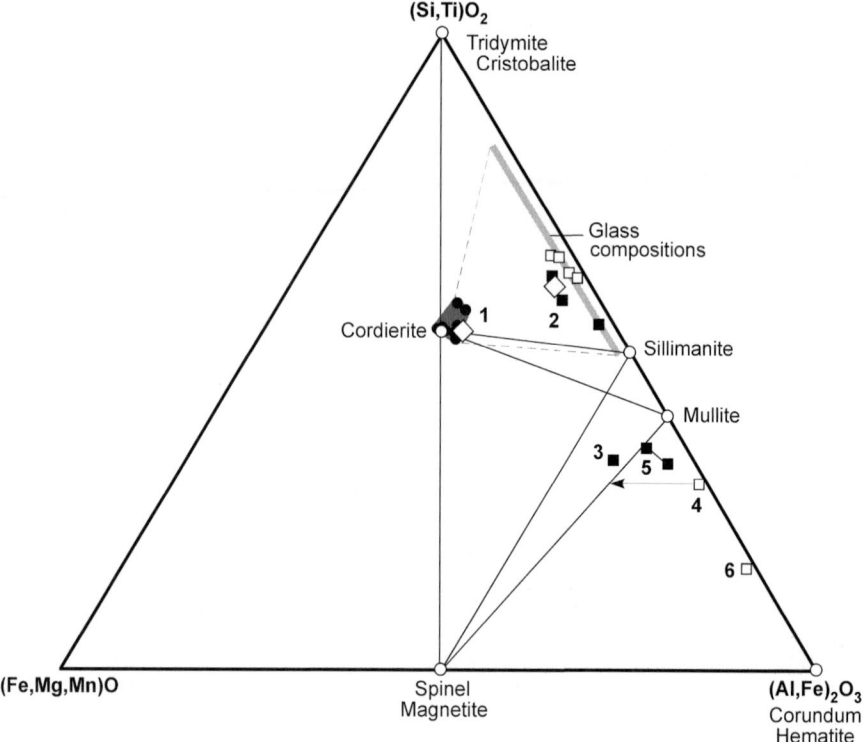

Fig. 5.13. $(Si,Ti)O_2$-$(Al,Fe)_2O_3$-$(Fe,Mg,Mn)O$ plot of pyrometamorphosed lithomarge-Fe bauxite rocks (xenoliths [*filled circles in shaded area*]; porcellanite), glass in xenoliths (compositional range shown by *thick grey line*), and unmetamorphosed lithomarge. *Arrow* from sample 4 (Oregon) indicates probable position shift to mullite-spinel tie line if FeO/Fe_2O_3 ratio is known

grains contain numerous cracks suggesting transition from cristobalite (or high-tridymite) to low tridymite. Of interest is the wide variation in the glass composition ranging from 73–35 wt.% SiO_2 and 25–59 wt.% Al_2O_3 (Fig. 5.13) with silica-rich glass occurring near tridymite and less aluminous glass near hercynite. This variation mimics compositional profiles typical for quenched margins of actively growing phases. Crystallisation of the glass would presumably have resulted in assemblages of variable amounts of tridymite, mullite and sillimanite as implied by the chemographic relations shown in Fig. 5.13.

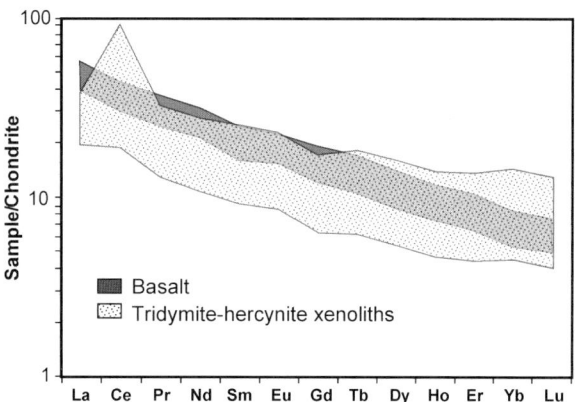

Fig. 5.14.
Chondrite-normalised REE abundance plot of tridymite-hercynite xenoliths and host tholeiitic basalt, SE China (redrawn and combined from Figs. 5A and B of Zou et al. 2004)

With reference to the system $FeO-Al_2O_3-SiO_2$ investigated by Schairer (1942) without the intervening narrow field of Fe-cordierite determined in the same system by Schairer and Yagi (1952), metastable co-precipitation of hercynite and tridymite in the xenoliths could have occurred between 1205 °C (where mullite would form) and 1073 °C (where fayalite would form) (Fig. 3.4). As shown in Fig. 5.13, xenolith compositions cluster near Fe-cordierite although no cordierite occurs in the xenoliths. This probably reflects nucleation difficulties without addition of cordierite seeds encountered in experimental work (Schairer and Yagi 1952), so that co-precipitation of hercynite and tridymite in the xenoliths with essentially bulk Fe-cordierite compositions implies metastable crystallisation.

Chondrite-normalised REE patterns of xenoliths and host basalt shown in Fig. 5.14 overlap and exhibit a positive Ce anomaly that is probably related to variable weathering of the soil profile where Ce and other LREE enrichment often occurs in the uppermost part of paleosols (Pan and Stauffer 2000). The presence of nearly pure hercynitic spinel and ilmenite in the xenoliths implies low fO_2 conditions during pyrometamorphism, a condition probably caused by organic matter in the paleosol although in amounts unable to facilitate the reduction of CeO_2 to Ce^{3+} at surface conditions giving rise to the positive Ce anomaly recorded in the xenoliths.

5.2
Aluminous Ultramafic Rocks

Examples of pyrometamorphosed aluminous magnesian-rich (ultramafic) rocks are rare. Except perhaps for the occurrence of cordierite they would be indistinguishable mineralogically from pyroxene hornfels facies assemblages, highlighting the importance of their occurrence, i.e. in typically narrow zones adjacent intrusions. Possible sanidinite facies assemblages for Ca-absent Al-ultramafic compositions (e.g. chloritite blackwall rocks) in terms of $MgO-Al_2O_3-SiO_2$ (Fig. 5.15) comprise:

1. Enstatite-tridymite (-cordierite)
2. Forsterite-enstatite (-cordierite)
3. Forsterite-cordierite (-spinel)
4. Forsterite-periclase (-spinel)

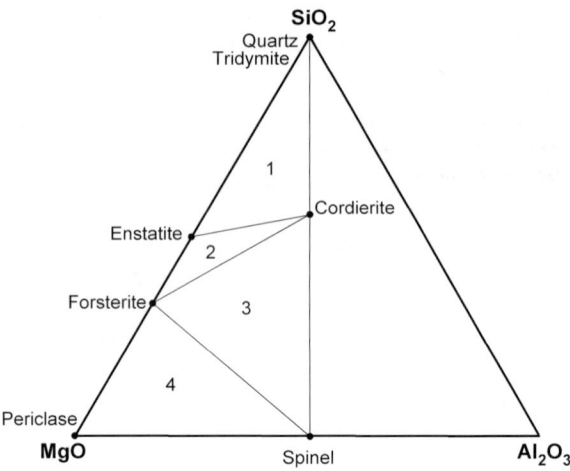

Fig. 5.15.
SiO_2-Al_2O_3-MgO plot showing possible sanidinite facies mineral assemblage fields (labelled 1 to 4 as detailed in text) in Al-rich ultramafic compositions

The reaction

anthophyllite = enstatite + quartz + H_2O

occurs at ~720–730 °C at < 1 kb (Greenwood 1963) and breakdown of Mg-chlorite according to the reaction

chlorite = forsterite + cordierite + spinel + H_2O

occurs at even lower temperatures (< 700 °C at < 1 kb) (Fawcett and Yoder 1966) (Fig. 5.16). These reaction temperatures are clearly not high enough for pyrometamorphic conditions and merely provide temperature minima for the occurrence of these assemblages in nature. It would take overstepping of the anthophyllite breakdown reaction by ~200–270 °C (at 0–1 kb respectively) to form tridymite (Fig. 5.16) and no natural examples of an enstatite–tridymite association are known. Maximum pressures in the Fe-free system involving an anhydrous cordierite are limited by the reaction

10 enstatite + 2 spinel = 5 forsterite + cordierite

at ~3 kb (Fig. 5.17) although pressure is lowered by substitution of iron and chromium in spinel, e.g. XAl_{spinel} of 0.7 = ~1 kb (Frost 1975) and probably also by the H_2O content of natural cordierite resulting in significant restriction of the forsterite-cordierite stability field to the extent where it may not be stable at all (i.e. at $XAl_{spinel} \leq 0.6$).

Tari-Misaka Complex

Serpentinised dunite-harzburgite and chromitite of the Tari-Misaka complex, western Japan, is intruded by granite and converted to assemblages of olivine (Fo_{92-94}), orthopyroxene (En_{91-95}), chrome spinel, and olivine (Fo_{97}), orthopyroxene (En_{96}), cordierite ($XMg - 0.99$), Mg-Al spinel ($XAl = 1.00$) within ~500 m (map distance) of the contact

Fig. 5.16.
T-P diagram showing reaction curves for (*1*) Mg-Fe chlorite (McOnie et al. 1975), (*2*) Mg-chlorite (Fawcett and Yoder 1966), (*3*) anthophyllite (Greenwood 1973), and (*4*) forsterite + cordierite formation with XSpl = 1.0 (Frost 1975). Positions of reaction *4* at lower XSpl – Al values (*filled circles*) due to Cr-substitution at 730 °C are from Fig. 14 of Frost (1975) with respective reaction lines (*dashed*) assumed parallel to that of XSpl – Al = 1.0

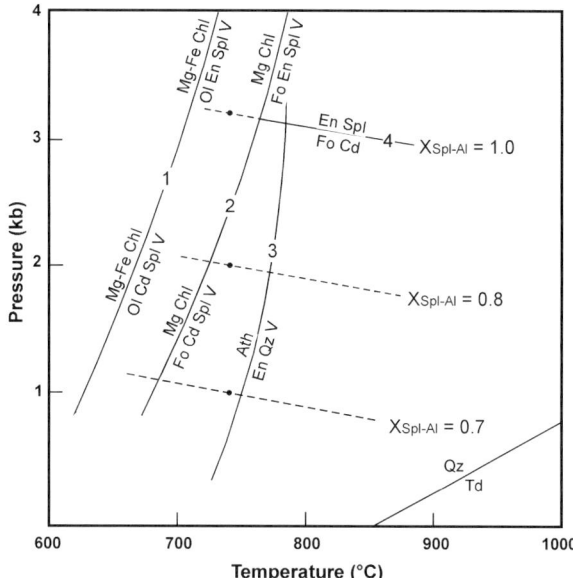

(Arai 1975). Comparison with experimental data in the system $MgO-Al_2O_3-SiO_2-H_2O$ (Fawcett and Yoder 1966; Seifert 1974) suggests a maximum temperature of ~770 °C at pressures of ~3 kb consistent with an XAl_{spinel} of 1.0 coexisting with olivine and cordierite (Fig. 5.16). Two pyroxene thermometry from a metagabbro associated with the ultramafic rocks near the granite contact yields a temperature of 890 °C although, as pointed out by Arai this seems to be too high for a granitoid magma and may reflect incomplete recrystallisation of primary pyroxenes.

5.3
Hydrothermally-Altered Andesite

Geothermal systems characterised by near-neutral NaCl or $NaCl-HCO_3-SO_4$ waters typically occur within the central vent region of andesitic volcanoes and give rise to complex mineral paragenesis when lava rises and pulses of high-temperature volatiles invade the hydrothermal envelope around these vents. Usually, mineral assemblages that form when a vent-hosted hydrothermal system is heated to near magmatic temperatures are not preserved because they are destroyed by large eruptions, but rare fragments from the magma-hydrothermal interface that may be ejected during phreatomagmatic eruptions contain sanidinite facies minerals.

White Island

Xenoliths ejected from the hydrothermal system above a shallow (< 1 km) degassing magma body below the crater of the White Island andesite volcano, New Zealand, contain mullite, high-sanidine, tridymite, cristobalite, rare indialite (the high temperature form of cordierite), wollastonite, plagioclase and magnetite-hematite (Wood

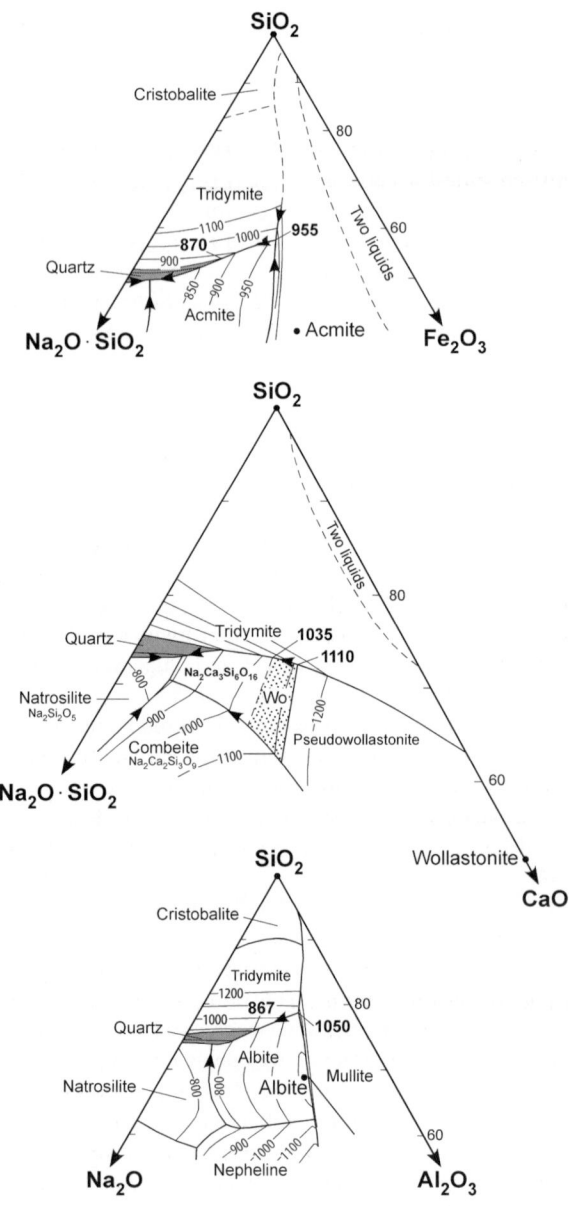

Fig. 5.17.
Phase equilibria relating to the occurrence of tridymite-acmite (system SiO_2-Fe_2O_3-$Na_2O \cdot SiO_2$ after Bowen et al. 1930), tridymite-wollastonite (system SiO_2-CaO-Na_2O after Morey and Bowen 1925), and tridymite-albite (system SiO_2-Al_2O_3-Na_2O after Schairer and Bowen 1947) in pyrometamorphic magma bombs, White Island volcano, New Zealand. See text

1994). The altered source rocks are crater lake-bed sediments that contain hydrothermal phases such as opaline silica, cristobalite, anhydrite, gypsum, natroalunite, pyrite and sulphur, with halite and silvite in cavities.

Under conditions of > 800 °C and PH_2O < 200 bar, mullite, sanidine and wollastonite formed by the reactions

2 natroalunite + 8 SiO$_2$ = mullite + 2 K-feldspar + 4 SO$_3$ + 6 H$_2$O

anhydrite + SiO$_2$ = wollastonite + SO$_3$

Quasi-pelitic sediments with natroalunite, kaolinite and pyrophyllite + partly alunitised andesitic glass are possible precursors of indialite that occurs in a partly fused porcellanite xenolith and in glass of an ejected block.

Wood (1994) and Wood and Brown (1995) describe an unusual example of *pyrometamorphic magma* (paralava) that occurs as bombs erupted with calc-alkaline basaltic andesite/andesite blocks, lapilli and ash from a ~30 × 50 m vent on White Island. The paralava consists of a highly peralkaline glass (Na$_2$O + K$_2$O > 10 wt.% with Na$_2$O ≫ K$_2$O; molar [Na$_2$O + K$_2$O]/Al$_2$O$_3$ = 1–22) with an average Cl content of 1.23 wt% that contains quench crystals of tridymite (in discrete mm-size aggregates), wollastonite and green clinopyroxene (inferred to be aegirine-augite or sodic hedenbergite) that may be locally intergrown. One pumiceous bomb contains mosaics of albite that intermingle with pale brown, less vesicular glass containing wollastonite and green pyroxene. The typically vesicular, green-black to olivine-green bombs often contain white siliceous ash and lapilli-size clasts and have been erupted in a molten state from small isolated melt pockets developed along the walls of the vent. Another 15 cm bomb contains a core of grey-black volcaniclastic metasediment surrounded by a zone of olivine-green weakly vesiculated material transected by expansion cracks. This, in turn, grades patchily into black, highly vesiculated paralava that contains the common white, silica-rich lithic clasts.

Likely protolith material of the paralava bombs is considered to be brine-soaked or halite-cemented crater lake sediments and acid-sulphate hydrothermally altered vent andesitic breccia and tuff. In some bombs, the aggregates of tridymite may represent unmelted residues of originally poorly crystalline opaline-silica/cristobalite acid altered material that has converted to tridymite. In the fusion zone, the temperature is inferred to have been between ~> 830 °C (temperature measured from the mouth of a fumerole) and 1020 °C (two-pyroxene thermometry from basaltic andesite) with pressures < 50 bar. As most of the glass compositions are silica-saturated, the occurrence of tridymite on the liquidus is likely. At atmospheric pressure, the inferred temperature range of fusion is consistent with cotectic formation of tridymite-aegirine between 870–955 °C in the system Fe$_2$O$_3$-SiO$_2$-Na$_2$O · SiO$_2$ (Bowen et al. 1930), a minimum temperature of 1035–1100 °C for tridymite-wollastonite in the system Na$_2$O-CaO-SiO$_2$ (Morey and Bowen 1925), and of albite-tridymite between 867–1050 °C in the system Na$_2$O-Al$_2$O$_3$-SiO$_2$ (Schairer and Bowen 1947) (Fig. 5.17). The environment at White Island may also be conducive to the formation of the compound Na$_2$Ca$_3$Si$_6$O$_{16}$ synthesised in the system Na$_2$O-CaO-SiO$_2$ (Fig. 5.17) that has not yet been found in nature.

Chemical heterogeneity of the paralava implies that fusion probably occurred in small, isolated areas and that the time between onset of melting and eruption was too short to allow compositional domains in a single pocket of melt to homogenise. Silicate melts that produced the paralava bombs were buffered by immiscible NaCl-saturated aqueous vapour and hydrosaline (NaCl-rich) melt with the high Na$_2$O and Cl of the melts a consequence of abundant halite in the volcaniclastic precursor rocks. Low sulphur contents of the paralava in comparison to unmetamorphosed sulphide and sulphate-bearing volcaniclastics with between 0.7-5.7 wt.% S, imply volatilisation and near minimum sulphur solubilities in the melts.

5.4
Vapour Phase Crystallisation

Ruapehu

Xenoliths of andesitic volcanic breccia erupted from Mt. Ruapehu, New Zealand, contain cavities in which SEM and petrographic observation indicates the following mineral paragenesis:

1. Osumilite + hedenbergite + labradorite, followed by,
2. hexagonal cordierite (0.18 wt.% K_2O) + Na-sanidine + low-Al (1.8 wt.% Al_2O_3) orthopyroxene + tridymite, and,
3. high-Al (8.7 wt.% Al_2O_3) orthopyroxene + pseudobrookite (Wood 1994).

Additional phases are titanite and apatite. The sequence of mineral assemblages implies sanidinite facies prograde and retrograde mineral formation where orthopyroxene rather than phlogopite is stable (Fig. 5.18). The early-formed osumilite + hedenbergite assemblage suggests a very low pressure of ~50 bar from the intersection of the osumilite + vapour-out curve and the stability curve of hedenbergite solid solution at ~960 °C in the field of tridymite (Fig. 5.18). Formation of the cordierite-bearing assemblage could have been facilitated by a small increase in temperature at higher PH_2O (Fig. 5.18) due to solidification and second-stage volatile release from the andesite magma generating conditions where Ptotal > Pload that resulted in eruption. Latest stage growth of pseudobrookite (Psb_{49} composition) in the cavities indicates crystallisation at ≥ 750 °C (Haggerty and Lindsley 1970). For the Ruapehu open vent environment, temperatures of vapour-phase mineralisation are ≤ 1050 °C, P_{load} of ≤ 80 bar, and a PH_2O of ~20 bar for the vapour-saturated hydrothermal system below the crater lake bed.

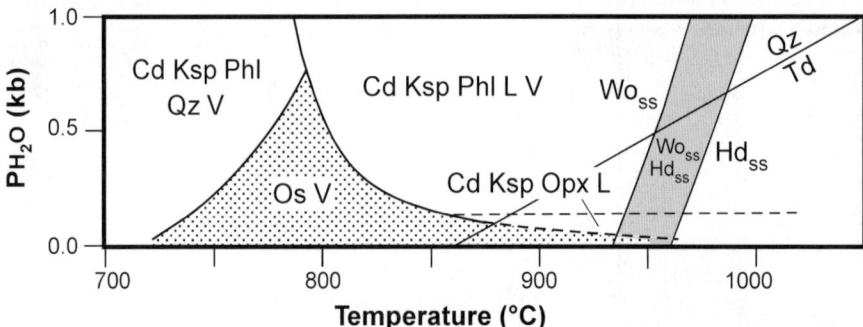

Fig. 5.18. Stability fields of osumilite- and cordierite-bearing assemblages (after Olesch and Siefert 1981) together with reaction curve for the inversion wollastonite$_{ss}$ to metastable hedenbergite$_{ss}$ (after Lindsley et al. 1969)

5.5
Lightning Strike Fusion

Adamello

Mountain top exposures of Adamello gabbro, Corone di Blumone, Italy, contain abundant evidence of lightning strike fusion resulting in the formation of glass-lined holes bored through outcrop edges and corners or shallow pits excavated on planar surfaces. One such occurrence has been examined where the lightning strike has melted an 0.8 cm diameter hole now lined by a black, vesicular glass up to 500 μm thick in hornblende gabbro (Fig. 5.19). The contact between the fused material and host rock is sharp and hornblende, plagioclase and magnetite of the gabbro adjacent the glass remain unaltered. A BSE image (Fig. 5.19) shows that the glass is crowded with dendritic quench crystals of spinel with the average mol% end member composition, ulvöspinel$_{9.0}$ hercynite$_{15.6}$ magnetite$_{15.4}$ hematite$_{31.2}$ jacobsite$_{1.0}$ magnesioferrite$_{24.7}$ fayalite$_{2.5}$. The hematite–magnesioferrite-rich composition of the quench spinel implies high oxidation conditions of crystallisation, i.e. > $-4\log \alpha O_2$ (Kang et al. 2000). Compositions of the host gabbro and glass (Fig. 5.20) indicate that the mafic composition of the glass was largely due to melting of hornblende and magnetite and only minor plagioclase. Deviation of glass compositions from the mafic-bulk-felsic tie line implies some alkali loss due to volatilisation.

It is interesting to note that the texture and quench mineralogy of the Adamello fulgarite is very similar to that of *ferrospheres* produced during industrial pulverised fuel firing of coal described by Sokol et al. (2000). The ferrospheres (from several up to 300 microns in diameter) result from the quenching of high-ferrous melt droplets consisting of dendritic or skeletal ferrispinels and complex ferrite aggregates within a Si-poor, Ca-Fe rich glass.

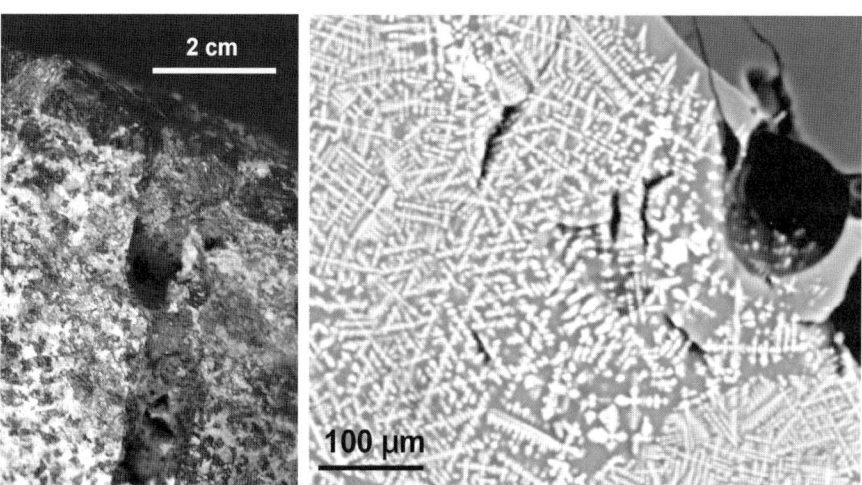

Fig. 5.19. Lightning strike fusion of Adamello hornblende gabbro, Corone di Blumone, Italy, that has created a glass-lined hole through the rock (*left*) and BSE image (*right*) of fusion product of glass crowded with quench crystals of magnetite

Fig. 5.20.
Plot of bulk rock composition, mafic and felsic fractions of Adamello gabbro and bulk rock composition of Darmstadt serpentinised peridotite in terms of wt.% $Al_2O_3 - [SiO_2,(Na,K)_2O] - [(Fe,Mg,Mn,Ca)O,TiO_2]$ showing relationship of glass composition fields (*shaded*) produced by lightning strike fusion. Darmstadt data from Frenzel and Stähle (1982)

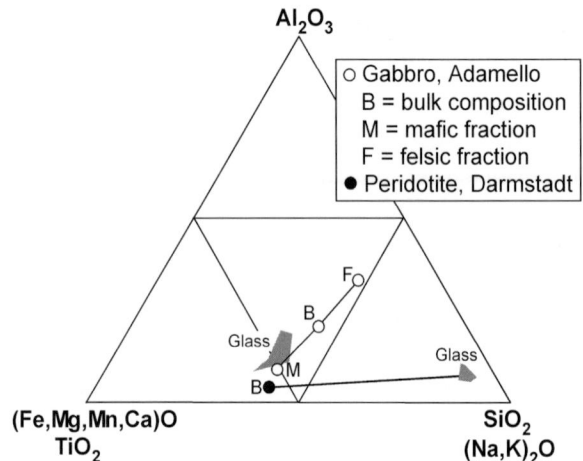

Frankenstein

Serpentinised peridotite on the mountain ridge of the Frankenstein, near Damstadt, Germany, has anomalous remnant magnetism and glassy fulgarites produced by lightning fusion (Frenzel and Stähle 1982). Compared to the bulk peridotite composition, the glass is strongly enriched in silica and depleted in MgO and FeO (Fig. 5.20). As the glass is crystal-free, fusion temperatures must have been > ~1850 °C (melting of olivine in lherzolite at low pressure inferred from the experimental data of Ito and Kennedy 1967), and clearly well above boiling points of the component minerals (olivine [Fo_{75-80}], diopside, brown hornblende, chrysotile, chlorite, tremolite, Fe-Ti oxides, spinel, pyrite) resulting in loss of elements more volatile than silica. Strike temperatures of 15 000 – 30 000°K are suggested by Frenzel and Stähle (1982). Although not related to fusion of ultramafic rocks in particular, the resultant highly siliceous glass is similar to moldavites.

Chapter 6
Anthropogenic Pyrometamorphism

6.1 Bricks/Ceramics

There is a close relationship between pyrometamorphism and brick/ceramic production in the sense that the two processes show a progression towards high temperature equilibria which is not ultimately attained and produce similar high temperature mineral assemblages (Fig. 6.1). In both cases, phase assemblages that persist after cooling are metastable although they indicate how far this progression has gone. Fired complex brick/ceramic bodies resemble buchites (both contain glass, new and residual minerals) and high temperature bloating of some ceramics is similar to the vesicularity commonly observed in buchites and paralavas. The most important differences between bricks/ceramics and pyrometamorphosed argillaceous and marl rocks is the fabric that is controlled by grain size and mineralogical heterogeneity. This affects variations in thermal conditions and heat transport. Fabrics of fired ceramic material are typically uniform in terms of grain size and random distribution of grains compared with argillaceous rocks in which there is usually a preferred orientation of clay minerals and detrital phyllosilicates. Bulk density and porosity are also very different.

Mineral and textural changes in non-carbonate- and carbonate-bearing mixtures of clay minerals, illite, muscovite, quartz and feldspar used in the manufacture of bricks and ceramics during firing at atmospheric pressure has been the subject of a number of studies, some of which are detailed below. These studies provide valuable evidence of temperatures, mineral reactions and textural transformations that must be closely analogous to those taking place during pyrometamorphism of clay-rich sediments and marls where only the end products are commonly preserved.

Non-carbonate Mixtures

Various mixtures of quartz together with kaolinite and muscovite fired for 2 hour periods at temperatures between 1000–1300 °C were performed by Brindley and Maroney (1960). New minerals produced are mullite, cristobalite, with minor corundum (in mica-rich mixtures only) and a poorly-crystallised spinel-type phase (only within the 1000–1100 °C temperature range), together with melt. In comparison with the system K_2O-Al_2O_3-SiO_2 of Schairer and Bowen (1947), there is a close approach to equilibrium in the highest temperature samples fired at 1300 °C as illustrated in Fig. 6.2.

The results imply that *mullite buchites* (mullite + glass) may form from sediments with a mica (illite/muscovite)/kaolinite ratio > 40/25 (at 35% quartz). With lower ratios of mica/kaolinite, *mullite-tridymite buchites* are produced. In quartz-poor (15%)

192 Chapter 6 · Anthropogenic Pyrometamorphism

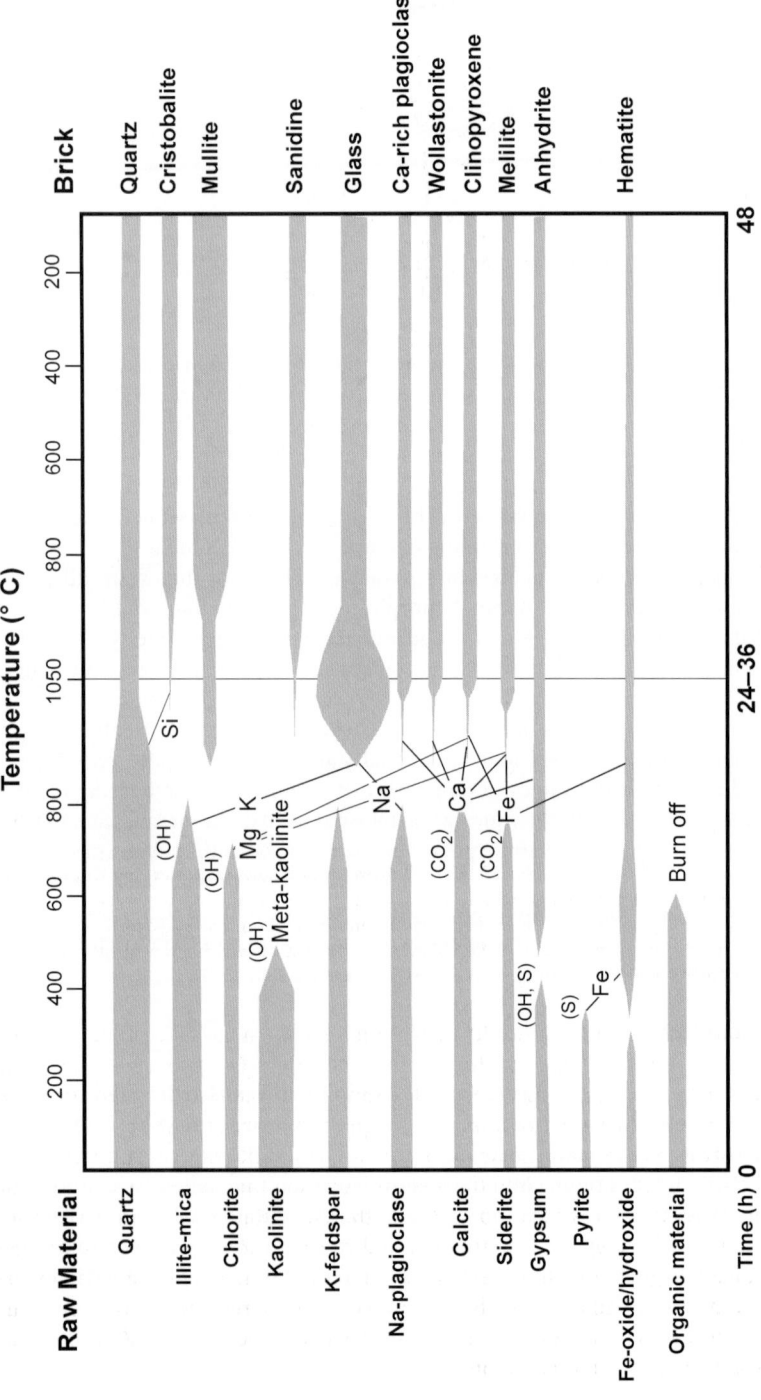

Fig. 6.1. Diagram showing mineral transformations that take place in typical brick raw material through the firing curve with a soaking temperature of 1050 °C (after Fig. 1 of Smith 2002)

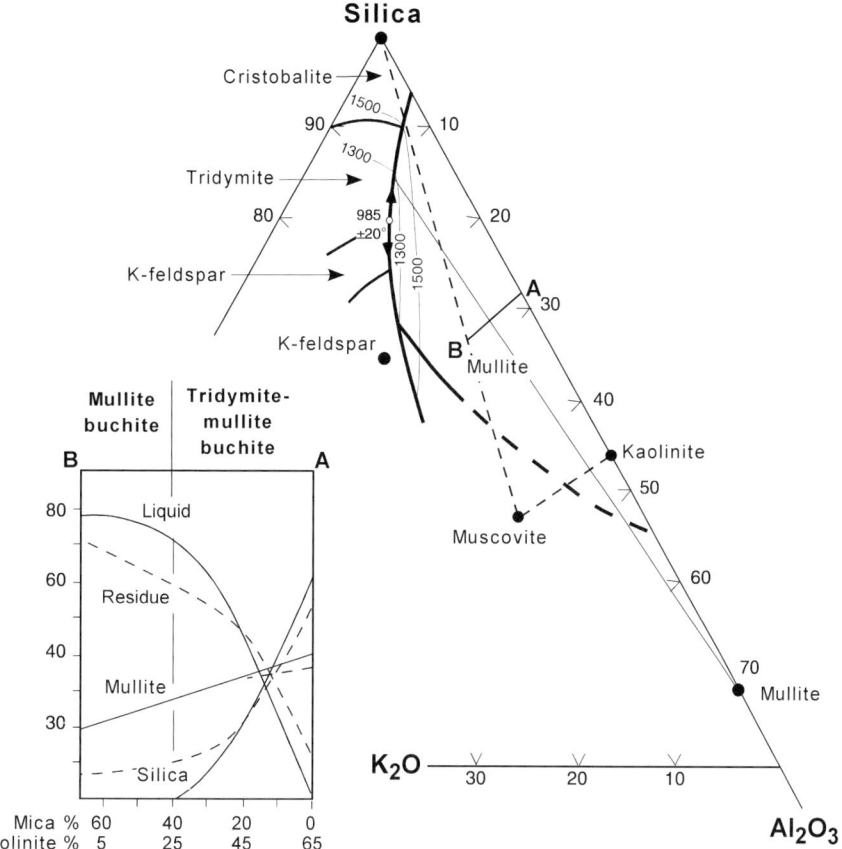

Fig. 6.2. Part of the system K_2O-Al_2O_3-SiO_2 (after Schairer and Bowen 1947) showing relationship of kaolinite-muscovite-quartz mixtures along *line A–B* in the *inset diagram* depicting relations between fired samples (2 h at 1300 °C) of mixtures that initially contained 35% quartz. *Solid lines* are calculated from the equilibrium phase diagram and dashed lines are derived from experimental data of Brindley and Maroney (1960). See text

compositions, mullite buchites would form by an increase in modal kaolinite, i.e. mica/kaolinite ratios < 40/45 (Fig. 6.2). "Impurities" such as Fe_2O_3 (2.04–4.70 wt.%), MgO (0.69 wt.%), TiO_2 (0.65 wt.%), CaO (0.50 wt.%) and Na_2O (0.80 wt.%) in the mica starting material of Brindley and Maroney (1960) appear to have little effect on the phase relations, with Fe^{3+}, Ti^{4+} and minor Mg entering the mullite (also corundum) structure, and Na, Ca, K partitioned into the melt because no sanidine was produced. Water disappears at temperatures above 1000 °C and therefore plays no part in the reaction. In Fig. 6.2, inverse discrepancies between liquid/residue and silica curves determined from the phase diagram and the experimental data result from the persistence of quartz reactant in the latter and also possibly because of an underestimation of the amount of cristobalite, i.e. by less than 50%, due to disordering which reduces X-ray intensity and thus the computed modal amount.

An example of changes in an industrial kaolin, mica, quartz red clay with firing between 750–1250 °C based on XRD and microscopic observation is shown in Fig. 6.3 (Cole and Segnit 1963) together with the starting composition in terms A'KF parameters. The following changes are observed with increasing temperature:

- 750 °C. The main clay mineral is brownish mica (Fe-rich) that is mixed with pale yellow patches of metakaolin. A strong red colour is developed.
- 950 °C. Mica anhydride is darker in colour than the starting mica and has a lower birefringence indicative of structural breakdown although the micaceous structure of the sample is still preserved. Former areas of metakaolin are transformed into amorphous Al-silicate.
- 1000 °C. Micaceous (celadonitic?) clays with elevated iron content begin to melt.
- 1050 °C. Clay mica residue is further darkened and is only slightly birefringent. A small amount of glass is present and feldspars have partially melted. A hard firing-skin (scratched with difficulty by a steel blade) forms.
- 1100 °C. Clay mica is decomposed to a dark brown, almost opaque mixture of Fe-oxide and glass. Feldspars have melted to an acidic glass. The red colour of the sample is darker.
- 1150–1200 °C. Abundant glass and Fe-oxides form a network throughout the sample. At 1200 °C mullite occurs within the glass and within areas of melted feldspar. Quartz

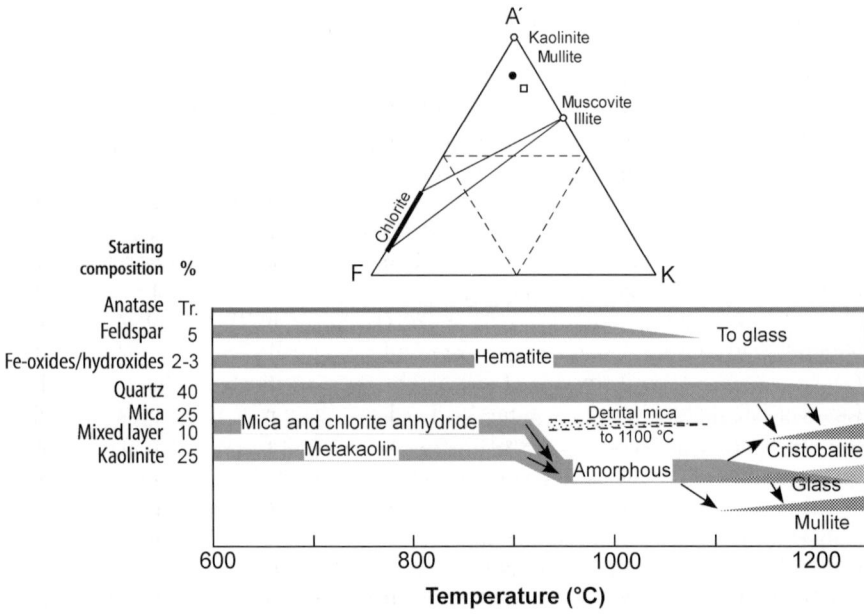

Fig. 6.3. Mineral composition of non-carbonate bearing clay and phases formed on firing to temperatures a little over 1200 °C (after Fig. 2 of Cole and Segnit 1963). *Above:* Clay starting composition (*filled circle*) is plotted in terms of A'FK parameters. Another fired non-carbonate clay starting composition (*open square*) discussed by Cultrone et al. (2001) is also plotted. See text

rarely shows incipient change to cristobalite (as indicated by a weakening of the XRD pattern). The sample is dark red-brown with spots of hematite.
- *1250 °C.* Magnetite is abundant in the darker brown glassy areas. Mullite occurs as well-formed crystals in the glass. Quartz grains show resorption at the margins and are partly converted to cristobalite. Bloating has occurred indicating extensive melting and the interior of the sample is porous and black.

SEM-EDX analyses of a similar fired clay composed of quartz, feldspar and phyllosilicates (smectite, illite/muscovite, kaolinite, paragonite) (Fig. 6.3) used in brick making is reported by Cultrone et al. (2001). Destruction of phyllosilicates occurs between 700–900 °C followed by vitrification which is significant at $T > 1000$ °C. At 700 °C muscovite begins to loose K as part of the dehydroxylation process prior to undergoing a solid-state phase change to mullite + K-feldspar (or melt) between 800–1000 °C (see Chapter 7) according to the reaction

$$3\,KAl_2(Si,Al)O_{10}(OH)_2 + 2\,SiO_2 = 3\,KAlSi_3O_8 + Al_6Si_2O_{13} + 3\,H_2O + melt$$
illite/muscovite   quartz   sanidine   mullite

Mullite first appears at 800 °C and by 1100 °C it is the second most abundant phase after quartz.

Carbonate-Bearing (Marl) Compositions

The high temperature mineralogy of fired carbonate brick pastes is very similar to that in pyrometamorphosed marls described in Chapter 4 except that in the fired samples the temperatures reached are ~75 °C lower than the wollastonite-pseudowollastonite inversion temperature at 1125 °C.

Peters and Iberg (1978) determined mineralogical changes during firing of marl compositions with modal mineralogy estimated by XRD of quartz [18–19%], albite [3–12%], kaolinite [0–17%], illite [21–35%], montmorillonite [0–16%], chlorite [3–12%], mixed-layer clays [0–14%], calcite [6–25%], dolomite [2–5%]). Newly-formed phases are gehlenite (with 15–20 mol% åkermanite), diopside (with < 20 mol% hedenbergite + some acmite), wollastonite, Ca-rich plagioclase and hexagonal anorthite, sodic sanidine ($K_{85}Na_{15}$–$K_{70}Na_{30}$), small amounts of hematite but no identifiable mullite, and with lime forming an intermediate phase. It can be noted that the typical pyroxene formed during the firing of Ca-silicate compositions is normally fassaite rather than a diopside-hedenbergite solid solution (Fig. 6.4) as commonly reported in this and other studies (Dondi et al. 1998). This is consistent with the occurrence of fassaitic pyroxene in pyrometamorphosed calc-silicate rocks (Chapter 4) and in slags with compositions plotting in the $CaO-MgO-Al_2O_3-SiO_2$ quaternary system (see below).

Mineralogical changes with increasing temperature in a marl composition are shown in Fig. 6.5a. Up to 500 °C there are no major mineralogical changes. Between 500–550 °C, kaolinite is destroyed. Dolomite decomposition begins at temperatures between 500–550 °C and is completed at 650 °C, whereas fine-grained calcite begins to decompose at 600 °C, accelerates at 650 °C and is complete at 700 °C. With the dis-

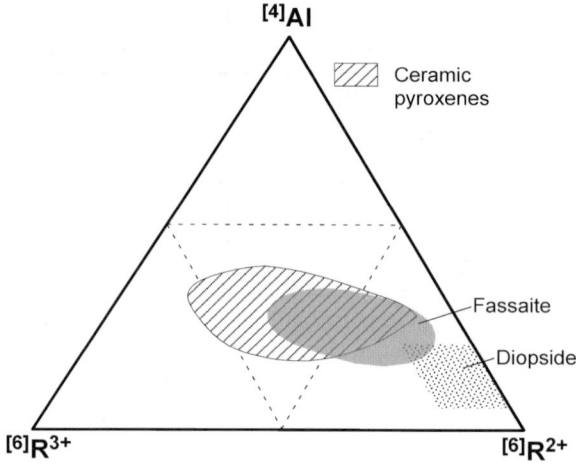

Fig. 6.4.
Ceramic clinopyroxene composition field in terms of $^{[4]}Al$-$^{[6]}R^{3+}$-$^{[6]}R^{2+}$ compared with the composition fields of natural fassaite and diopside pyroxenes (after Fig. 4 of Dondi et al. 1998)

appearance of calcite, crystallisation of lime occurs, reaching a maximum at ~825 °C before disappearing again between 900–950 °C. The major clay mineral constituents, illite and montmorillonite, begin to slowly decompose between 650–700 °C, their amounts rapidly decreasing between 800–850 °C. They disappear completely at 950 °C at which temperature sanidine appears. Chlorite begins to decompose at ~550 °C and disappears at 725 °C. At temperatures between 850–900 °C the formation of gehlenite, clinopyroxene, wollastonite and anorthite coincide with a decrease in quartz and lime.

The above mineral changes with increasing temperature are also accompanied by dimensional changes in some of the phases as shown in Fig. 6.5b. Above the α–β inversion in quartz at 575 °C, expansion occurs in calcite-rich marls above 700 °C due to release of CO_2. At temperatures > 820 °C however, there is a notable shrinkage resulting from dehydroxylation and compaction of amorphous decomposition products of illite and montmorillonite. The amount of shrinkage diminishes at temperatures higher than 900 °C when renewed expansion begins due to the formation of a framework of Ca-silicates which counter the shrinkage produced by amorphous material. Only after fusion begins at > 1050 °C does shrinkage again occur (Fig. 6.6).

Comparison of starting chemical compositions and those of their respective crystalline fractions after firing is shown in Fig. 6.7. As the compositions lie within the quartz-anorthite-wollastonite/pyroxene stability field, gehlenite is metastable as implied by its transformation to anorthite and wollastonite during the "soaking" period at 1050 °C (Fig. 6.5) according the reactions

$Ca_2Al_2SiO_7 + 2 SiO_2 = CaSiO_3 + CaAl_2Si_2O_8$
gehlenite   quartz   wollastonite   anorthite

$Ca_2Al_2SiO_7 + 3 SiO_2 \cdot Al_2O_3 = 2 CaAl_2Si_2O_8$
gehlenite   amorphous material   anorthite

with the åkermanite component in gehlenite and possibly also Fe, Mg in the amorphous material contributing to the formation of pyroxene.

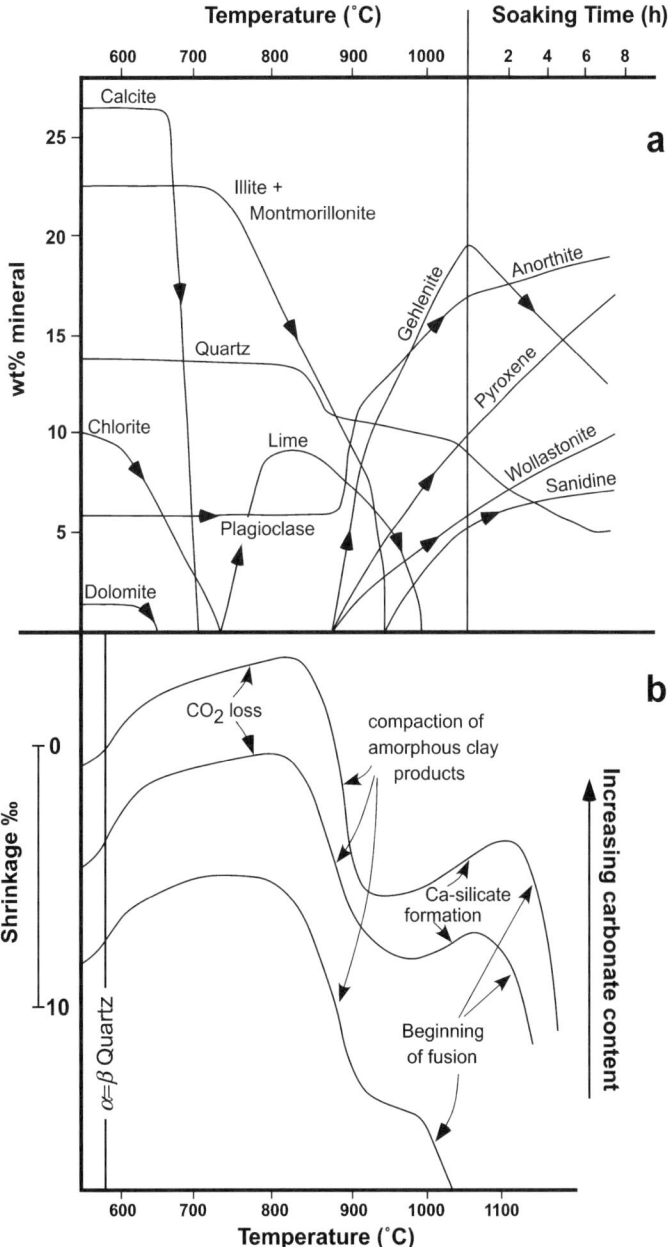

Fig. 6.5. a Changes in mineral composition during firing of carbonate-rich raw brick material at a heating rate of 1 °C min^{-1} and a soaking temperature of 1050 °C (after Fig. 1 of Peters and Iberg 1977). **b** Dimensional changes on firing of raw brick materials with variable carbonate content. In carbonate-poor material, shrinkage caused by vitrification passes directly into shrinkage caused by melting (after Fig. 2 of Peters and Iberg 1977)

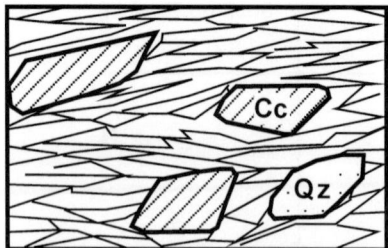

>650 °C
Claystone containing grains of calcite and quartz. Fine interstices occur between the clay.

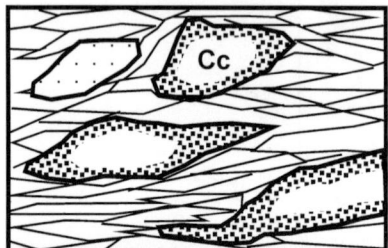

700–800 °C
Rims of calcite grains decompose to CaO. Lime crystals grown on walls of pores due to release of CO_2. In carbonate-poor areas, lime grows on the clay minerals.

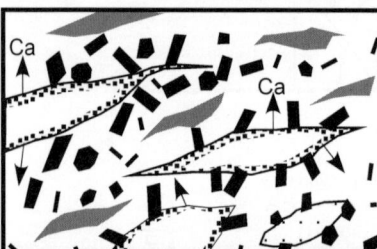

820–850 °C
 Ca and Ca-Al silicates

Clays mostly decomposed to amorphous material (grey areas) that reacts with CaO and partly quartz to form Ca and Ca-Al silicates. Formation of amorphous material causes strong shrinkage.

950–1000 °C
Ca and Ca-Al silicates form a framework preventing further shrinkage. Their continued growth leads to expansion. Lime is used up together with soe quartz. Crystal size in comparison to size of interstices is greatly exaggerated.

<1300 °C
Melt

Formation of melt causes strong shrinkage of interstices that begin to become rounded.

Fig. 6.6. Textural/mineralogical changes in marl brick material with firing between 600–1300 °C (after Peters and Jenni 1973, p. 56)

Fig. 6.7.
Compositions of carbonate-rich tile and brick raw materials (open circles) and crystallised fractions of fired material (filled circles) in terms of mol% $(Ca,Fe,Mg)O - Al_2O_3 - SiO_2$ (after Fig. 3 of Peters and Iberg 1977)

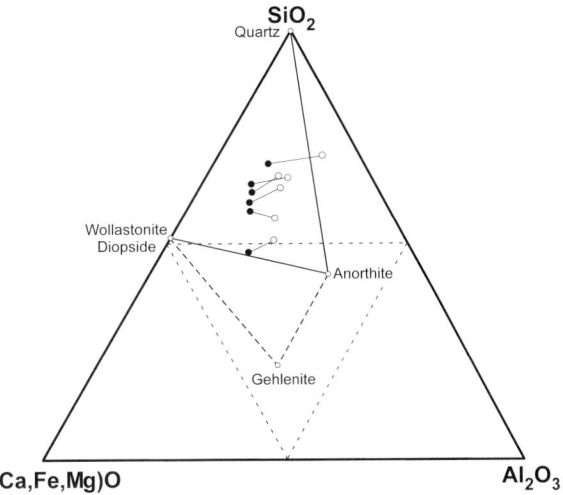

Metastable mineral and textural transformations in a brick paste with additional calcite and dolomite, fired at temperatures between 700–1100 °C are further detailed by Cultrone et al. (2001) using optical, XRD, SEM-EDX and EPMA methods. Important features of the transformation are as follows:

- ~700–800 °C. At ~700 °C dolomite starts to decompose according to the reaction

  $$CaMg(CO_3)_2 = CaO + MgO + 2CO_2$$
  dolomite   lime   periclase

  followed at 830–870 °C by calcite decomposition

  $$CaCO_3 = CaO + CO_2$$
  calcite   lime

  and leading to cracking of the brick as a result of shrinkage.
- *800–900 °C.* At temperatures > 800 °C, melting begins and calc-silicates appear. Gehlenite forms at the boundaries between carbonates and dehydroxylated phyllosilicates through the reaction

  $$KAl_2(Si_3Al)O_{10}(OH)_2 + 6CaCO_3 = 3Ca_2Al_2SiO_7 + 6CO_2 + 2H_2O + K_2O + 3SiO_2$$
  illite   calcite   gehlenite

  with wollastonite forming by the usual reaction between lime and quartz.
- *900 °C.* At 900 °C clinopyroxene occurs at the interface between dolomite–quartz according to the reaction

  $$CaMg(CO_3)_2 + 2SiO_2 = CaMgSi_2O_6 + 2CO_2$$
  dolomite   quartz   diopside

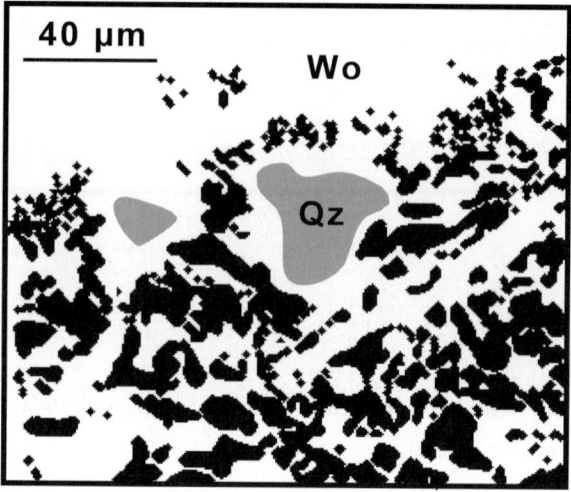

Fig. 6.8.
Complex "fingered" texture developed at the outer contact of wollastonite reaction rims around dolomite. See text. (Drawn from BSE image photo Fig. 5g of Cultrone et al. 2001)

and at the same time anorthite + sanidine form at the expense of illite, calcite and quartz

$$\underset{\text{illite}}{KAl_2(SiAl)O_{10}(OH)_2} + \underset{\text{calcite}}{CaCO_3} + SiO_2 = 2\underset{\text{sanidine}}{KAlSi_3O_8} + 2\underset{\text{anorthite}}{Ca_2Al_2SiO_8} + 2CO_2 + H_2O$$

- 1000–1100 °C. There is a significant reduction in the modal amount of gehlenite at 1000 °C that is probably associated with the formation of anorthite and wollastonite as peak intensities of both minerals increase at 1100 °C, similar to the results of Peters and Iberg (1977). At temperatures > 1000 °C, dolomite/silicate (i.e. quartz or phyllosilicate) contacts are commonly lined by wollastonite. The outer parts of the reaction rims are characterised by a complex texture consisting of the protrusion of "fingers" of ?siliceous melt within the wollastonite (Fig. 6.8). The formation of this texture may be the result of partial melting at the carbonate-silicate interface with viscous flow (mass transport) taking place through the porous clay-quartz matrix. The thick (up to 250 µm) wollastonite rims the surround former carbonate grains and the presence of unreacted inclusions such as quartz (Fig. 6.8) reflect high reaction and diffusion transport rates at ~1100 °C.

6.2
Spoil Heaps

Chelyabinsk

A large number of high temperature minerals have formed in an aggressive gaseous media of O_2 from the atmosphere and S, F, Cl within naturally burned coal-bearing spoil heaps of the Chelyabinsk brown coal basin, South Urals, Russia (Chesnokov and Shcherbakova 1991; Sokol et al. 1998, 2002a, 2002b; Sharygin et al. 1999). There are some 50 such pit-heaps located in the vicinity of Chelyabinsk that range from 40 to

Fig. 6.9. Burnt spoil heaps, Chelyabinsk coal basin, southern Urals, Russia. (Colour photo supplied by Dr. E.V. Sokol)

70 m in height with volumes reaching 1 000 000 m^3 (Fig. 6.9). In addition to coal, the spoil heaps also contain mudstone, siltstone, sandstone, marl, siderite concretions, and fragments of petrified wood. Spontaneous combustion of the coal material began with smoldering, infrequently passing through a phase of flame combustion generating temperatures of 1000–1200 °C. This process continued over a twenty year period from 1960 to 1980 and resulted in variable extents of pyrometamorphic alteration to form clinker, including so-called "black blocks" and nodules of cordierite, and paralava (Fig. 6.10).

High temperature red and cherry-coloured fine grained mudstone *clinker* is the main constituent of the heaps and contains cordierite, mullite, tridymite (rare quartz and α-cristobalite), hematite and magnetite. Lower temperature pink mudstones lack cordierite and contain quartz, hercynite and mullite, while in the lowest temperature yellow-coloured examples, primary muscovite is partly altered to mullite.

Grey-violet and grey crystalline *cordierite nodules* less than 20 cm in size, also referred to as fluorine-rich (up to 1.6% F) paralava, are associated with the high temperature clinkers and are commonly developed adjacent channels of hot gas jets (Fig. 6.11). In addition to minerals in the associated clinker, the nodules also contain anorthite and sometimes biotite. Walls of gas vesicles in the nodules are covered by small (< 1 mm) crystals of hematite, magnetite, pseudobrookite, mullite, cordierite, anorthite and biotite, accessory topaz, apatite and periclase, while surfaces of annealed mudstone fragments in cavities have crusts of anorthite, wollastonite and esseneite. Intergranular spaces in the nodules contain fluorite, sellaite (MgF$_2$) or a K-Al acid glass. The cordierite nodules appear to have resulted from the melting of a mixture of siderite concretions, mudstone and carbonaceous mudstone (Fig. 6.10) under oxidizing conditions. The annealing process proceeded with gas-transport reactions as indicated from reaction zones at the contact of the nodules enclosed in mudstone, the common proximity of the nodules to gas vents and the abundance of pores and cavities in them.

Black layered and porcellanous burnt mudstone contaminated by "soot" occurring in "*black blocks*" contain cordierite, quartz, tridymite, mullite, K-Mg osumilite, peri-

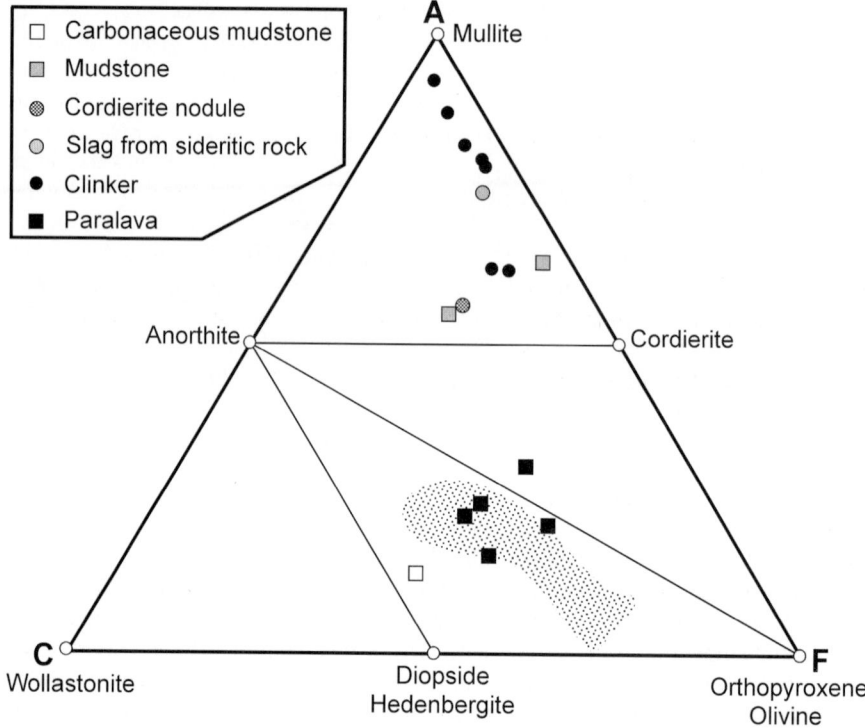

Fig. 6.10. ACF plot of unburnt calcareous mudstone and mudstone, burnt yellow, pink and red mudstone clinker, cordierite nodule, slag after sideritic mudstone, and basaltic paralava from burnt spoil heaps, Chelyabinsk coal basin, southern Urals, Russia (data from Chesnokov and Scherbakova 1991; Sokol et al. 1998; Sharygin et al. 1999)

Fig. 6.11.
Red clinker with holes created by burning gas jets. The walls of the holes are coated with glass that contains abundant quench crystals of cordierite and fayalite (× 4). Pyrometamorphosed spoil material, Chelyabinsk coal basin, southern Urals, Russia (photo supplied by Dr. E.V. Sokol)

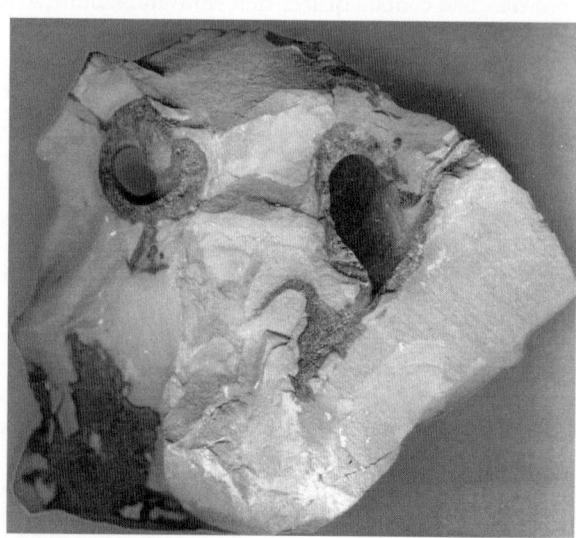

clase, sanidine, graphite, iron carbides and native iron. On the walls of cracks in fragments of woody coal within the "black blocks", vapour-phase crystallisation of orthorhombic (svyatoslavite) and hexagonal (dmisteinbergite) forms of $CaAl_2Si_2O_8$ occur together with anorthite, troilite, kogenite, chondrodite, norbergite, fluor-phlogopite, fayalite, titanite, spinel and graphite. Hexagonal anorthite as a high temperature synthetic phase was first reported by Davis and Tuttle (1952) and its occurrence at Chelyabinsk together with orthorhombic $CaAl_2Si_2O_8$, appears to be the first recorded natural occurrence.

An unusual mineral assemblage is found in fragments of pyrometamorphosed *petrified wood* with the development of nut-like concentrically-zoned aggregates having a dense anhydrite shell enclosing a friable core (Fig. 6.12). The anhydrite shell contains apatite, chondrodite, fluorite, forsterite, anorthite and wollastonite. Cores consist of portlandite (after lime), sometimes with larnite and spurrite in calcitic varieties or hematite, magnesioferrite and magnetite in sideritic varieties.

Paralava (parabasalt) is only found in the largest of the intensely burning spoil heaps and results from melting of a finely crushed mixture of mudstone, calcareous claystone, siderite concretions and carbonaceous matter. Areas of melt generation are associated with a system of channels that provided passage for incandescent gases. Separate parts of the melt flowed to the bottom of the spoil heaps and collected in "chambers" to form *massive* greenish-grey paralava, up to a few cubic metres in size. This type of paralava is fine-grained, holocrystalline with a doleritic texture between olivine, clinopyroxene, anorthite, pleonaste spinel and leucite. Interstices are occupied by fayalite, Ti-magnetite, K-feldspar, pyrrhotite and rarely devitrified opaque glass. Dark brown *stalactitic* paralava has formed during the downward flow of iron silicate melt in the spoil heap and consists of plagioclase, clinopyroxene, spinel, skeletal Fe-olivine and leucite and devitrified opaque glass with Ti-magnetite, pyrrhotite and kirschsteinite filling interstices. An intricate branch-like arrangement of 2–3 cm thick *veins* of parabasalt are related to localisation of numerous gas channels that heated carbonate-clay host rock to melting point. Interesting reaction minerals have developed between the parabasalt veins and their wallrocks (Sharygin et al. 1999):

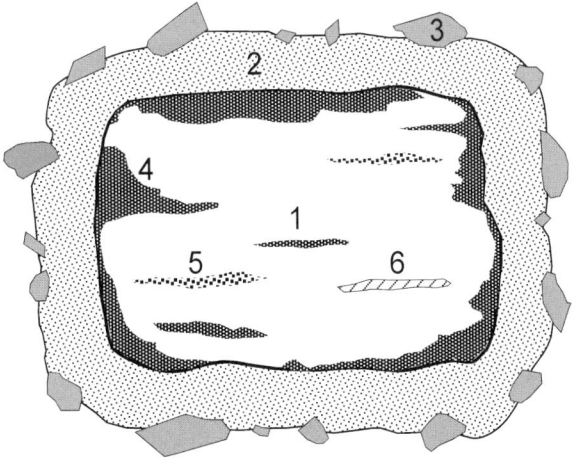

Fig. 6.12.
Cross-section of a concentrically zoned nut-like aggregate formed during the burning of petrified wood, pyrometamorphosed spoil material, Chelyabinsk coal basin, southern Urals, Russia. *1* = friable core of largely portlandite after lime; *2* = dense anhydrite mantle; *3* = rock fragments; *4* = srebrodolskite; *5* = fluorallestadite (wilkeite; $Ca_5(SiO_4,PO_4,SO_4)_3$ (F,OH,Cl)); *6* = spurrite, larnite etc (after Fig. 7 of Chesnokov and Tsherbakova 1991; scale not given in original diagram)

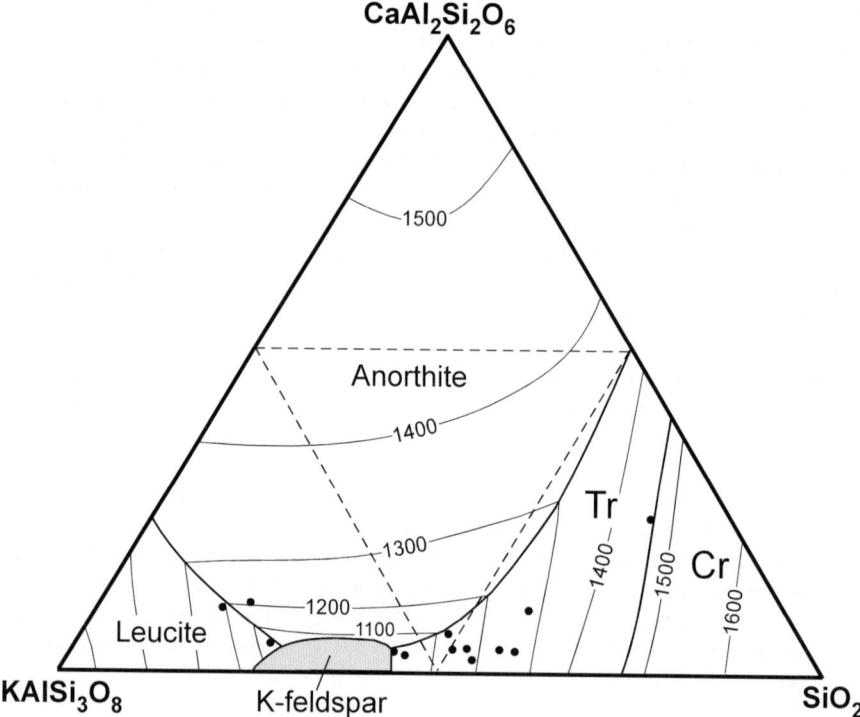

Fig. 6.13. Normative anorthite-leucite-silica composition diagram of acidic glasses from veined parabasalts, pyrometamorphosed spoil material, Chelyabinsk coal basin, southern Urals, Russia (after Fig. 4 of Sharygin et al. 1999)

1. Reaction with annealed mudstone has formed a chilled zone (to 0.5 cm thick) consisting of cordierite, ilmenite (± rutile) and acidic K-Al glass.
2. Reaction with claystone has produced corundum, cordierite, a silica polymorph and acidic glass.
3. Mineralogical zoning is well developed at the contact with a fragment of roasted petrified wood of dolomite-ankerite composition. The core of the fragment consists of fine-grained aggregates of oldhamite, ferropericlase, lime (reacted to portlandite) and Cl-bearing mayenite. Towards the parabasalt, magnesioferrite, P-bearing larnite and gehlenite occur while at the contact coarse grained low Al-melilite has formed.
4. Reaction with sideritic material has produced an outward zonation of magnesioferrite + Ca-rich ferrites + periclase, periclase + melilite, and melilite.

Like those from East Kazakhstan (see Chapter 3), the parabasalts are rich in iron with 11.3–16.9 wt.% FeO. They differ in composition from natural basalts in containing slightly higher Al_2O_3 (14.6–18.9 wt.%), and lower MgO (4.6–6.5 wt.%) and Na_2O (0.05–0.33 wt.%). The first melts appear to originate in areas of abundant siderite due to the low temperature dissociation of $Fe(CO)_3$ with subsequent melting of annealed pelitic rocks, Ca-Mg and Fe-carbonates with rising temperature. Homogenisation temperatures of primary inclusions in anorthite, olivine, augite, leucite and apatite in the veined parabasalt provide minimum temperatures of host mineral formation with: Al-spinel → anorthite (1250–1125 °C) → Mg-Fe olivine (≫ 1140 °C) → augite (1225–1145 °C) → leucite (> 1180 °C) → apatite, Ti-magnetite → K-rich anorthite (1125 °C) → fayalite, kirschsteinite, hedenbergite → K-Ba feldspar (1060 °C). The composition of the parabasalt melt evolved from Fe-rich basalt (1250–1200 °C) to rhyolite (1100–1000 °C) although in one parabasalt, coexisting basic and acid glass implies immiscibility. In terms of normative An-Or-silica content, the acid glass compositions indicate temperatures between 1250–1000 °C (Fig. 6.13) and span the temperature range of mineral formation given above.

The occurrence of oldhamite, schungite, sulphides (mainly pyrrhotite), native iron, the common coexistence of fayalite with titanomagnetite, and substitution of P in olivine and kirschsteinite, indicates that pyrometamorphism occurred under fO_2 conditions not exceeding that of the QFM buffer. Although the "black block" mineral assemblages indicate the most reducing conditions, the initial high porosity of the waste heaps insured the presence of a S, C and H-containing gas network system and good aeration, so that the majority of mineral assemblages formed under more oxidizing conditions. Reaction of F, Cl and S gases in the coal heaps has resulted in many of the OH-bearing minerals becoming Cl- and F-substituted analogues, e.g. fluor-silicates (humite group minerals, amphiboles, cuspidine-chlorocuspidine, ellestadite), phosphates (fluorapatite, wagnerite) and fluorborate that are considered to be the products of gas-transport reactions (Sokol et al. 2001a).

6.3
In-situ Gasification

In-situ gasification, or underground coal gasification (UCG), is an unconventional technique of coal utilisation associated with large-scale firing of rocks adjacent to a coal seam to create the simplified basic reaction

$$\text{coal [macerals + minerals] + oxidant } [O_2/\text{air}] + H_2O$$
$$= \text{gas } [CO_2 + CO + H_2 + CH_4 + N_2 \ldots] + \text{ash + heat}$$

Centralia

Pyrometamorphic features and mineralogy in and around an underground gasifier near Centralia, Washington, USA, have been detailed by Kühnel and Scarlett (1987), McCarthy et al. (1989) and Kühnel et al. (1993) and are similar to those associated with the burning of coal seams described in Chapters 2 and 3. A reconstruction of the temperature regime of the gasifier is shown in Fig. 6.14 and mineral transformations on heating of argillaceous overburden rocks and crystallisation of new minerals from cooling of gasification residues are summarised in Fig. 6.15. Close to the horizontal injection well, temperatures were high enough to melt the upper part of the steel tubing.

In the upper part of the gasifier, cavities of ~0.5 m size have formed in which stalactites and stalagmites of molten argillaceous rock of the overburden occur. This area

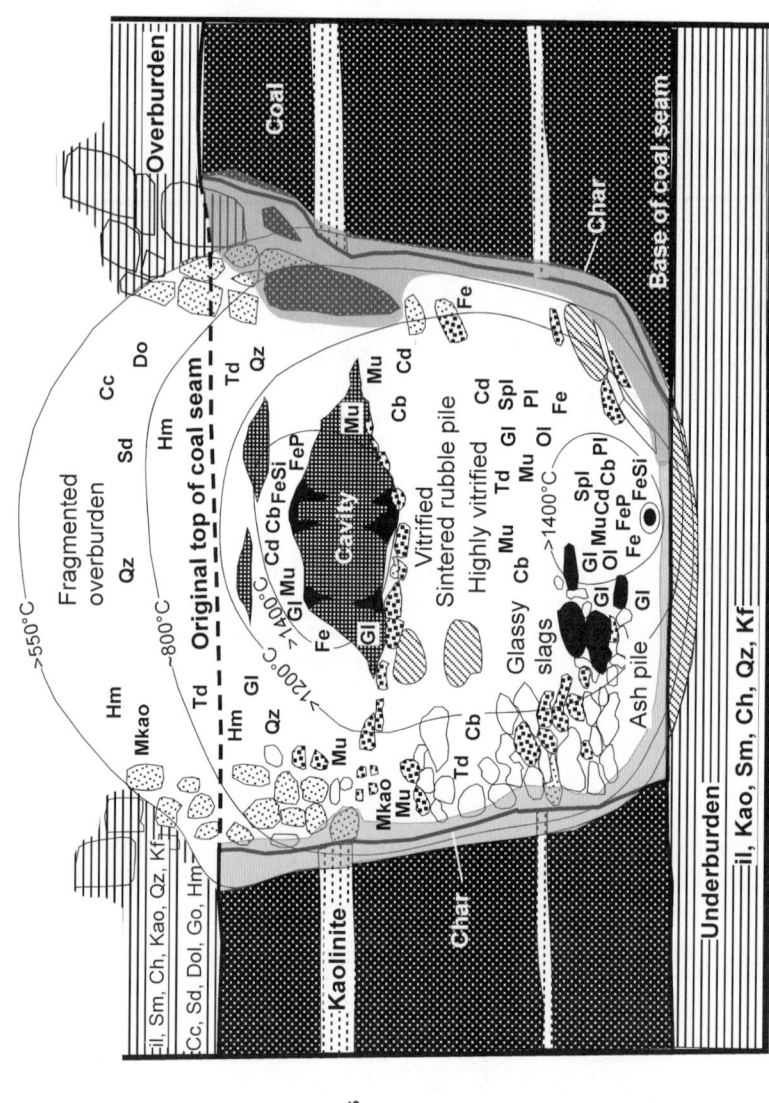

Fig. 6.14. Cross section showing reconstruction of temperatures in the underground coal gasifier (UCG) reactor, Centralia, Washington, USA, based on formation of high temperature minerals from argillaceous rocks in the overburden and from minerals in the burnt coal seam (see Fig. 6.15) (redrawn from Fig. 4 of Kühnel et al. 1993)

Cb	cristobalite
Cc	calcite
Cd	cordierite
Ch	chlorite
Do	dolomite
Gl	glass
Go	goethite
Hm	hematite
Fe	native iron
FeP	iron phosphides
FeSi	iron silicides
il	illite
Kao	kaolinite
Kf	K-feldspar
Mkao	metakaolinite
Mul	mullite
Ol	olivine
Pl	plagioclase
Py	pyrite
Qz	quartz
Sd	siderite
Sm	smectite
Spl	spinel
Td	tridymite
	vitrified clay

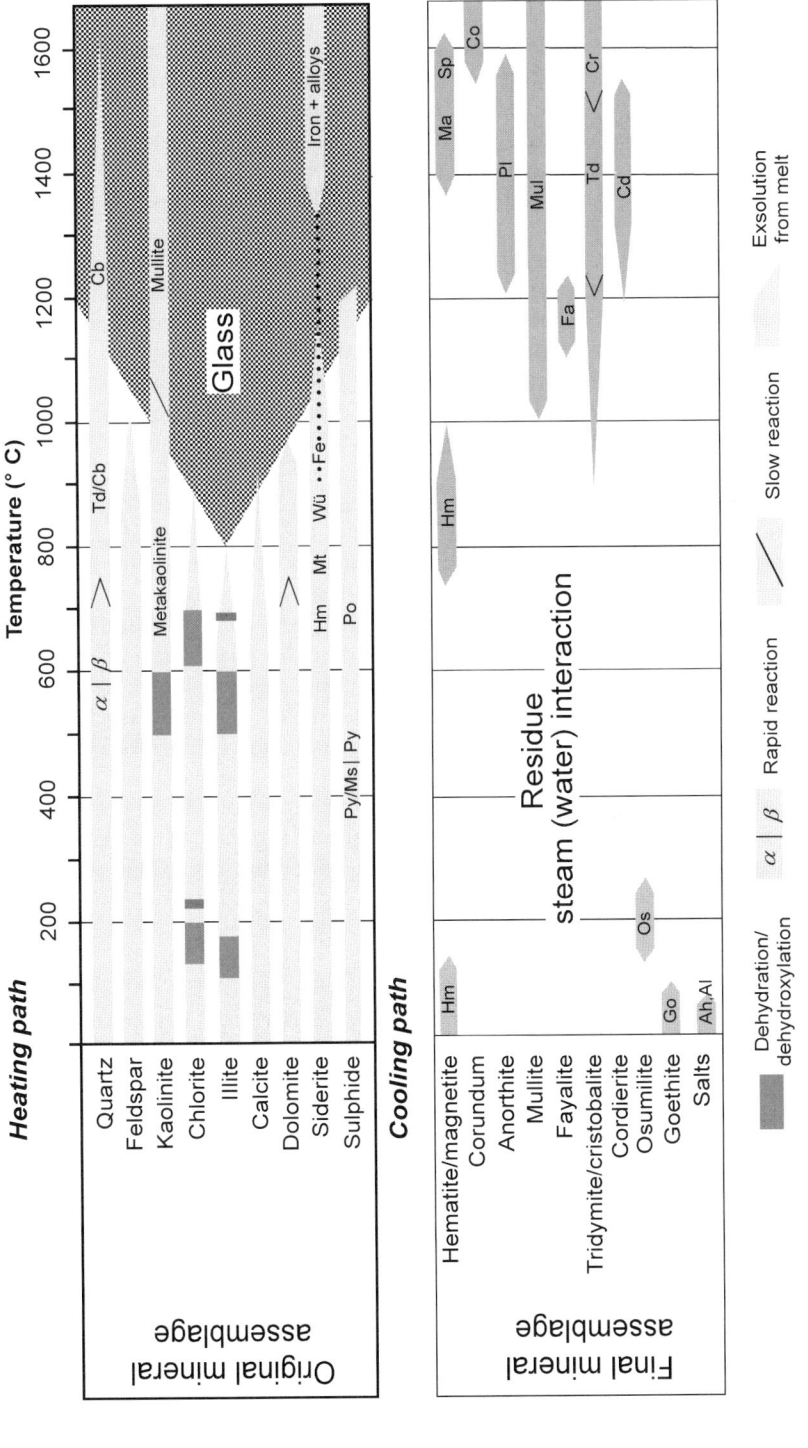

Fig. 6.15. Mineral transformations and the development of mineral assemblages on heating argillaceous overburden rocks and cooling of underground coal gasifier residues, Centralia, Washington, USA (after Fig. 1 of Kühnel et al. 1993). Note the surprisingly low temperature of osumilite formation

represents the upper hot spot in the gasifier. Melt derived from illite-rich clay having a viscosity at 1400 °C similar to honey has flowed down onto a glowing pile of mullite-rich ash derived from kaolinitie intercalations in the coal seam resulting in the formation of a breccia of sintered mullite-rich rocks in a glassy matrix.

At 800 °C droplets of melt begin to form along the grain boundaries of illite and smectite and with increasing temperature feldspars and quartz begin to melt. When the temperature exceeds 1200 °C the argillaceous rocks are > 80% melted and they are completely molten at temperatures above 1350 °C. As cooling in the gasifier is rapid, much of the melt is preserved as glass and crystalline phases are characterised by dendritic, skeletal, hollow and needle-like crystals of cristobalite, tridymite, fayalite, cordierite, mullite, anorthite, corundum, hercynite and magnetite, with droplets of metallic iron and Fe + FeS microspheres. Gases released from gasification of the coal, dehydration and dehydroxylation of clays, thermal dissociation of sulphides and carbonates are explosively released when fragments or overburden fall into the melt causing vesiculation, foaming and bloating.

Cracking and fracturing of the heated roof rock of the coal seam caused by an increase in pore pressure created by water released from clay minerals, the orientation of which controls the orientation of fractures, eventually causes disintegration of the roof rocks by spalling and collapse into the cavity created by the gasification process to form a vitrified and sintered rubble pile with associated glassy slag.

Prevailing reducing conditions of the gasifier and the development of local hotspots causes metallisation (beginning at ~700 °C) of ferruginous phases such as goethite, hematite, siderite and Fe-bearing silicates such as chlorite, together with the formation of ferroalloys such as iron silicides. Sulphur generated from thermal dissociation of sulphides may react with native iron to form troilite that contains exsolved Fe_3P (Kühnel et al. 1993).

Influx of oxygen and steam into the gasifier through fractures during heating and influx of groundwater during cooling causes hydrothermal alteration of the warm rocks of the gasifier to form osumilite, hematite, goethite and various salts such as alum and anhydrite at temperatures < 300 °C (Fig. 6.15).

Thulin

During an underground coal gasification experiment at Thulin, 15 km east of Mons, Belgium, shale and sandstone within 5 m of the top of the burnt coal seam were pyrometamorphosed at temperatures ranging from 1083 °C to 1500 °C (Nzali et al. 1999). Melting/crystallisation produced parabasalt consisting of anorthite, interstitial clinopyroxene and magnetite with minor K-feldspar and biotite. Heterogeneous, vesicular glassy enclaves consist of domains of mullite-sillimanite-spinel-corundum and corundum-anorthite occur within the parabasalt. Textures indicate that sillimanite is replaced by mullite and mullite is replaced by corundum suggesting the reactions

$$3\,\text{sillimanite} = \text{mullite} + SiO_2$$

$$\text{mullite} = 3\,\text{corundum} + SiO_2$$

with SiO$_2$ presumably dissolved in the acidic melt and temperatures of between 900–1000 °C. Vesicles (20–100 mm diameter) around which a mineral zonation is evident occur in micaceous sandstone, with the amount of fusion (glass) increasing towards the vesicle (Fig. 6.16) implying that it was a burning gas vent. Siderite lenses elongated parallel to stratification in shale are converted to calcite and andradite together with wollastonite, monticellite, magnetite, pyrrhotite and chalcopyrite.

6.4
Drilling

California

Fused core ends from well drilling in arkose with fine shale interlayers, California, described by Bowen and Aurousseau (1923), contain small amounts of tridymite, sillimanite (possibly mullite) in glass with fayalite developed near the iron drilling pipe. Relic grains within the glass are quartz, plagioclase, microcline and unidentified clay material, together with small metallic fragments from the drill. Experimental determination of fusion temperatures of core material under reducing conditions found that slight sintering occurred at 1050 °C and extensive melting occurred at 1100 °C, while at 1150 °C a proportion of glass similar to that in the fused core was produced within 1 hour.

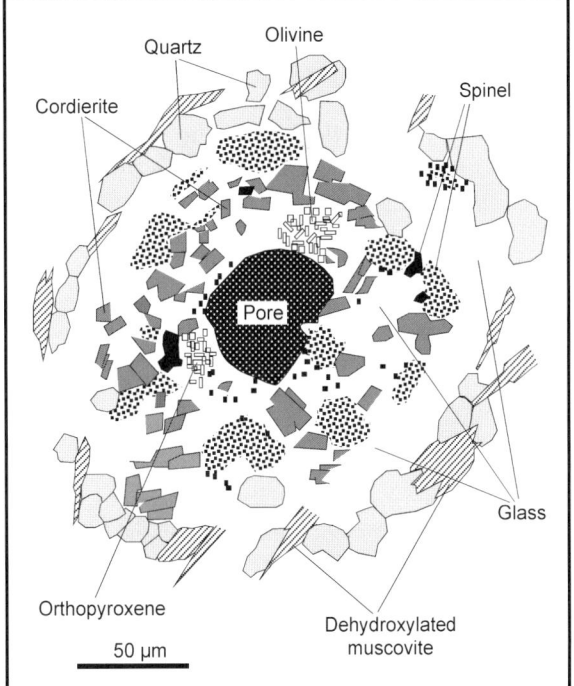

Fig. 6.16.
A melt pocket developed around a burning gas vent in micaceous sandstone, underground coal gasifier at Thulin, Belgium (redrawn from Fig. 3 of Nzali et al. 1999)

Denmark

Melting of clastic sediment during coring (Lavø-1 well, Denmark) described by Pedersen et al. (1992), appears to have been caused by restricted circulation of the mud lubricant with friction induced by rotation heating the sticky silicate mass up to that of partial melting. The thermally affected rock occurs over a interval of about 30 cm from the base of the drill hole, of which ~21 cm is glassy. This grades upward into a mylonite in which siltstone fragments and mineral grains + vitrinite are embedded in a dark matrix of mylonitised mica-rich silt- or mudstone characterised by the presence of illite and metakaolinite which disappear within a 5 mm thick zone in contact with the melted rock. The glassy part of the core is vesicular (vesicles < 0.5 mm) and is characterised by radial and concentric fissures that formed during thermal contraction. The glassy portion is fairly homogeneous on a cm-scale and displays flow texture caused by variable amounts of minute opaque and semi-opaque particles. It is widespread in extremely

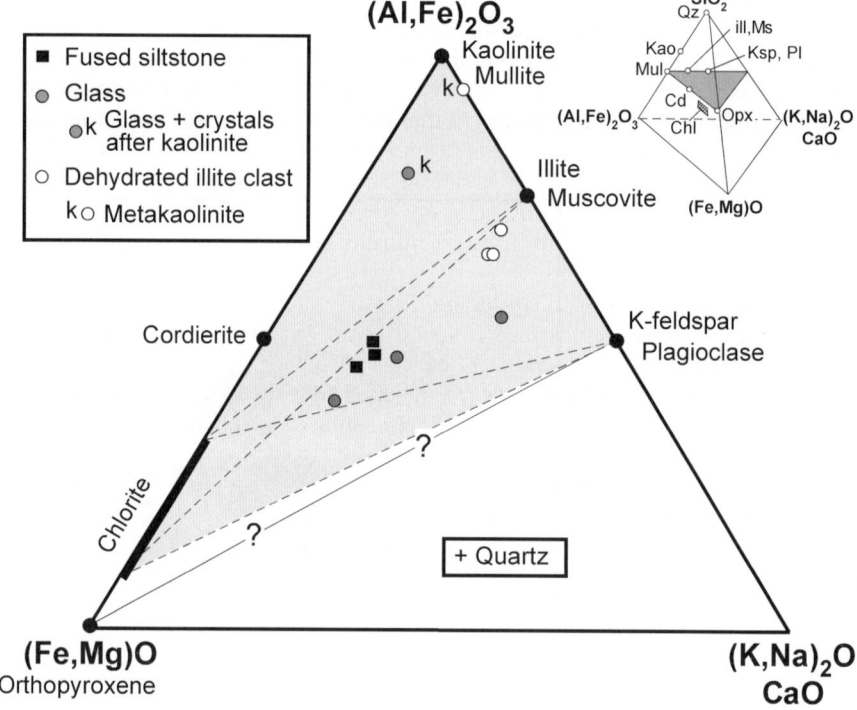

Fig. 6.17. Primary kaolinite, illite, fused siltstone and glass compositions of partly melted clastic sediment from the base of the Lavø-1 well, Denmark, plotted in the 50 mol% $(Al,Fe)_2O_3 - (Fe,Mg)O - [(K,Na)_2O + CaO] \cdot (+ H_2O)$ composition plane projected from SiO_2 (*inset diagram*). Kaolinite and chlorite are projected from above and below the composition plane respectively. All iron in fused rock and glass is computed as FeO; in illite and metakaolinite as Fe_2O_3 (analyses from Pedersen et al. 1992). Shaded area defined by detrital/authigenic assemblage of kaolinite, illite, muscovite, microcline, plagioclase, probable chlorite/smectite and quartz in original siltstone from which mullite, cordierite and glass formed during pyrometamorphism. See text

fine-grained interstitial areas and is colourless to dark brown with a variable composition often on the scale of a few microns. Included in the glass are relic grains of quartz, K-feldspar (transformed microcline), mica, Fe-bearing illite and metakaolinite together with subordinate zircon, Ti-oxides, vitrinite, and rare chromite and tourmaline. Pyrometamorphic minerals, identified by XRD, comprise cordierite (indialite) and a very small amount of mullite.

The chemographic relationship between primary and pyrometamorphic minerals is shown in Fig. 6.17 terms of a 50 mol% $(Al,Fe)_2O_3 - (Fe,Mg)O - [(K,Na)_2O + CaO]$ composition plane projected from SiO_2. The plot suggests that chlorite and/or smectite were also part of the primary assemblage although these minerals are not detected by XRD suggesting that they may have melted, e.g. between ~700–900 °C (Cultrone et al. 2001). Glass compositions indicate that in addition to cordierite and mullite, sanidine and possibly orthopyroxene may have formed if crystallisation had occurred.

From an Fe^{3+}/Fe^{2+} ratio of less than 0.02 in the glass determined by Mössbauer-spectroscopy, the prevailing oxygen fugacity at the time of pyrometamorphism, assuming a temperature of 1000–1100 °C, could have been between $\log fO_2 = -15.2$ to -13.7, i.e. from slightly below to around that of iron-wüstite buffer conditions, respectively. Very low redox conditions are also reflected in the Fe-Ti oxide compositions and the presence of troilite. Clastic grains (< 0.01–0.08 mm) comprise rutile and composite Fe-Ti oxides. These are rimmed and replaced by high temperature armalcolite and pseudobrookite which also occur as spongy aggregates with rutile. Significant intergrain compositional variation implies different equilibrium domains with different fO_2, e.g. Fe-poor armalcolite indicates high-temperature reduction caused by carbon compounds. Globular troilite occurs in the glass and is also associated with vitrinite, rutile and armalcolite.

Small fragments of metallic iron within the glassy rock derived from the low-alloy steel drilling bit and tubing have a bainitic structure indicating rapid cooling of between 10–100 minutes from high temperature from a parent austenite structure developed when the steel particles were briefly reheated. Vitrinite as 0.005–0.04 mm sized grains, has a mean reflectance in oil of $R_o = 4.89\%$, and a maximum reflectance of $R_{max} = 6.55\%$ indicting that it is meta-anthracite (i.e. > 98% fixed carbon). These high values contrast with vitrinite values of $\sim R_o = 0.5\%$ only 50 cm away from the melted rock and underscore the extremely high thermal gradient.

6.5
Slag

Oil Shale

Slag derived from fused oil shale subjected to a furnace temperature of 760 °C in the presence of steam is described by Phemister (1942) as consisting of cordierite, calcic plagioclase (An_{80-90}), anorthite, fayalitic olivine ($\sim Fa_{75}$), clinopyroxene (with a yellow colour and most probably fassaitic-rich), hercynite and magnetite in colourless to brown, turbid and opacised glass. The composition of the slag is variable. Parts are almost holocrystalline whereas other parts are glass-rich. Adjacent the firebrick of the retort a ~1 mm wide porcellanous zone is developed consisting of plum-coloured

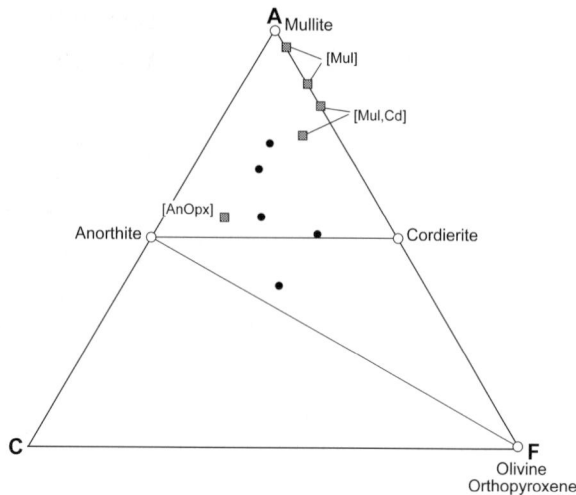

Fig. 6.18.
ACF plot of ash from fused oil shales (data from Phemister 1942) (*filled circles*) compared with shale compositions (*filled squares*) used in melting experiments by Wyllie and Tuttle (1961), the fused products of which contain newly-formed mullite, mullite + cordierite or anorthite + orthopyroxene as labelled

glass containing crystals of bytownite and with cordierite concentrated along the firebrick interface and in glass veins extending into the unrecrystallised brick. Similar cordierite-glass coronas surround fragments of unfused shale in the slag.

Phemister (1942) considers the cordierite-rich areas may have resulted from selective diffusion of oxides within the heated shale, i.e. diffusion of CaO and FeO into areas where melting has begun, leaving areas relatively enriched in MgO and Al_2O_3 that crystallised as a sinter composed of concentrations of cordierite cemented by a small amount of glass. In this respect it is interesting to note that in the experimental study of melting of natural shales by Wyllie and Tuttle (1961), cordierite only formed in bulk compositions with > 1.64 wt.% MgO (19–21 wt.% Al_2O_3) and anorthite formed in a shale with high CaO (6.93 wt.%). Chemographic relationships in terms of ACF parameters of the fused oil shales and those of the shale compositions used in the melting experiments of Wyllie and Tuttle (1961) are shown in Fig. 6.18.

Blast Furnace Slag

Non-metallic blast furnace slag is developed during iron production where iron ore or scrap iron is reduced to a molten state by burning coke fuel fluxed with limestone and/or dolomite. The slag is used as a supplementary additive in cement production. Slag compositions are essentially Fe-poor marls and can be depicted within the system CaO-SiO_2-Al_2O_3-MgO. Pelletised and granulated slags examined by Scott et al. (1986) are characterised by the crystallisation of melilite ($Åk_{17-73}$ Ge_{27-83}) or merwinite, with minor amounts of early-formed oldhamite and iron from compositions of Al_2O_3 (10.2–17.9 wt.%), CaO (39–41 wt.%), MgO (5–10 wt.%). Glass compositions are similar to those of the slag material. Whether melilite or merwinite occurs depends on the CaO content of the slag, i.e. if CaO > 42 wt.% merwinite forms (Fig. 6.19). This is probably the reason why merwinite is not found in natural pyrometamorphosed marls which typically have bulk CaO contents < 37 wt.%. Other slags with higher Al_2O_3

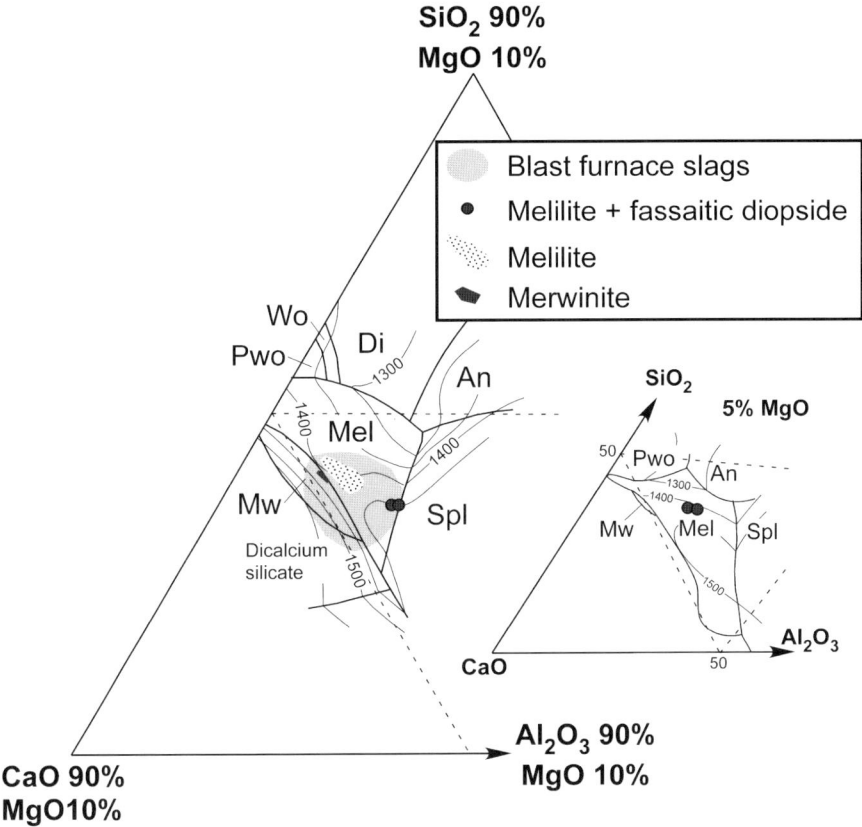

Fig. 6.19. Melilite, merwinite and melilite + fassaitic pyroxene-bearing slag compositions plotted in a section through the system CaO-Al_2O_3-SiO_2-MgO at 10 wt.% MgO (data from Scott et al. 1986, Fig. 4; and Butler 1977). Composition field of blast furnace slags from Fig. 11.1 of Moranville-Regourd (1998). *Inset* shows plot of melilite + fassaite slag in the system CaO-Al_2O_3-SiO_2-MgO at 5% MgO (after Verein Deutscher Eisenhüttenleute 1995; Fig. 3.324, p. 160)

(21–23 wt.%), less CaO (~33.9 wt.%), and similar MgO (5–6 wt.%) contain fassaitic pyroxene instead of merwinite, melilite ($Åk_{28-51}$ Ge_{43-69} Na-Ge_{3-6}), oldhamite (with up to 25 wt.% Mn and 8 wt.%. K) and pale yellow silica-poor, Al, Ca, K-rich glass that represents the residual liquid after crystallisation of melilite and pyroxene with a composition near a eutectic in the system melilite–anorthite–leucite representing the final liquidus temperature of the slag (Butler 1977). In Fig. 6.19, the melilite + fassaite slag compositions (in which fassaite crystallised after melilite) plot on the 1450 °C isotherm in the Al-rich part of the melilite field and with falling temperature would be expected to produce a final liquid on the melilite-diopside cotectic at around 1250 °C. Additional components, particularly alkalis, in the slag would lower this temperature.

Except for low iron content (maximum of 1.53 wt.% FeO) and the presence of merwinite, the silicate composition and mineralogy of the melilite-bearing blast furnace

Fig. 6.20.
Plot of glass compositions in slag derived from fusion of bog iron in terms of the system $FeO-Al_2O_3-SiO_2$ (Schairer and Yagi 1952). Data from Müller et al. (1988)

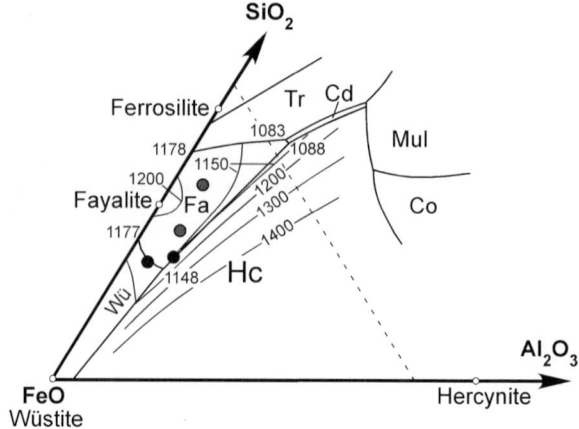

slags is comparable to marls that have been fused by combustion of organic matter, e.g. the Si-poor, Ca-rich marl compositions in Iran and the Central Apennines described in Chapter 4. To form wollastonite in the slags would require a higher silica content.

Bog Iron

Iron-rich slag (61–69 wt.% FeO) derived from melting of bog iron from medieval ironworks near Schieder, Germany, are characterised by fayalite (48–98 vol.%) displaying spinifex textures, wüstite, native iron and magnetite with accessory hematite, "iddingsite", vivianite, rutile, chalcopyrite and pyrite (Müller et al. 1988). The bog iron is essentially "limonite" (= goethite) (61.9–66.2 wt.% FeO), together with minor amounts of detrital quartz, plagioclase, orthoclase, muscovite, hornblende, chlorite, epidote, zircon and tourmaline. Bulk rock and slag compositions have SiO_2, Al_2O_3, FeO making up 98 wt.% of total oxides and in the system $FeO-SiO_2-Al_2O_3$ plot in the fayalite field with slag compositions clustering near the fayalite–wüstite cotectic indicating temperatures of ~1150 °C (Fig. 6.20). Such compositions are extreme but might be encountered in nature from pyrometamorphism of goethite–hematite-rich bauxite (lithomarge).

Solid Waste

High temperature processes developed for the treatment of solid waste and its incinerated residues result in the formation of glass and crystalline silicates that resemble natural paralava formed through the action of burning coal seams, oil and gas. One example from the municipal solid waste incinerator at Basel, Switzerland, has been investigated by Traber et al. (2002). At kiln temperatures of 1200–1400 °C, waste samples undergo extensive melting to form a dark, porous slag consisting of anorthite (An_{90-95} $Ab_{4.3-9.0}$ $Or_{0.6-1.9}$), melilite (Ge_{20-34} $Åk_{33-43}$ Na-Mel_{20-27} Fe-$Åk_{4-6}$ and with the Zn end member *hardystonite* varying between 4–6) and a glassy to microcrystalline interstitial SiO_2-Al_2O_3-CaO-rich phase (Fig. 6.21). Additional minerals are corundum

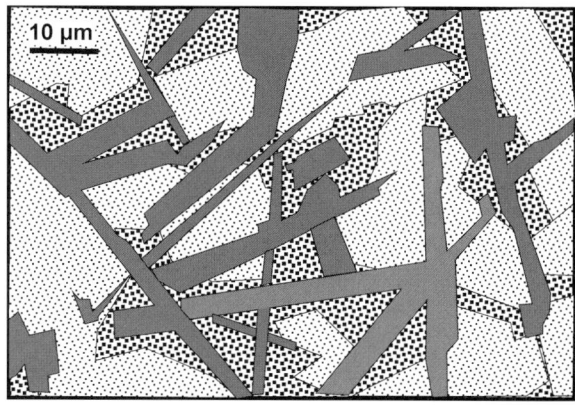

Fig. 6.21.
Texture between anorthite (*dark grey*), melilite (*light stipple*) and glassy to microcrystalline interstitial material (*coarse stipple*) formed from melting and cooling of solid waste, Basel, Switzerland (drawn from a BSE image photo Fig. 3 of Taber et al. 2002)

and minor (< 1 vol.%) metallic inclusions. The metallic inclusions are preferentially associated with gas pores and include Fe-P alloy, pure Fe, Al and Si. No sulphides are present. Relic phases include quartz, calcite and lime.

About 80 wt.% of the bulk slag composition consists of SiO_2, Al_2O_3 and CaO (av. 42.4%; 20.5%; 21.6% respectively, i.e. a marl composition) with Al relatively constant and the CaO/SiO_2 ratio varying from 0.3–0.7. Samples with high CaO/SiO_2 (> 0.54) have undergone extensive melting, crystallisation of anorthite and melilite, and a greater degree of homogenisation in comparison with those with low ratios (< 0.52) that contain more glass. The glassy varieties are characterised by incipient crystallisation of plagioclase or of SiO_2-rich molten scrap glass. Compositions of the waste product plot within the primary field of anorthite in the system Al_2O_3–CaO–SiO_2 with 5 wt.% MgO (Verein Deutscher Eisenhüttenleute 1995, p. 160; Fig. 3.6b) and indicate that crystalline/glass-poor samples plot closer to the cotectic line with gehlenitic melilite. Within the system gehlenite–åkermanite–Na-melilite, melilite compositions indicate temperatures of between ~1400–1350 °C in agreement with those recorded in the kiln (Fig. 6.22).

The solid waste of this study is very heterogeneous in terms of composition and particle size and with melting resulting a large variety of reactions. Short incineration time of 30–60 minutes insures disequilibrium reaction products, variable mineral compositions, preservation of glass, partial devitrification of glass, locally variable redox and temperature conditions – all characteristics of the fusion of natural marl and other protoliths during pyrometamorphism.

6.6
Ritual Burning, Vitrified Forts

Ötz Valley

Partially molten biotite gneiss from a locality in the Ötz Valley, Tyrol, Austria, where immolation of ritual animal offerings took place between 450–15 BC, is described by Tropper et al. (2004). The unmelted gneiss consists of quartz, oligoclase, biotite, with accessory apatite and zircon. Partial melting has caused the formation of foamy patches of dark glassy material at the surface of the rock and as internal layers. The glassy

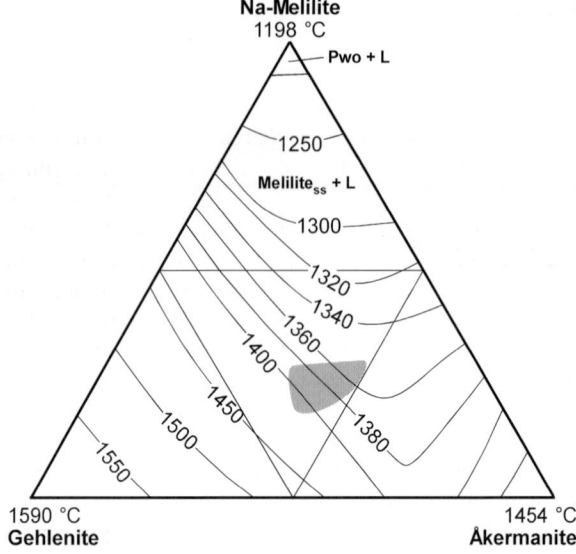

Fig. 6.22.
Melilite compositions from waste slag projected in the system gehlenite-åkermanite-Na-melilite (after Fig. 6 of Taber et al. 2002). See text

material consists of a "basaltic" assemblage of olivine (Fo_{49-71}; 0.30–8.9 wt.% P_2O_5), augite (XMg 0.35–0.76; 0.6–3.5 wt.% Al_2O_3), plagioclase (An_{40-60}), Ti-magnetite within colourless to dark brown glass (wt.% 63.5–69.1% SiO_2; 5.0–16.1% FeO; 0.5–6.0% CaO; 1.5–3.4% Na_2O; 1.3–5.5% K_2O). Near the contact between glass and gneiss whitlockite [$Ca_9(Mg,Fe)(PO_4)_6(PO_3OH)$] occurs together with olivine, augite, plagioclase and glass. The occurrence of unusual P-bearing olivine and whitlockite reflects the presence of animal bone fragments at the burning site.

Melting experiments between 800–1300 °C at 1 bar on slabs of gneiss with interlayers of animal bone in graphite crucibles resulted in the formation of P-rich olivine, whitlockite and plagioclase and indicated that the reaction

biotite + quartz = olivine + Ti-magnetite + K-rich melt

exceeded 1000 °C at fO_2 conditions near the CCO buffer. Glass compositions indicate that feldspar and possibly clinozoisite alteration of plagioclase was also involved with biotite during the melting to produce clinopyroxene and more Ca-rich plagioclase according to the reactions

biotite + feldspar(1) = olivine + feldspar(2) + K-rich melt

biotite + plagioclase + quartz = clinopyroxene + K-Na-rich melt

biotite + clinozoisite + quartz = anorthite + K-rich melt

the gave rise to the "basaltic" plagioclase, clinopyroxene, olivine, Ti-magnetite, glass assemblage in the melted areas (see Chapter 7).

El Gasco

A quartz-rich pumiceous rock that crops out on a hill top near the village of El Gasco, Cáceres, Spain, contains the only known occurrence of ferroan ringwoodite [$(Mg,Fe^{3+})_2SiO_4$] on Earth (Diaz-Martínez and Ormö 2003). The pumice has resulted from partial melting of quartzite and in addition to ringwoodite, the silica glass (lechatelierite) contains hercynite and iron droplets. The occurrence may be compared with process of constructing Bronze- and Iron-age vitrified forts in northern and western Europe through the firing of wood and stone structures where temperatures reached 1235 °C. Metallic iron spherules and droplets are found in many of these vitrified forts indicating strongly reducing conditions during melting. Wood casts and silica glaze on the surface of some of the pumiceous clasts at El Gasco and highly variable phosphorous content that may be due to melting of apatite, incorporation of wood-ash and addition of bones, strengthens the possibility of an origin by anthropogenic burning. The problem is the presence of the high pressure phase, ringwoodite which forms at ~1200 °C and 60 kb for more Fe-rich compositions (Agree 1998). Diaz-Martínez and Ormö (2003) explain the occurrence as follows. Melting of a chloritic matrix in the quartzite caused localised high volatile pressures within the closed, confining 3-D quartz framework. With increasing temperature, and as soon as the discrete internal pressure reached the limit of quartz tenacity, the framework structure broke and vesicles expanded by a process of hydrodynamic cavitation (Spray 1999) releasing the accumulated presssure. Ringwoodite, hercynite and metallic iron droplets occur within glass adjacent these vesicles. Expansion ended when the pressure equalled the high viscosity of the surrounding clast-rich glass. The vitrocrystalline rock at El Gasco may thus be an example of short time development of high pressures during melting and vapourisation of matrix minerals in quartzite that resulted in the formation of quench crystals of ringwoodite in glass. Reducing conditions and high temperatures of the vitrification process may also have allowed ferroan ringwoodite to form at lower pressure than that under which it normally forms.

6.7
Artificial Fulgurite

SE Otago

Raeside (1968) describes an interesting example of artificial fulgurites produced by high voltage electrical discharge in a soil, SE Otago, New Zealand. The slag-like fulgurites were found in a shallow furrow approximately 83 m long in topsoil running roughly parallel to an 11 000-volt power line that had fallen and discharged into the soil over period of about 40 minutes. The fused material occurred as irregular tube-like fragments up to 1 cm diameter and 2 cm long, and irregular, non-tubular fragments up to 9.5 cm long and 7 cm wide. Inner surfaces of the hollow tube-like fragments and concave surfaces of the irregular fragments are coated with glass the colour of which changes outwards from pale grey, through a brick red partly fused vesicular layer to an almost black weakly fused vesicular outer layer coated with a loose layer of charred soil. The vesicular nature of the glass can be attributed to a high soil moisture content. Quartz, zircon, epidote with minor actinolite and albite along with an abundance of small opaque inclusions of possible carbon and Fe-oxide occur throughout the glass.

The soil protolith is developed on loess derived from weathered low grade greywacke and schist. It is characterised by ~40–45 % each of quartz and albite, with accessory white mica, chlorite, epidote, rare actinolite, zircon, titanite, Fe-oxide and ~15% clay content containing numerous "limonite" mottles and soft Fe and Mn concretions. As quartz within the glass is angular and shows no evidence of fusion, the glass was probably formed by the melting of albite, mica, chlorite and clay with the humus fraction transformed to carbon. The change from dark grey to the black outer zone of the fulgurites may reflect stages in the transformation of humus material and hydrated ferric hydroxides together with localised variation in oxidation conditions.

Chapter 7

Metastable Mineral Reactions

Metastable melting and high temperature disequilibrium reaction mechanisms are important processes in pyrometamorphism. Because of kinetic factors such as low diffusion rates, low fluid pressure and short term heating, reaction textures in pyrometamorphic rocks do not generally achieve thermodynamic equilibrium and disequilibrium mineral assemblages arrested in various stages of up-temperature reaction are typically preserved. It is only with a coarsening of grain size during annealing at high temperatures, that thermodynamic equilibrium is approached during pyrometamorphism. Using light optics, the initial stages of mineral reactions can rarely be resolved because they occur over very small distances and the reaction products are typically extremely fine grained. Consequently until the advent of the electron probe microanalyser (EPMA), scanning electron microscope (SEM) and the transmission electron microscope (TEM), important details of such reactions were essentially unknown until the mid 1980's and terms such as "amorphous", "cloudy", or "altered", were commonly used to describe the optically unresolvable nature of fine-grained reaction assemblages. The advanced techniques have made it relatively easy to establish fine scale reaction relationships in terms of composition, crystallography, and mechanisms of transformation between reactants and products.

In rocks, individual minerals generally do not react as isolated chemical systems as shown, for example, by possible reactions between a quartz-albite-phengite- chlorite assemblage in a metasediment undergoing pyrometamorphism (Fig. 7.1). Except perhaps in the central areas of larger grains, breakdown reactions will involve diffusive interaction with adjacent minerals by intercrystalline diffusion towards grain boundaries and then across grain boundaries by way of a fluid or melt (Fig. 7.1). Also, grain boundaries are probably never actually "dry" at the stage when minerals begin to react in response to increasing temperature. Initial stage dehydroxylation of phyllosilicates would ensure the presence of water molecules along their grain boundaries as well as along those of associated anhydrous phases such as quartz and feldspars. Volatiles such as H_2O, CO_2 and SO_2 released during heating can act as "mineralisers" with catalytic effects and they may modify the path of thermal transformations, change the temperatures and rates of reactions and even the reaction products themselves.

When mineral reactions occur under conditions where there is a large temperature overstep of equilibrium conditions, reaction rates can be very fast because ΔG_r is large. Metastable phases form in place of stable assemblages as predicted by the Ostwald step rule where mineral transformation occurs via a sequence of steps in-

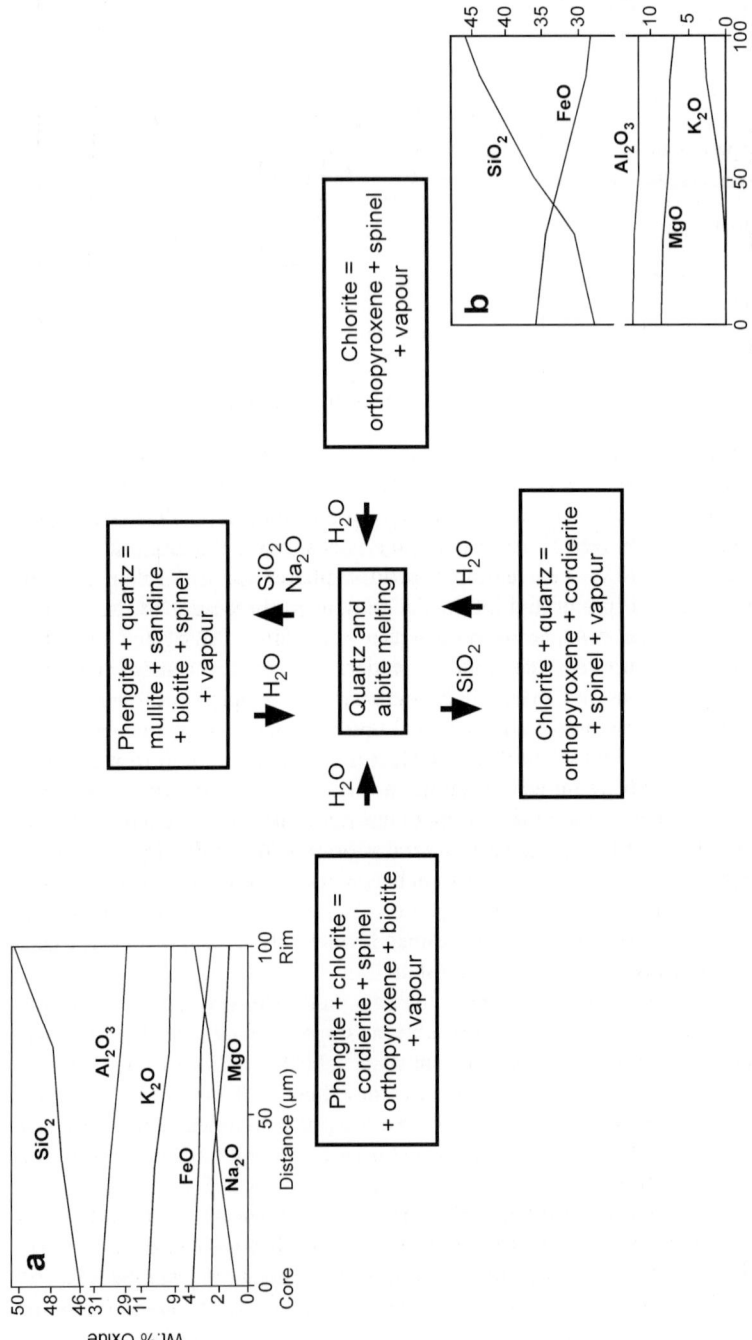

Fig. 7.1. Relationship of four reactions involving chlorite, phengite, ± quartz to adjacent melting of albite + quartz. Water released during dehydration promotes melting and aids diffusive interaction between grains and within grains as shown by core to margin composition variation across chlorite (**a**) and phengite (**b**) adjacent quartz (after Figs. 2, 3 and 18 of Worden et al. 1987)

Fig. 7.2.
Diagram illustrating the Ostwald step rule. Direct transformation from state 1 to state 4 involves a large activation energy (ΔG_α^I) and may be very sluggish in comparison to transformation by way of a sequence of steps, each represented by metastable phases and each involving smaller activation energies that may be kinetically more favourable (after Fig. 5.9 of Putnis and McConnell 1980)

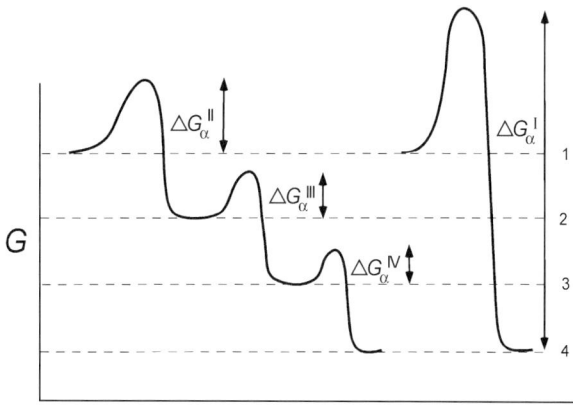

volving kinetically more favourable activation energies relative to direct transformation that involves a large activation energy (Putnis and McConnell 1980) (Fig. 7.2), and once formed the thermodynamic driving force to produce a stable assemblage is lowered by a reduction in free energy. Rubie (1998) considers the amount of temperature overstepping required for nucleation of product phases may differ depending on the degree of structural (strain and interfacial energies) mismatch between reactant and product. For the most part, however, nucleation during melting reactions that typify many pyrometamorphic reactions are expected to require considerable temperature overstepping if several product phases form. For example, experimental work on the breakdown of muscovite + quartz by Rubie and Brearley (1990) has shown that the product phases (e.g. mullite, biotite, spinel) form sequentially rather than simultaneously and this can have important effects on the kinetics of melting which proceeds in a series of steps controlled by punctuated supersaturation of required components needed to form the successive mineral products. While this is typically the case for dehydration reactions, minerals such as those crystallising from melts in buchites require an ordered separation of chemical components by diffusion through a relatively homogeneous medium (melt) to specific mineral nucleation sites, and such phases are typically present as quench crystals.

The presence of defects such as cleavage planes and twin boundaries are important loci for the commencement of high temperature transformation because such boundaries provide low activation energy sites for nucleation to occur and pathways for mass transfer of material to and from the reaction site. In hydrous minerals, such defect sites might be expected to be characterised by higher water activity than in more ordered parts of the grain and this also contributes to a lowering of the activation energy for nucleation.

Pyrometamorphism is often associated with the formation of melt. For this to occur the temperature must exceed that of mineral solidi by a substantial amount (perhaps 100–200 °C) before melting begins. Once melting starts, then the excess heat (per unit mass of rock) is available to contribute to the latent heat of melting according to the equation

$$\Delta q = (T - T_s)C_p$$

where T_s is the solidus temperature and C_p is the heat capacity of the rock undergoing melting (Rubie and Brearley 1990). In pyrometamorphism, the most important reactions (except in calc-silicate rocks) are incongruent melting ones as they require a substantial overstepping of the solidus temperature in order for new although often metastable crystalline phases to nucleate.

In most cases, steady state conditions of recrystallisation are seldom attained except perhaps in the formation of some sanidinites (Chapter 3). With a rapid increase in temperature, the rate of mineral nucleation will lag behind the change in temperature until maximum temperatures are reached as shown in Fig. 7.3. During pyrometamorphism, peak temperatures are also unlikely to be maintained for any substantial time before cooling commences. As the rate of cooling is generally significantly less than the rate of heating, except in cases of quenching on eruption, it seems likely that the majority of nuclei produced will be formed during cooling under a low temperature gradient (Fig. 7.3). Thus the period of time over which effective nucleation can occur will be a substantial fraction of the total available time for mineral transformation and this has the effect of preserving various stages of the reaction within individual crystals.

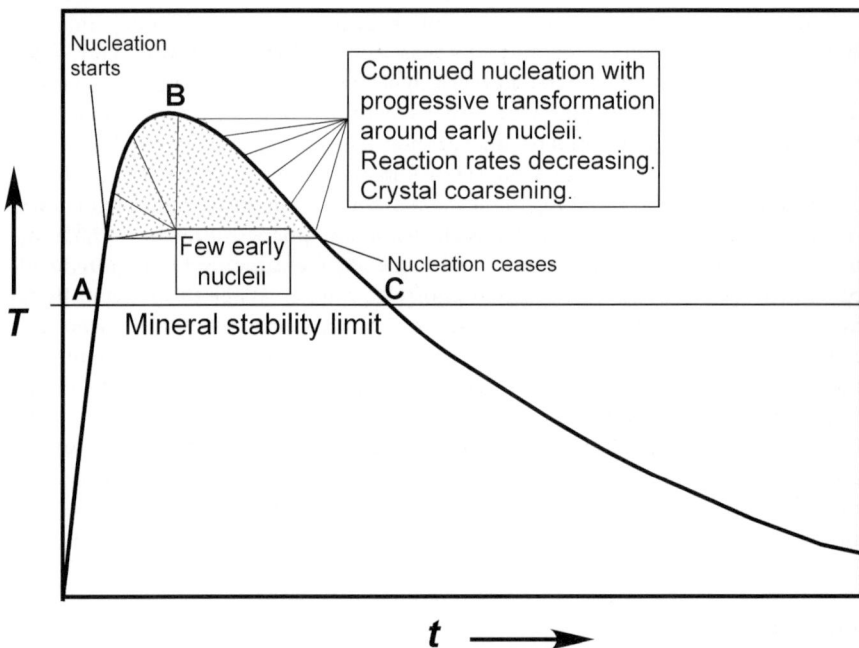

Fig. 7.3. Hypothetical T-t path for a rock undergoing pyrometamorphism. During rapid heating the thermal stability of a mineral is overstepped until nucleation of reaction phases begins at A. Few nucleii form with increasing temperature between A and B (thermal maximum) because of the rapid temperature rise. Most mineral nucleii form during the longer cooling period whilst earlier-formed phases continue to react and coarsen. Nucleation ceases at C or possibly before if the rock is a xenolith quenched on eruption (after Fig. 15 of Brearley 1986)

For example, energetically favourable areas such as grain boundaries, dislocations, cleavages, twin planes, which begin to react during rising temperature experience a longer period of time for nucleation than areas where nucleation begins during slower cooling. These latter nuclei will also tend to decrease rapidly as temperature drops and the rate of transformation becomes increasingly sluggish.

Nucleation rate changes rapidly as a function of temperature. If the rate of increase of temperature (dT/dt) is fast then the number of nuclei formed is likely to be limited. Also, the nucleation rate of each reactant phase in a mineral breakdown will differ because in each case the free energy involved in forming nuclei will vary for each phase. In some cases the nucleation of certain phases depends on the prior nucleation of others in order to create a suitable low energy nucleation site, such as for example, an interface between two minerals. Once nucleation has occurred, growth will be the rate-determining step and the overall reaction will be controlled by diffusion of the slowest element.

In this chapter, up-temperature breakdown products, crystallisation of new products from liquid, and textures of some common rock-forming silicates are described.

7.1
Quartz

In the absence of water and at low pressure, pure silica begins to melt at 1700 °C and in nature this is only likely to occur during lightning fusion. From the many examples described above, resorbed quartz grains in partially melted rocks are often surrounded by clear glass. The glass adjacent melting quartz grains is never 100% SiO_2 but contains considerable amounts of Al_2O_3 and alkalis with totals less than 100% reflecting the presence of water, indicating that quartz melts in conjunction with feldspar and/or a phyllosilicate at considerably lower temperatures approaching those of minimum melt production. Nevertheless, there are many examples (Chapter 3) where melt must be sufficiently saturated with SiO_2 in order for tridymite to nucleate on partly melted quartz grains. An example of this in a quartz-rich (porcellanite) buchite xenolith in basalt from the Three Kings islands, New Zealand, is shown in Fig. 7.4 that implies melting of quartz + Na-K feldspars until the feldspars were completely melted (no feldspars remain) followed by nucleation of tridymite needles around unmelted quartz on cooling indicating (at 100 bar) a minimum temperature of ca. 960 °C for both wet and dry melting, i.e. ~80 °C above the quartz-tridymite inversion (Fig. 7.5). With quenching on eruption, perlitic cracking developed around the relic quartz grains presumably due to the inversion from high to low quartz at 573 °C.

7.2
Plagioclase

The beginning of melting of plagioclase feldspar is typically indicated by the development of a "fingerprint" texture as described in several examples cited in Chapter 3, where those parts undergoing fusion consist of a fine intergrowth of feldspar and glass (e.g. Guppy and Hawkes 1925; McDonald and Katsura 1965; Sigurdsson 1971; Tsuchiyama and Takahashi 1983). The "finger print" fusion texture has been experimentally pro-

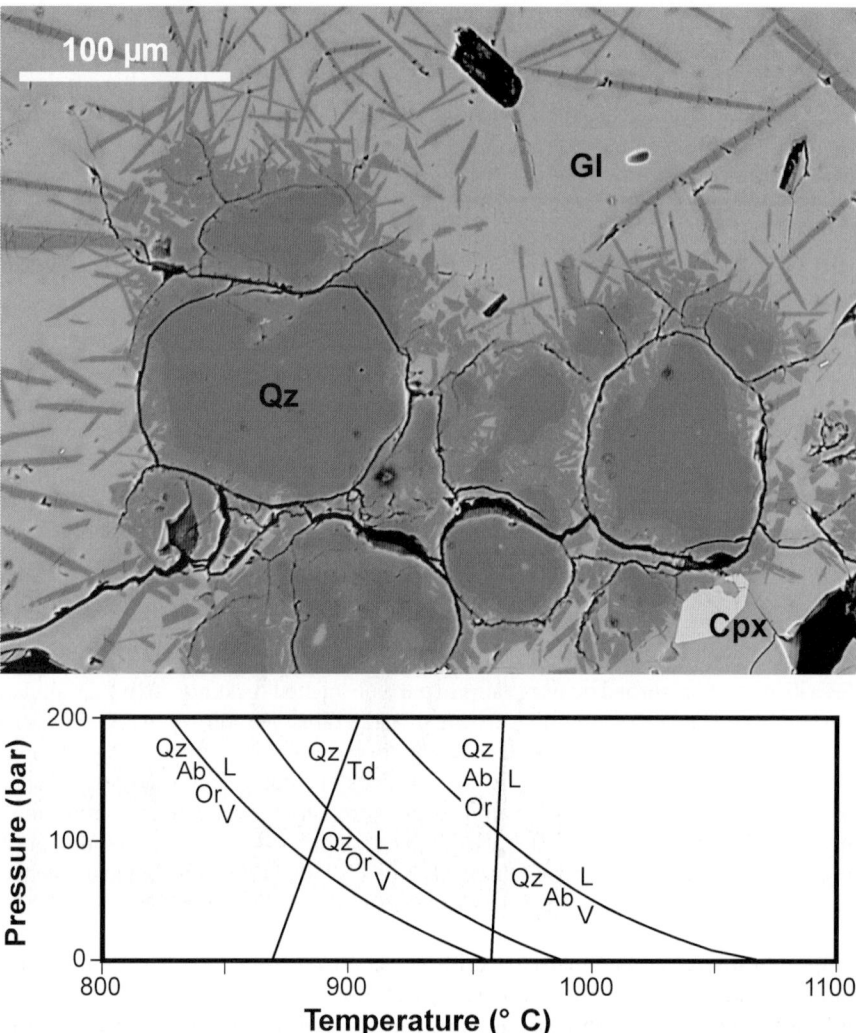

Fig. 7.4. BSE image photo showing needles of tridymite nucleated on partially melted quartz and occurring within surrounding glass, xenolith in basalt, Three Kings islands, New Zealand. Note the distribution of perlitic cracks developed around the relic quartz grains. *Below* is a *T-P* diagram showing wet and dry feldspar + quartz melting curves (after Tuttle and Bowen 1958; Shaw 1963) in relation to the quartz-tridymite inversion. See text

duced by Johannes (1989) using single crystals of plagioclase (~An_{60}) surrounded by quartz powder + H_2O at temperatures between 820–880 °C. Partial melting results in the development of "fingerprint" reaction zones of different thickness parallel to (001) and (010) as shown in Fig. 7.5a within which newly-formed An-rich plagioclase ($An_{82–85}$) mantles relic "islands" of the original plagioclase. The Or-component in the Ca-rich plagioclase rims is $Or_{0.6–0.8}$, significantly less than that of the starting plagioclase with

Fig. 7.5.
a Drawing of "fingerprint" texture developed around the margin of part of a crystal of plagioclase (An_{60}) experimentally reacted with quartz (after Fig. 2 of Johannes 1989).
b Drawing from a BSE image photo (Fig. 4 of Johannes 1989) showing the texture of new high temperature plagioclase (An_{86}) developed in the marginal reaction zone parallel to (010) of the plagioclase crystal shown in (**a**). *Black* = glass

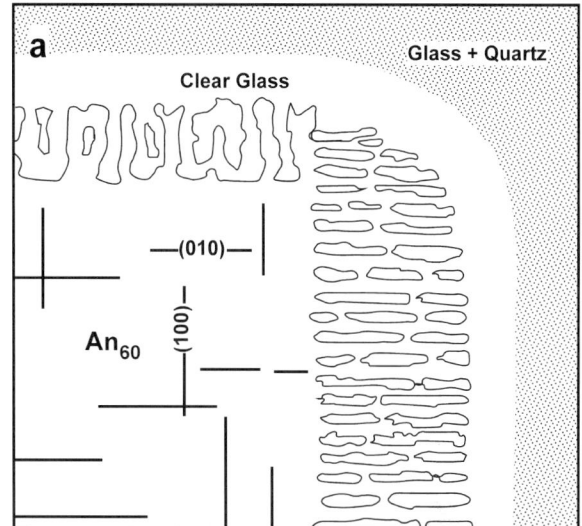

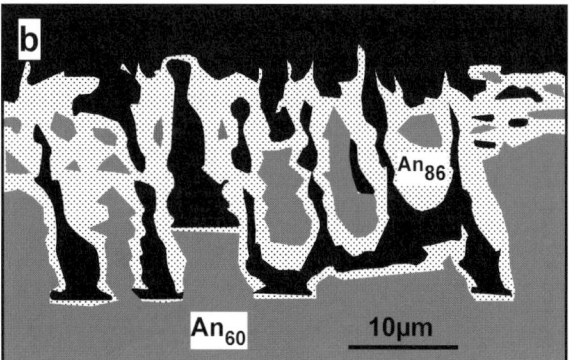

$Or_{1.8-1.9}$. Contacts between the relic and new plagioclase are sharp (Fig. 7.5b) and diffusion profiles are not developed. This suggests that the plagioclase reaction is probably surface controlled, with the geometry of the melt corrosion surface shown in Fig. 7.5b indicating slow internal diffusion of NaSi-CaAl components compared to the rate of melting. In comparison to K_2O in the new Ca-rich plagioclase (0.03–0.22 wt.%), there is significant enrichment of K_2O (up to 2.24 wt.%) in the melt.

The An-content of newly-formed plagioclase increases with time until a stable composition is reached depending on temperature. In the example shown in Fig. 7.6a from a partially melted argillite, this would be achieved after 700 hours. In Fig. 7.6b, newly-formed plagioclase compositions (in 0.5 and 1.0 kb PH_2O runs > 750 hours) are plotted with respect to temperature. Slopes of the solidus curves for partially melted argillite and greywacke compositions are similar to those of Winker and von Platen (1961) and Winkler (1979) for plagioclase in experimentally melted amphibolite-grade greywacke compositions, and are significantly steeper than the 2 kb solidus in the system Qz-Or-Ab-An-H_2O (Johannes 1984). The difference can be ascribed to the

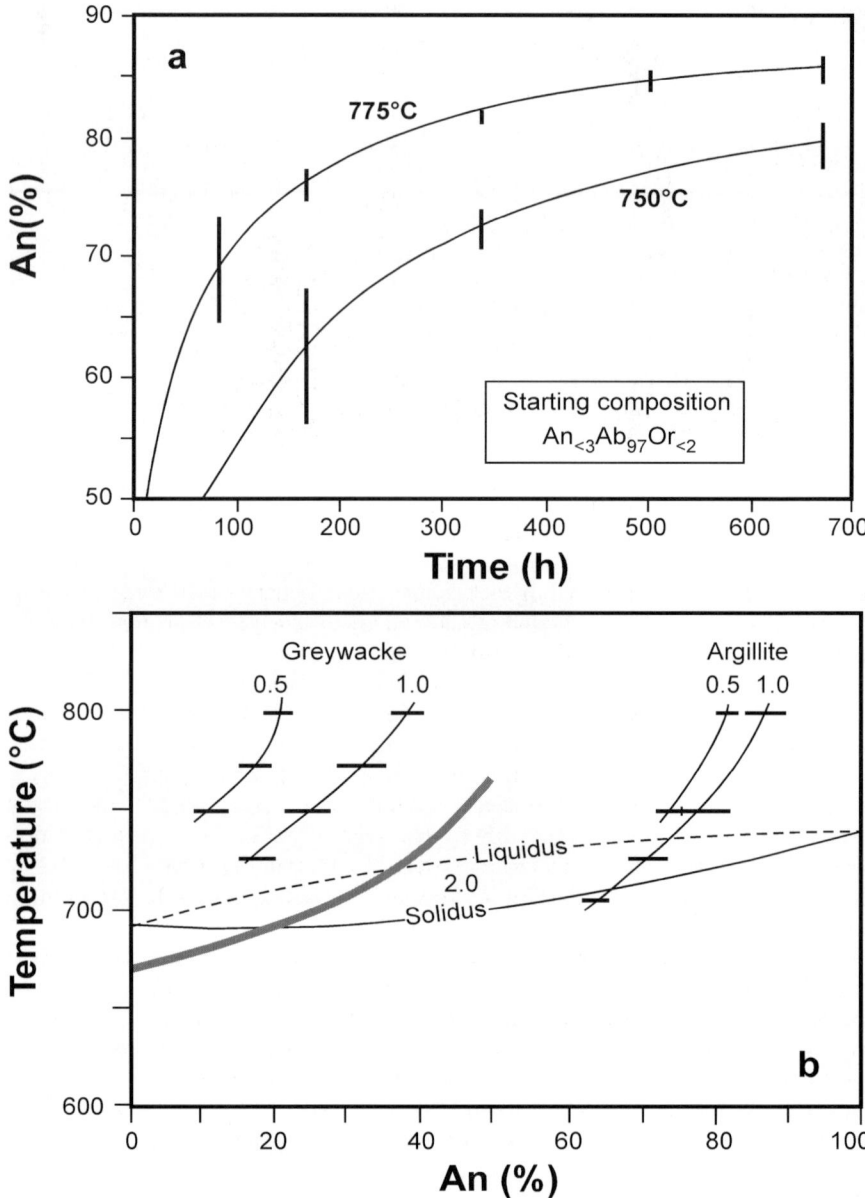

Fig. 7.6. a Compositional changes of plagioclase (Ab_{97}) with time at 750 and 775 °C and 1 kb (unpublished experimental data of Kifle 1992 using a prehnite-pumpellyite facies argillite starting composition). **b** T-An% diagram showing changes in plagioclase compositions coexisting with melt with increasing temperature at 0.5 and 1.0 kb PH_2O in low grade greenschist facies greywacke and argillite bulk compositions (unpublished experimental data of Kifle 1992). *Thick grey curve* represents change in plagioclase composition coexisting with cotectic melt in amphibolite grade metagreywacke (Winkler and von Platen 1961; Winkler 1979). Solidus-liquidus curves for plagioclase at 2 kb after Johannes (1984). See text

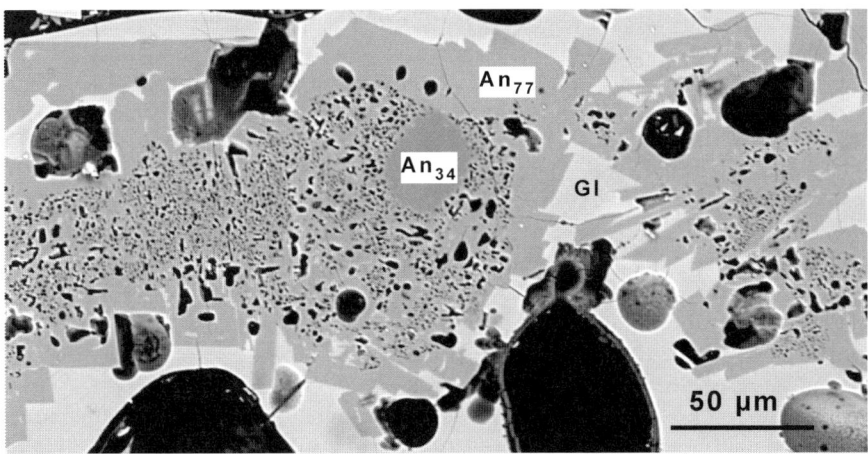

Fig. 7.7. BSE image photo showing partially melted (An_{34}) and newly-formed plagioclase (An_{77}) in a pelitic xenolith, Eifel, Germany. Note the abundance of holes in the reacting plagioclase and the euhedral habit of newly-formed plagioclase in contact with glass. See text

complex mineralogy of the starting rock compositions (detrital and lower grade greenschist facies neometamorphic mineral assemblages for the greywacke/argillite example) with respect to a synthetic mixture of quartz and feldspars.

Figure 7.7 shows relations between the breakdown of plagioclase ($An_{34.2} Ab_{64.4} Or_{1.4}$) and formation of new plagioclase ($An_{77.7} Ab_{21.9} Or_{0.4}$) in a pyrometamorphosed amphibolite xenolith from the Eifel area, Germany. The old plagioclase contains an unreacted core but for the most part the grain (or more probably grain cluster) is characterised by a sieve-like texture between relic (darker grey areas), newly-formed plagioclase (lighter grey areas) and holes that represent vesiculation of melt within the plagioclase when the xenolith was erupted. The reacted plagioclase is overgrown by euhedral, homogeneous Ca-plagioclase extending outwards into the surrounding glass indicating that it has grown from the melt.

From the above experimental and natural data, it can be concluded that the melting of plagioclase is a solution/reprecipitation process where the original plagioclase is progressively dissolved and at the same time more Ca-rich plagioclase is formed with the melt enriched in the Ab (and Or) component(s). Subsequently, new Ca-rich plagioclase may grow epitactically on the surfaces of both initial and reacted plagioclase by crystallisation from the melt. After cessation of the melting process, plagioclase compositions may become equilibrated by solid state diffusion.

7.3
Muscovite

The beginning of muscovite breakdown without general destruction of the crystal lattice involves dehydroxylation (e.g. Gaines and Vedder 1964; Guggenheim et al. 1987). This initially involves the formation of water molecules from structural OH groups, followed by diffusion through the crystal lattice. The temperature at which dehydroxylation begins is largely controlled by grain size, finely divided muscovite beginning to loose its structural water at temperatures as low as 400 °C. Large well-crystallised muscovite may be resistant to thermal decomposition at temperatures > 500 °C and Gaines and Vedder (1964) report that a thin sheet of muscovite heated at 600 °C for a long enough time shows a loss of OH that is accelerated between 700–800 °C. Guggenheim et al. (1987) demonstrate the presence of two overlapping, but poorly defined dehydration peaks at ~550 and 750 °C suggesting that muscovite dehydroxylation occurs over a considerable temperature interval and involves some hydroxyls being lost before others. Using Pauling bond-strength summation calculations, they show that the strength of the Al-OH bond is greatly affected by the coordination number of neighboring polyhedra. When the polyhedra are in octahedral coordination, OH is lost at lower temperatures than when the polyhedra are in 5-fold coordination (after dehydroxylation has been initiated).

Dehydroxylation is accompanied by delamination of the muscovite resulting in an increase in surface area due to a pressure increase caused by the concentration of water molecules along the K-ion cleavage planes, e.g. Grapes (1986). Sanchez-Navas and Galindo-Zaldivar (1993) and Sanchez-Navas (1999) demonstrate that during heating at $T > 500$ °C, there is a high diffusivity of K along muscovite (001) planes induced by water adsorbed along the basal planes resulting from dehydroxylation. At 700 °C, muscovite shows a slight depletion in K_2O which becomes more pronounced at 1100 °C coupled with an increase in CaO and SiO_2 and an oxide total of ~100% indicating complete dehydroxylation (Cultrone et al. 2001). The release of H_2O and K_2O during dehydroxylation is clearly an important factor in promoting the melting of psammitic and pelitic rocks.

High temperature metastable muscovite breakdown reactions recognised in pyrometamorphosed rocks are

muscovite + quartz = mullite/sillimanite + peraluminous melt

muscovite = corundum/Al-mullite + peraluminous melt

(e.g. Grapes 1986) with the K-feldspar component typically dissolved in the melt. If a significant phengite component ($[Fe,Mg]^{2+} + Si^{4+} = 2\,Al^{3+}$) is present in the muscovite (i.e. > 2 wt.% FeO + MgO), pleonaste spinel and/or biotite are also possible reaction products, e.g.

phengitic muscovite + quartz ($\pm\,H_2O$)
= biotite ± pleonaste ± mullite/sillimanite + K-feldspar/melt

phengitic muscovite = biotite + pleonaste + corundum + K-feldspar/melt

phengitic muscovite = biotite + mullite + K-feldspar + K-feldspar/melt

(e.g. Brearley 1986; Brearley and Ruby 1990). Where biotite and/or spinel do not form, the sillimanite, mullite, corundum reaction products may be enriched in Fe^{3+}, e.g. up to 5.0 and 6.8 wt.% Fe_2O_3 in mullite and corundum respectively indicating oxidizing conditions (Grapes 1986). An example of muscovite breakdown in a partially melted pelitic xenolith from the Eifel area, Germany, is shown in Fig. 7.8a and b in which the high temperature reaction products are mullite, biotite, hercynitic spinel, sanidine and peraluminous glass.

Rubie and Brearley (1987) and Brearley and Ruby (1990) studied dry and H_2O-added disequilibrium breakdown of muscovite in contact with quartz using rock cores of quartz-muscovite schist between 680–775 °C at 1 kb over a period of 5 months. In dry runs (757 °C; 5 months), *larger* muscovite grains (> 0.3 mm in length) are replaced by a fine-grained aggregate of K-feldspar + sillimanite + minor biotite + rare hercynitic spinel, together with narrow (< 0.1 micron wide) rims of melt along some grain boundaries. Smaller muscovite grains (< 0.3 mm long) are pseudomorphed by an aggregate of mullite + sillimanite + biotite + melt surrounded by a rim of clear melt. In H_2O-added experiments (757 °C; 5 months), muscovite reacted to mullite (and possibly also sillimanite) + biotite + melt.

From their isothermal experimental kinetic data at 1 kb, Rubie and Brearley (1987) construct a schematic time-temperature-transformation (TTT) diagram for the breakdown of muscovite + quartz and muscovite alone in the system K_2O-Al_2O_3-SiO_2-H_2O (KASH) involving the equilibrium reactions

muscovite + quartz = K-feldspar + sillimanite + H_2O (1)

muscovite = K-feldspar + corundum + H_2O (2)

and their higher temperature metastable reaction equivalents where the K-feldspar and H_2O components are dissolved in a peraluminous melt

muscovite + quartz = sillimanite/mullite + melt

muscovite = corundum/Al-mullite + melt

Five possible reaction pathways of muscovite breakdown in the *quartz-present system* at 1 kb and labeled A, B, C, D, E, F are shown in Fig. 7.9:

1. Path A represents conditions close to the equilibrium reaction (1). Provided that nucleation occurs due to sufficient overstepping of the equilibrium temperature (555 °C), Ms-Qz will react directly to Ksp-AS over a time interval between the "start" and "finish" curves that define the Ms-Qz-Ksp-AS field.
2. Path B occurs at a higher temperature than the quartz-absent (in this system metastable) reaction (2) at 610 °C. Before this reaction reaches completion, the "start" curve of reaction 1 is intersected with nucleation of Ksp-AS. Muscovite disappears to form the metastable assemblage of Ksp-Co-AS. The final assemblage of Qz-Ksp-AS thus eventuates after a longer time because of the formation of intermediate metastable Co.
3. Path C occurs at higher temperatures where muscovite breaks down according to reaction (2) to produce, first the metastable assemblage Qz-Ksp-Co, then Qz-Ksp-Co-AS and finally Ksp-AS by way of the reaction, Co + Qz = AS.

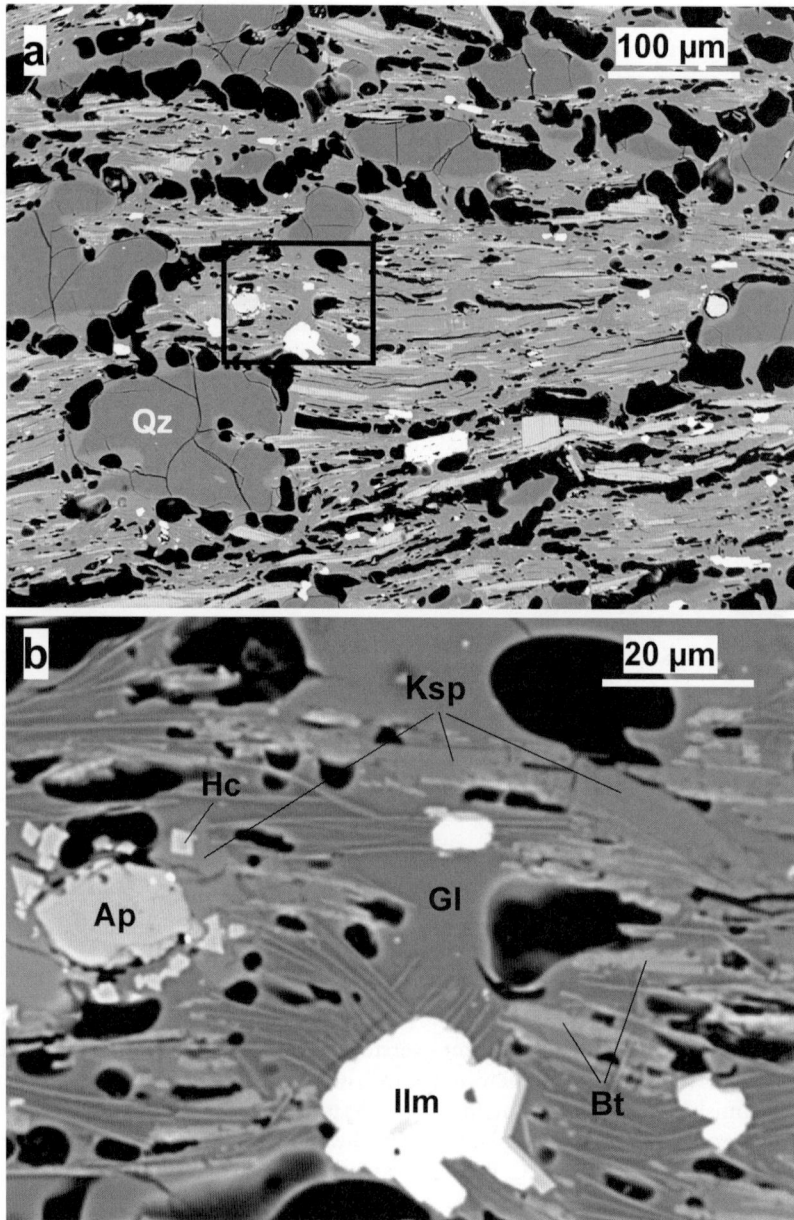

Fig. 7.8. BSE image photos showing melting reaction between muscovite and quartz in pyrometamorphosed pelitic schist xenolith in phonolite, Eifel, Germany. **a** Texture of partially melted quartz and muscovite. Holes indicate vesiculation of melt due to volatile loss on eruption. *Square* denotes enlarged area shown in (**b**). **b** Detail of muscovite breakdown reaction to biotite, hercynite (clustered around apatite and as overgrowths on ilmenite), mullite (acicular crystals), K-feldspar (sanidine) and melt (glass). Inclusions of apatite are unreacted and ilmenite contains exsolved rutile

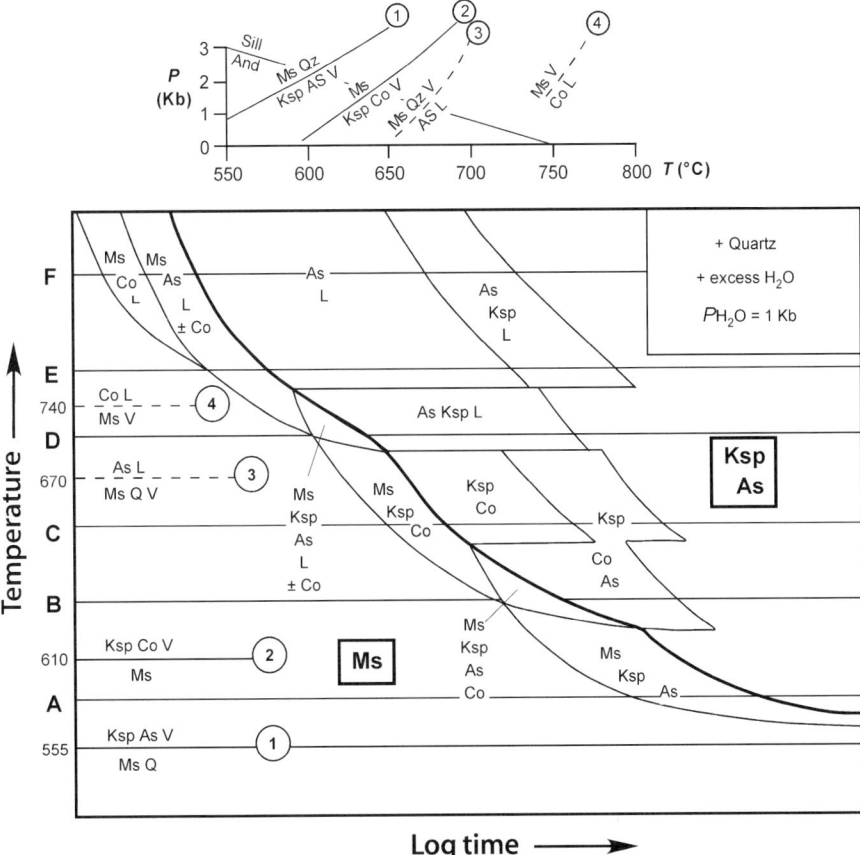

Fig. 7.9. Schematic Time-Temperature-Transformation (TTT) diagram for the equilibrium reaction Ms Qz = AS Ksp V (at 555 °C) with excess Qz and PH_2O of 1 kb. Five different isothermal reaction pathways are labelled A–F that occur at temperatures above those of their respective stable (1, 2) and metastable (3, 4) reactions (*dashed lines*) as shown in the T-P diagram above. In 757 °C H_2O-added experiments, muscovite reacts along the initial stages of reaction pathway F and a subsequent stage would involve formation of K-feldspar. It should be noted that only one Al-silicate phase is considered in the diagram. At 1 kb PH_2O, sillimanite and mullite can form above ~680 °C and andalusite is stable at lower temperatures (after Fig. 11 of Rubie and Brearley 1987). See text

4. The higher temperature reaction paths D, E and F involve formation of melt and the number of intermediate metastable assemblage steps increases with increasing temperature overstepping of equilibrium reaction (1).

It is well known that the presence of significant amounts of H_2O facilitates faster reaction rates than when fluid is absent or only present in small amounts. However, the presence of fluid catalyses metastable reactions including the formation of a melt phase in accordance with the Ostwald step rule and with reference to the TTT diagram (Fig. 7.9), significantly increases the time required to form an equilibrium Ksp-AS assemblage.

7.4
Chlorite

High temperature end products (+ H_2O) of the breakdown of chlorite have been determined from many studies as (Fig. 7.10):

1. Forsterite, enstatite, Mg-cordierite, spinel after Mg-rich chlorite (clinochlore) (Yoder 1952; Roy and Roy 1955; Nelson and Roy 1958, Fawcett and Yoder 1966; Chernosky 1974; McOne et al. 1975; Jenkins and Chernosky 1986; Cho and Fawcett 1986; Barlow et al. 2000).
2. Olivine$_{ss}$, orthopyroxene$_{ss}$, cordierite$_{ss}$, quartz, spinel$_{ss}$, magnetite$_{ss}$, corundum after Fe-Mg chlorite ± O_2 (McOnie et al. 1975; Worden et al. 1987).
3. Fayalite, Fe-cordierite, Fe-gedrite, quartz, mullite, pyrophyllite, corundum, hercynite, magnetite, hematite, after Fe-chlorite (chamosite) (Turnock 1960; James et al. 1976).

Firing of Mg-rich chlorite (clinochlore) (XMg = 0.94) by Barlow et al. (2000) shows that prior to the metastable formation of olivine, orthopyroxene and spinel at ~867 °C,

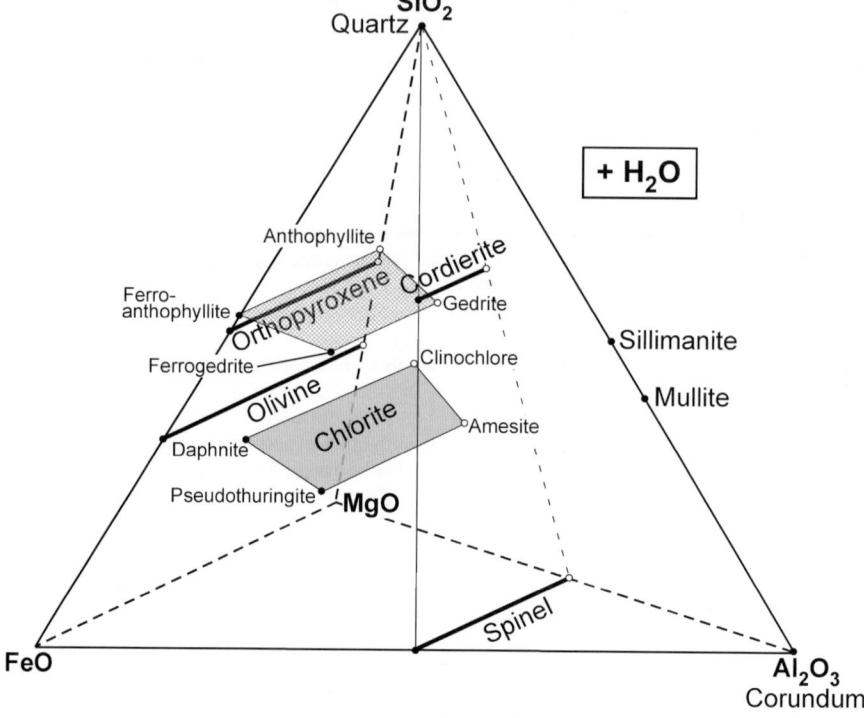

Fig. 7.10. Range of chlorite compositions and compositions of possible high temperature breakdown phases of chlorite projected from H_2O in terms of mol% FeO – MgO – Al_2O_3 – SiO_2

dehydroxylation of first the brucite layers (at ~590 °C) and then the talc layers (complete at ~870 °C) of the chlorite structure occurs (Fig. 7.11). Barlow et al. (2000) suggest that dehydroxylation of the brucite layers fails to produce a suitable environment for the formation of new phases, although a long-range ordering is developed caused by the MgO in the brucite layers being "forced" into a rigid and planar alignment giving rise to a well-ordered 14 Å spacing. When the talc layers loose water, SiO_2 and Al_2O_3 become available to react with MgO remaining from the brucite layers. Initially, epitaxial growth of olivine occurs along the {001} planes and as the reaction proceeds with further disruption of the SiO_2-rich tetrahedral layers orthopyroxene forms and may consume olivine according to the reaction

$$Mg_2SiO_4 + 2 SiO_2 = Mg_2Si_2O_6$$
forsterite + silica = enstatite

with tetrahedral Al largely used to form spinel (some Al may also enter the orthopyroxene). The absence of cordierite in this example may be explained by the lack of common crystallographic planes within the clinochlore structure along which it could nucleate.

A TEM and EPMA study of the disequilibrium breakdown of Fe-rich chlorite ($XMg = 0.33$) in the contact aureole of the dolerite intrusion at Sithean Sluaigh, Scotland, described in Chapter 3 was made by Worden et al. (1987). There is no evidence that chlorite has undergone melting. Three kinds of reactant intergrowths (individual phases < 2 microns) in chlorite are identified and can be characterised by the following reactions:

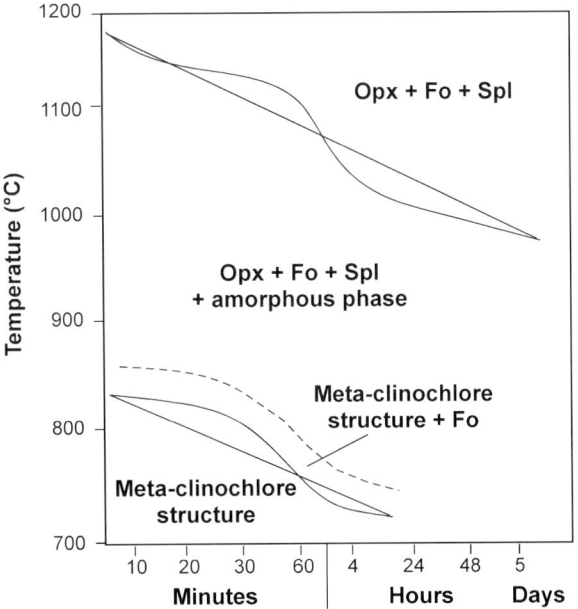

Fig. 7.11.
Time-Temperature-Transformation (TTT) diagram for clinochlore (after Fig. 2 of Barlow et al. 2000). See text

chlorite = spinel + orthopyroxene + H_2O

chlorite + quartz + O_2 = magnetite + spinel + orthopyroxene + cordierite + H_2O

chlorite + phengite = orthopyroxene + cordierite + spinel + biotite + H_2O

The first reaction occurs in the central areas of large chlorite grains (2–3 micron longest dimension). This is effectively an isochemical environment whereby the reaction site is isolated from surrounding minerals. The reactions involving quartz and phengite occur at the margins of chlorite grains. The marginal sites are clearly not isochemical and involve diffusive interaction with the surrounding minerals or with a melt that is developed along albite-quartz contacts in the sample studied (Fig. 7.1). In contrast to the internal parts of reacting grains, chlorite is completely consumed at the marginal reaction sites.

There is no relative orientation of the reactant phases in the chlorite partly due to distortion of the (001) planes. However, there are the following relationships between the breakdown products:

$[111]_{spinel}$ /// $[001]_{biotite}$ /// $[010]_{orthopyroxene}$

$[010]_{spinel}$ /// $[010]_{biotite}$ /// $[100]_{orthopyroxene}$

A modulated structure indicated by TEM and slight asterism shown by the diffraction pattern indicate that the orthopyroxene is disordered. It contains significant Al_2O_3 (4–8 wt.%) in accordance with increasing substitution of Al in orthopyroxene with increasing temperature at low pressure in the orthopyroxene, spinel, forsterite system (Fujii 1976). The lower XFe ratios of orthopyroxene (0.5), cordierite (0.4–0.5) and biotite (0.3–0.4) compared to that of the chlorite (~0.6) is balanced by the presence of spinel in all the reaction assemblages (Fig. 7.1). Worden et al. (1987) infer a temperature of about 700 °C for the chlorite breakdown reactions.

Laser-heating of chloritised biotite by Viti et al. (2003) shows that at 810 °C spinel, olivine and amorphous silica begin to crystallize as irregular 10–100 nm size domains. At 940 °C the olivine and spinel form larger grains typically elongated parallel to chlorite/biotite (001) with spinel [111] and olivine (100) parallel to chlorite/biotite [001].

7.5
Biotite

Petrographic observations described in Chapter 3 indicate that with heating biotite commonly becomes darker in colour and eventually appears black because of the formation of (in the main) Fe-oxide (e.g. Grapes 1986; Brearley 1987a). This is the result of oxidation of Fe^{2+} during dehydroxylation by way of a reaction like

$Fe^{2+} + OH^- = O^{2-} + Fe^{3+} + H$

(Addison et al. 1962; Vedder and Wilkins 1969) resulting in the formation of Fe-oxides. In most biotites, however, some of the iron present is almost certainly present as Fe^{3+} (as an oxyannite component ($K[Fe^{2+}Fe^{3+}_2]AlSi_3O_{12}$) so that the development of magnetite could also occur without any oxidation because of hydrogen loss. Vedder and Wilkins (1969) find that hydroxyl ions near vacancies in octahedral layers of biotite sheets are entirely lost at temperatures below 850 °C over periods of < 24 hours. As with muscovite dehydroxylation, expansion along (0001) planes occurs and potassium may also be lost, e.g. Grapes (1986).

Melting reactions involving biotite ± quartz ± plagioclase are described by Le Maitre (1974), Grapes (1986) and Brearley (1987b) as

biotite + quartz = olivine + Ti-magnetite + melt (Fig. 7.12)

biotite + Ab comp. plagioclase ± quartz
   = pleonaste + Al-magnetite + Na-sanidine + peraluminous melt

Fe-Mg biotite = Mg-Al biotite + magnetite + pleonaste + K-feldspar/melt + vapour

A further reaction occurring in a quartzofeldspathic xenolith from the Eifel, Germany, and shown in Fig. 7.13 is

biotite + quartz + plagioclase
   = orthopyroxene + Ti-Al magnetite ± pleonaste + peraluminous melt

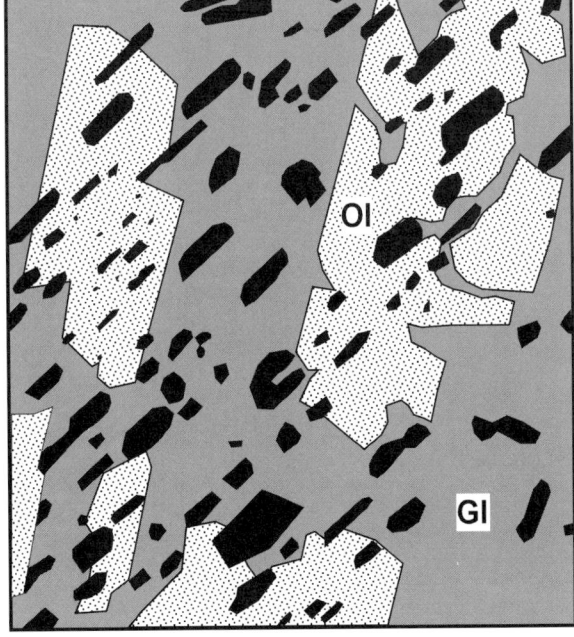

Fig. 7.12.
Drawing from BSE image photo (Plate 2a of Le Maitre 1974) of olivine ($Fo_{75\pm5}$), Ti-magnetite (*black*) and glass after biotite in a granite xenolith in basalt, Mt. Elephant, Victoria, Australia. Width of drawing = 60 μm

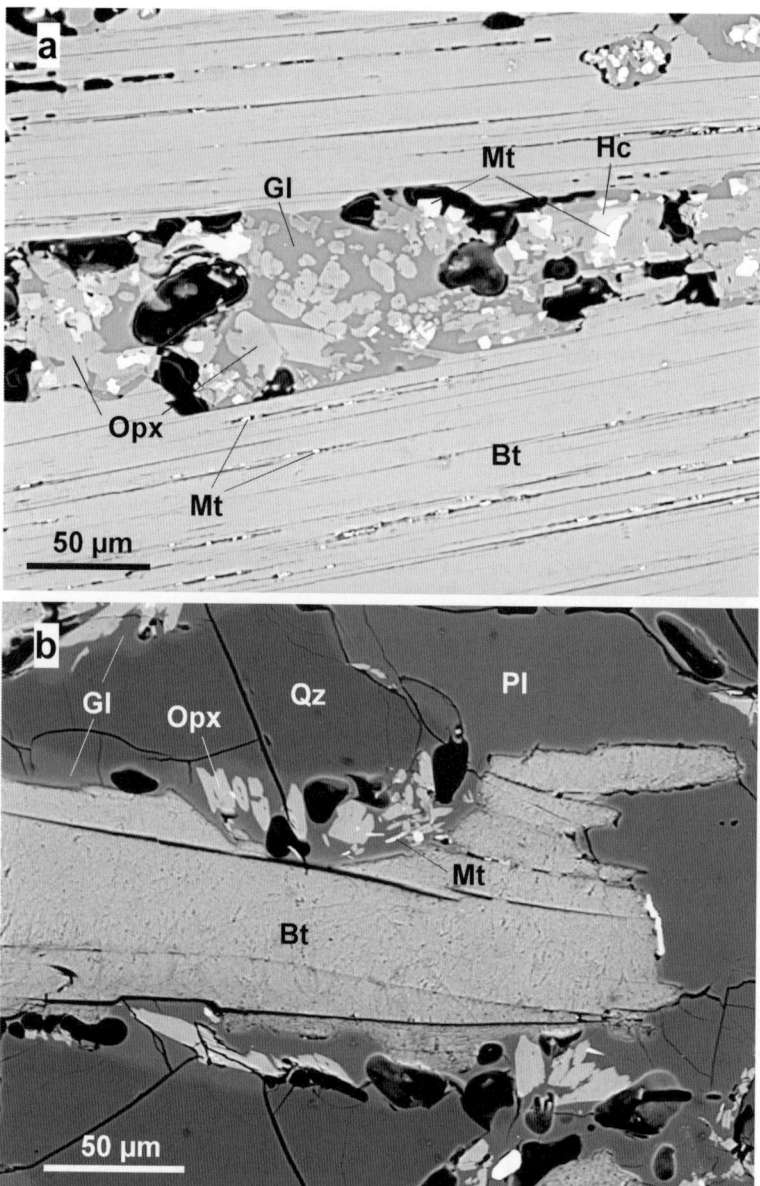

Fig. 7.13. BSE image photos of the breakdown of biotite in a partially melted pelitic xenolith, Eifel, Germany. **a** Biotite showing initial stage of breakdown to magnetite ± glass along cleavage planes and internal area of complete melting to orthopyroxene, hercynitic spinel, Ti-Al magnetite and glass. **b** Biotite rimmed by glass in contact with quartz and sodic plagioclase with newly formed orthopyroxene and magnetite in glass adjacent embayed (melted) areas of the biotite

A balanced isochemical reaction (except for H_2) for the reaction involving the formation of more Mg-rich biotite above is given by Brearley (1987b) as

$$K_{1.71}(Mg_{1.79}Fe^{2+}_{2.93}Al_{0.7}Ti_{0.28})Al_{2.45}Si_{5.55}O_{20}(OH)_4$$
$$= 0.788\,K_{1.73}(Mg_{2.22}Fe_{2.44}Ti_{0.36}Al_{0.64})Al_{2.26}Si_{5.74})_{20}(OH)_4 + 0.34\,KAlSi_3O_8$$
$$+ 0.26\,Mg_{0.15}Fe_{0.85}Al_2O_4 + 0.261\,Fe_3O_4 + 0.288\,H_2O + 0.136\,H_2$$

The reaction assumes that Fe^{3+} is zero in reactant and product biotite, although some oxidation undoubtedly occurred through loss of H_2 and in producing magnetite (also an Fe_3O_4 component in pleonaste of 6.6–33.1%). It would be expected that with increasing temperature the new biotite is more Ti-rich than the reactant biotite because no rutile is produced in the reaction although a small amount of Ti enters the spinel (e.g. 0.2–1.9 wt.% TiO_2 in pleonaste). Also, as the Ab content of the K-feldspar ranges up to Ab_{15}, plagioclase may have been involved in the reaction.

Experimentally-induced disequilibrium breakdown of aluminous iron-rich biotite ($XMg = 0.42$–0.47, 800 °C, 1 kbar, 2 days to 8 weeks duration) according to the overall reaction

Al-biotite = Al-orthopyroxene + hercynitic spinel + magnetite + peraluminous melt

is reported by Brearley (1987a). Notable is the considerable variation in the compositions of the spinel and pyroxene reaction products that is dependent on reaction time and location, i.e. whether they develop along cleavage planes or within ordered biotite. As shown in Fig. 7.14a, compositional ranges of the orthopyroxene reactant become more restricted with increasing run time and in the case of pleonaste spinel (Fig. 7.14b), less magnetite-rich. This implies that metastable (and therefore variable) compositions move towards stable equilibrium compositions as a function of time at the same temperature. These characteristics and the production of melt rather than K-feldspar as predicted from a stable breakdown reaction of biotite, underscore the metastable nature of the reaction.

The experiments show that two different reaction pathways occur within a single biotite crystal:

1. Along prominent defects such as cleavage planes there is nucleation of abundant spinel with melt and with the biotite composition becoming less aluminous.
2. In areas of ordered biotite, initial spinel is magnetite-rich and then becomes more aluminous (higher hercynite and spinel components) which shifts the biotite to a more Mg-rich, Al-poor composition. A delayed nucleation of orthopyroxene maybe because of the need to first establish high energy nucleation sites by the formation of interfaces between spinel and biotite. Orthopyroxene equilibrates to higher Al and Fe contents with time and the overall reaction can be described as

Mg-poor/Al-rich biotite
= Mg-rich/Al-poor biotite + spinel + orthopyroxene + melt

Fig. 7.14.
Plots of orthopyroxene (Al versus Mg/Mg + Fe) (**a**) (after Fig. 8b of Brearley 1987a) and pleonaste spinel ($FeAl_2O_4$-$MgAl_2O_4$-Fe_3O_4) (**b**) (after Fig. 7 of Brearley 1987a). Composition fields in terms of experimental run time at 800 °C/1 kb. *Dotted field* = 48 h; *horizontal-lined field* = 161 h; *grey-shaded field* = 309 h; *black field* = 1460 h

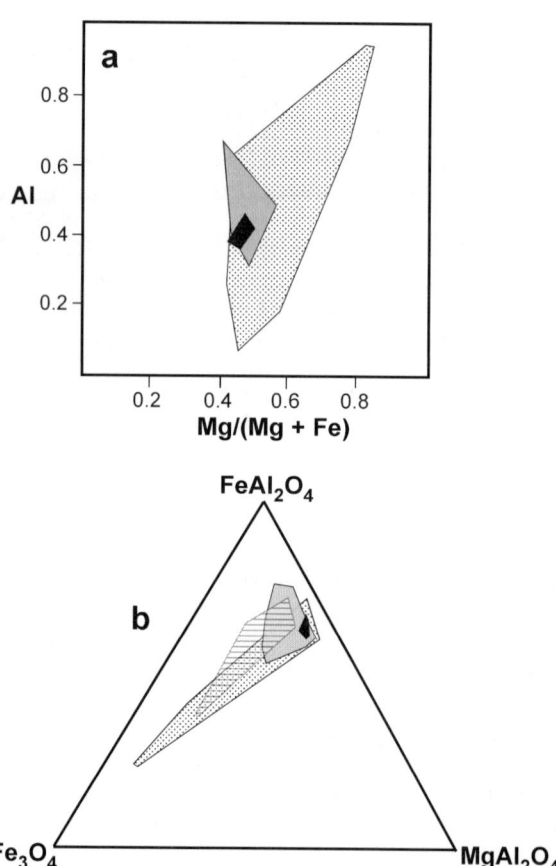

7.6
Calcic Amphibole

Thermal decomposition of amphiboles to pyroxenes and silica under non-oxidising conditions, and to pyroxene, silica and Fe-oxides under oxidising conditions occurs at temperatures between 700–800 °C (e.g. Ghose and Weidner 1971; Xu et al. 1996). Xu et al. (1996) show that the breakdown of tremolite is characterised by the formation of (010) clinopyroxene slabs (18 and 36Å in thickness) along the *b* axis of the tremolite, the clinopyroxne slab thicknesses corresponding to two or four tremolite chains along the *b* axis. Thus, transformation to clinopyroxene involves the breaking of every two tremolite chains to form four pyroxene chains. This decomposition occurs by rearrangement of Si-O tetrahedral and interdiffusion of Ca and Mg atoms within (100) octahedral bands with diffusion enhanced by H_2O released during dehydroxylation of the amphibole.

The breakdown of more complex (hornblende) compositions (av. $[K,Na]_{0.77}Ca_{1.86}$ $[Ti_{0.14}Mg_{2.89}Fe^{2+}_{1.09}Fe^{3+}_{0.62}Al_{0.26}][Si_{6.39}Al_{1.61}]O_{22}[OH]_2$) without melting between ~720–850 °C occurs in arkose within the contact aureole of the Rhum intrusion, Scot-

Fig. 7.15.
BSE image photo showing a breakdown assemblage of olivine, clinopyroxene, magnetite, glass after hornblende in a pyrometamorphosed amphibolite xenolith, Eifel, Germany. The replaced hornblende is in contact with clinopyroxene that has reacted along the contact to magnetite and a more Mg-rich clinopyroxene (*darker grey tone*). *Black areas* = holes in glass caused by exsolution of H_2O from melt on eruption

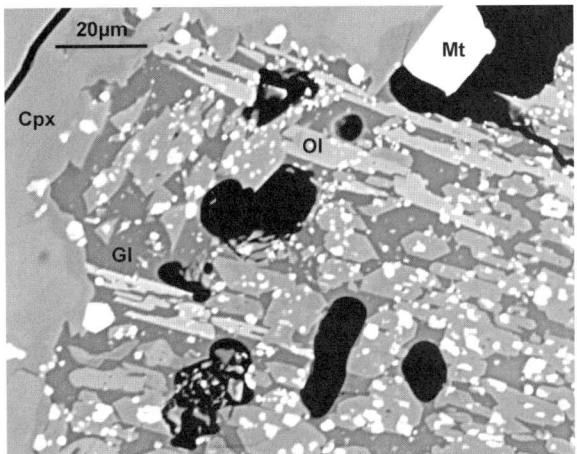

land (Holness and Isherwood 2003). The onset of hornblende breakdown is indicated by the formation of fine grained ilmenite-magnetite along cleavage planes that eventually develops into a pseudomorphic aggregate of clinopyroxene ($Ca_{46-44}Mg_{45-36}Fe_{10-19}$; < 1.4 wt.% Al_2O_3), orthopyroxene (XMg = 68–52), magnetite and biotite, together with plagioclase ($Ab_{62}An_{35}Or_3$) that may have originally formed an intergrowth with the amphibole. The amphibole pseudomorphs are typically zoned with a coarse-grained outer rim of clinopyroxene (developed where amphibole is in contact with plagioclase), an inner rim of orthopyroxene and a fine-grained central area of pyroxene, plagioclase, Fe-Ti oxides, ± biotite. With increasing temperature, the grain size of the reaction minerals becomes coarser and the clinopyroxene rim becomes less prominent. The proposed overall amphibole breakdown reaction is

hornblende + anorthite ± Fe_{fluid}
  = clinopyroxene + orthopyroxene + biotite + magnetite

Isocon analysis (after the method of Grant 1986) suggests that Si, Al, Ca, Na, K remain immobile during the reaction, Fe was either gained (in Fe-poor amphibole) or lost (in Fe-rich amphibole), and Mg was removed from the core areas of the pseudomorphs to form the monomineralic clinopyroxene rims.

A pyrometamorphosed amphibolite xenolith, west Eifel, Germany, provides an example where hornblende of unknown composition has been replaced by olivine (instead of orthopyroxene), clinopyroxene, magnetite and an Al-rich mafic melt as shown in Fig. 7.15. Olivine is Fo_{69}, clinopyroxene is $Ca_{50}Fe_{21}Mg_{29}$ (6.97 wt.% Al_2O_3; 0.94 wt.% TiO_2), spinel is Al-Ti magnetite (~ wt.% 4.4 TiO_2; 15.3 Al_2O_3; 5.7 MgO) and the breakdown reaction can be described as

hornblende = clinopyroxene + olivine + Fe-Ti oxide + melt

where the average wt.% melt composition is 43.6% SiO_2; 1.2% TiO_2; 18.4% Al_2O_3; 12.1% FeO; 1.0% MgO; 13.0% CaO; 7.0% Na_2O; 1.7% K_2O.

7.7
Clinopyroxene

Experimental work using volcanogenic greywacke starting material by Kifle (1992) at low pressure ($PH_2O < 1$ kb) and temperatures between 775–800 °C, indicates that clinopyroxene (detrital grains of $Ca_{37-43}Mg_{37-43}Fe_{14-21}$; 1.7–3.6 wt.% Al_2O_3; 0.4–0.7 wt.% TiO_2) reacts to orthopyroxene ($XMg = 0.55$–0.62) and possibly minor ilmenite in the presence of a peraluminous melt (Fig. 7.16). As the glass contains a maximum CaO content of 1.2 wt.%, it is apparent that Ca and excess Si needed to form orthopyroxene from the clinopyroxene is partitioned into the melt resulting in the crystallisation of new plagioclase ($An_{34-39}Ab_{59-64}Or_2$) around the boundary of the relic clinopyroxene, i.e.

$$Ca(Mg,Fe)Si_2O_6 = (Mg,Fe)SiO_3 + [CaO + SiO_2]_{in\ melt}$$
clinopyroxene   orthopyroxene

At 500 bar/800 °C the reaction is complete after 1450 hours.

Nano-petrographic evidence of high temperature pyroxene transformation with decreasing pressure in a partially melted microgabbro xenolith (several cubic centimeters in size) erupted from the Beaunit maar, Massif Central, France, is described by Faure et al. (2001). The xenolith consists of fine grained plagioclase, clinopyroxene and orthopyroxene with abundant interstitial pockets of brown glass and is inferred to have been derived from the granulitic lower crust (870–970 °C/7–8 kbar). Incorporation into basaltic magma and subsequent rapid journey to the surface resulted in the following sequence of reactions involving clinopyroxene as shown in Fig. 7.17:

1. Topotactic transformation of orthopyroxene to clinopyroxene (Cpx A in Fig. 7.22) with higher Al, Ti than primary clinopyroxene in the gabbro at temperatures ~1200 °C when the xenolith was incorporated into basalt. Calcium for clinopyroxene formation was probably derived from melting of plagioclase.

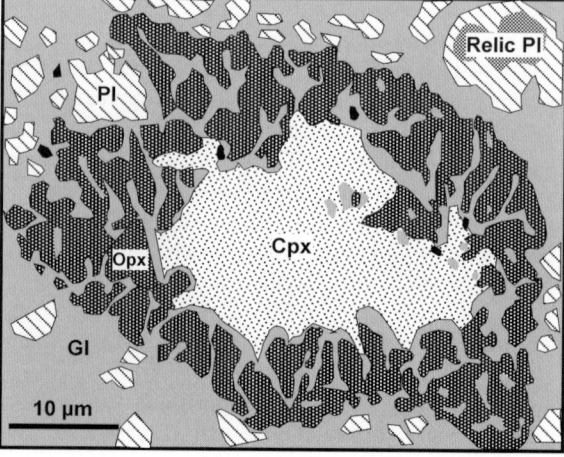

Fig. 7.16.
Drawing from BSE image photo (Plate 5.21B of Kifle 1992) showing the breakdown of clinopyroxene (augite) in the presence of melt (*Gl* = glass) to orthopyroxene and plagioclase (An_{34-39}). The relic plagioclase core has the composition (An_{15}). *Black* = ilmenite

2. Phase separation in Cpx (A) to form thick (50 nm) lamellae of $C2/c$ high pigeonite and $C2/c$ augite (Cpx B) during nearly isothermal decompression on ascent. Exsolution could have occurred over a minimum of 7 hours based on calculation using a Ca diffusion coefficient of $D = 10^{-15}$ cm^2 s^{-1} and $t = h^2/D$ where h is the distance (cm) between centres of adjacent lamellae and t is time in seconds.
3. Eutectic melting of pigeonite + augite (Cpx B) to form another, more Mg-rich, clinopyroxene (Cpx C) + ilmenite due to further isothermal decompression over a possible period between 30 minutes to 9 hours estimated from the thickness of melt films that vary between 500 nm to 2 μm.
4. Rapid cooling on eruption of non-melted pyroxene (Cpx B) produced spinoidal decomposition and martensitic transformation from $C2/c$ high pigeonite to $P2_1/c$ low pigeonite.

Although highly sensitive to kinetic factors, calculation of the time required for each of the above pyroxene transformations gives a minimum residence time of the xenolith in the basalt of 16 hours, a magma ascent velocity from a depth of 30 km of 1.8 km h^{-1}, and with the two exsolution episodes caused by pressure decrease.

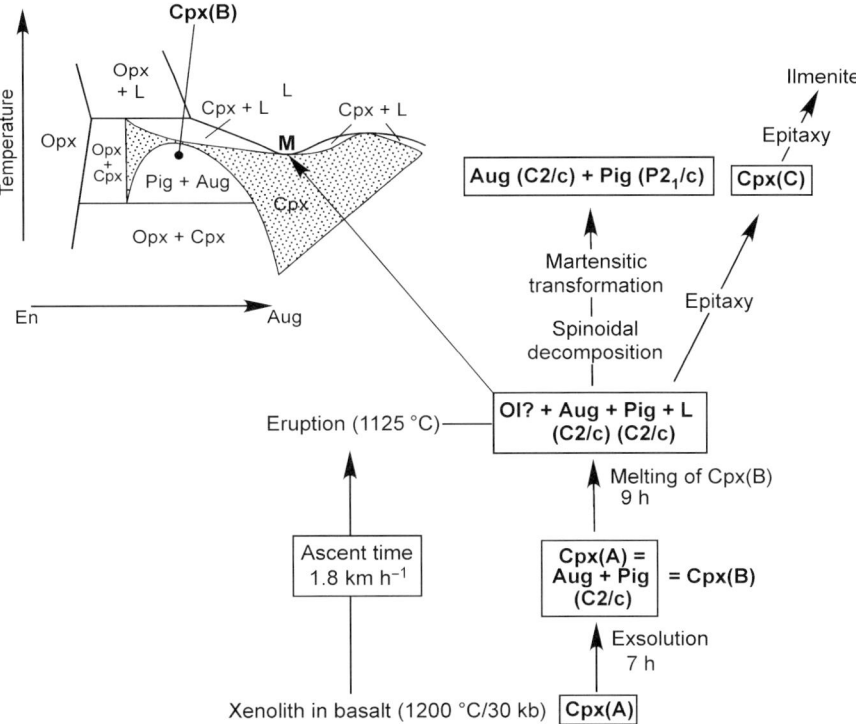

Fig. 7.17. Diagram illustrating clinopyroxene reactions during nearly isothermal decompression of a gabbro xenolith in basalt, maar de Beaunit, Massif Central, France, together with a schematic isobaric temperature–composition diagram relevant to the reactions (after Longhi and Bertka 1996) (redrawn from Figs. 7b and 8 of Faure et al. 2001). See text

Switzer and Melson (1969) document the pyrometamorphic breakdown of omphacitic pyroxene that occurs with kyanite in an eclogite xenolith entrained in kimberlite magma. The replacement assemblage of the omphacite is a submicroscopic mixture of stellate (quenched) plagioclase, Na-poor clinopyroxene and glass. In comparison with residual liquid compositions in the system nepheline-diopside-silica (Schairer and Yoder 1967), high normative albite (40%) and nepheline (23%) of the glass would be expected from incongruent melting of omphacite. Melting experiments using omphacite from the eclogite at atmospheric pressure by Switzer and Melson (1969) indicate that it begins to melt at 1030 °C and is completely molten at 1260 °C.

7.8
Al-Silicates

The reaction of *andalusite* porphyroblasts to corundum and K-feldspar under pyrometamorphic conditions has been noted by Lacroix (1893) (Fig. 7.18), Brauns (1912a) and Sassi et al. (2004) implying that andalusite breaks down in the presence of muscovite according to the reactions

andalusite = corundum + SiO_2

muscovite = corundum + K-feldspar + H_2O

Quartz is not found with the corundum + K-feldspar reaction products and silica is presumed to have been either been removed in an aqueous phase (lower temperature) or partitioned into melt (higher temperature). The common presence of hercynitic spinel with corundum (Fig. 7.18) suggests that andalusite contains some iron and that the reaction occurred under reducing conditions. In the pure Al_2O_3-SiO_2 system, the andalusite breakdown reaction at low pressures (< 1000 bar) occurs at a maximum temperature and log silica activity of ~700 °C and −0.1 respectively (Fig. 7.19a).

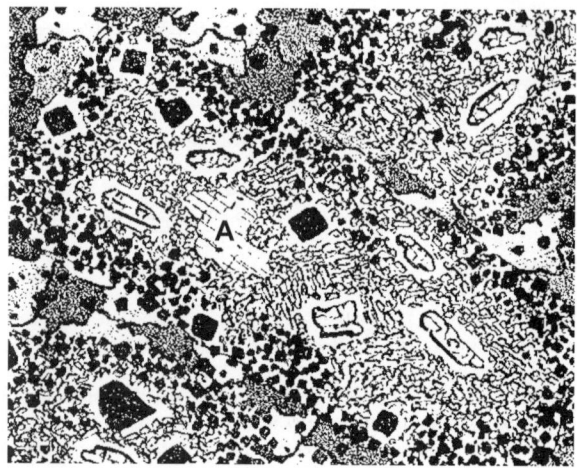

Fig. 7.18.
Breakdown of porphyroblasts of andalusite (relic core labelled '*A*') to corundum (small and some large crystals) and spinel (*black*) within sanidine. Interstitial *stippled grains* = biotite (Fig. 14 of Lacroix 1893). No scale given in original. Micaceous schist enclave in biotite trachyte, Cantal, Massif Central, France

Fig. 7.19.
T-aSiO$_2$ diagram at 650 bar for sillimanite-andalusite-corundum equilibria after Fig. 10 of Markl (2005). *Numbers* refer to aAl$_2$SiO$_5$ values. Note that the added Fe^{3+}-bearing andalusite-sillimanite transition lies at a higher temperature than that of the Fe^{3+}-free transition (data from Holdaway 1971)

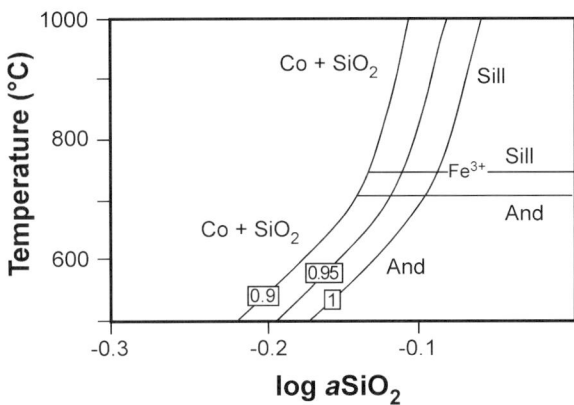

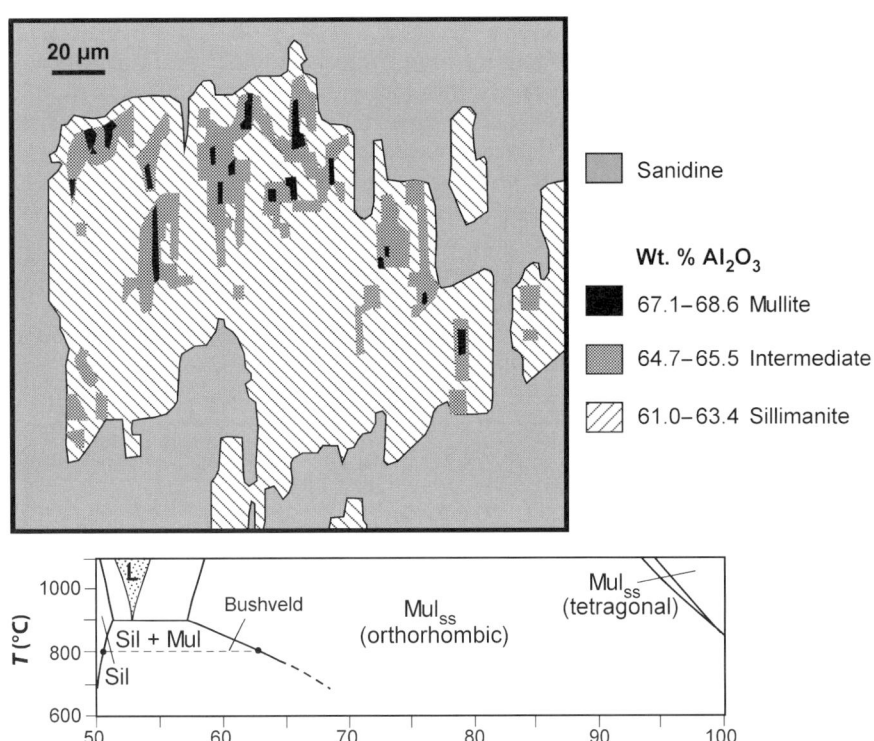

Fig. 7.20. Drawing from a BSE image photo showing the distribution of Al$_2$O$_3$ in a heterogeneous sillimanite-mullite grain enclosed by sanidine, buchite xenolith, Eifel, Germany (after Fig. 24 of Grapes 1986). *Below* is a diagram showing possible phase relations between sillimanite and corundum at temperatures < 1100 °C (after Fig. 6 of Cameron 1977). Stable coexisting sillimanite-mullite compositions from a sillimanite-mullite-corundum-sapphirine xenolith (Bushveld Intrusion, South Africa) described by Cameron (1977) help delineate a Sil + Mul field

Addition of Fe_2O_3 resulting in andalusite-corundum activities of < 0.9, would increase the maximum temperature and lower $aSiO_2$ of the reaction.

An example of the apparent transformation of *sillimanite to mullite* is described by Grapes (1986) in a buchitic xenolith from the Eifel, Germany. As shown in Fig. 7.20, the bulk of the grain is sillimanite (~50 mol% SiO_2) that contains blebs of mullite (44 mol% SiO_2). Another phase of intermediate composition (45–47 mol% SiO_2) occurs mainly with mullite. The texture suggests the possibility that mullite has exsolved from sillimanite and that the areas of intermediate composition represent a quenched concentration gradient due to sluggish diffusion of Al and Si between sillimanite and mullite. The "exsolved" mullite is more Si-rich compared to other homogeneous mullite grains with ~42 mol% SiO_2 in the buchite suggesting that the sluggish nature of the reaction

sillimanite = mullite + SiO_2

(Fig. 7.21) might explain the absence of a silica phase that should have formed if equilibrium was maintained or that excess silica was removed from the reaction site and dissolved in the buchite melt. Newly-formed quartz does occur in graphic intergrowth with cordierite in contact with glass in this xenolith and provides a lower P of ~1.3 kb and an upper T range of 710–1060 °C for the exsolution of mullite depending on which data set in Fig. 7.21 is adopted for this reaction. Cameron (1976a) describes an example of the reverse reaction – mullite with exsolved sillimanite in a corundum-sillimanite-mullite xenolith from the Bushveld Intrusion, South Africa (Chapter 3), that he attributes to slow cooling (Fig. 7.20).

The above observations are illustrated by a phase diagram at temperatures < 1100 °C for the system Al_2SiO_5-Al_2O_3 (after Cameron 1977) (Fig. 7.20) that indicates a two-

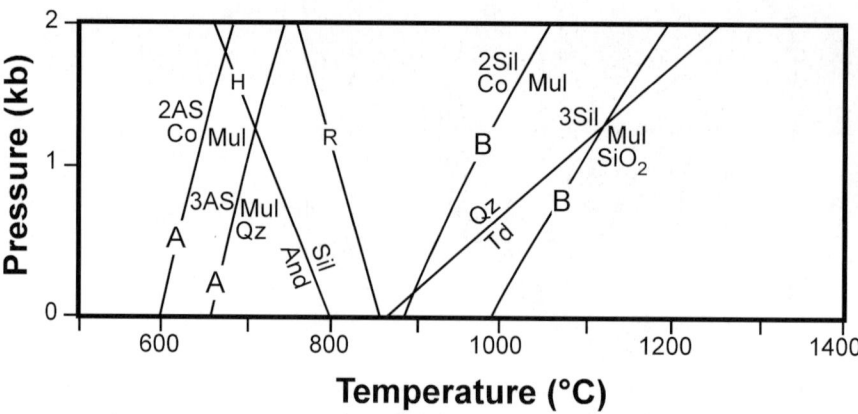

Fig. 7.21. *T-P* diagram of phase equilibria in the Al_2O_3-SiO_2 system (after Fig. 13 of Markl 2005). The shift in the position of reaction curves labelled *A* and *B* results from using different values of heat capacity (*cp*-function coefficients) in calculating the reactions; *A* = data set of Robie and Hemingway (1995); *B* = modified data set to attain agreement between calculations and natural constraints (see Markl for details). Andalusite-sillimanite reaction curves after Holdaway (1971) (*H*) and Richardson et al. (1969) (*R*)

phase field of sillimanite + mullite between sillimanite and orthorhombic mullite$_{ss}$ below 900 °C within which compositions are probably metastable.

An interesting example of inferred pyrometamorphic breakdown of *kyanite* in eclogite xenoliths within kimberlite, Roberts Victor Mine, South Africa referred to above, is documented by Switzer and Melson (1969). Kyanite is marginally melted (usually at contacts with garnet) to corundum and mullite with additional sapphirine in the glass reported by Chinner and Cornell (1974). It is suggested that the melting of kyanite was caused by a sudden decrease in pressure at high temperature followed by quenching in the rapidly ascending, expanding gas-rich kimberlite magma.

7.9
Garnet

There are few natural examples of the pyrometamorphic breakdown reaction of garnet (see sections on Mull, Sithean Sluaigh, Traigh Bhàn na Sgúrra in Chapter 3) Brauns (1912a,b) shows examples of euhedral almandine-rich garnet (Alm$_{75}$) in sanidinite xenoliths from the Eifel area, Germany, that are surrounded by glass and remain as stable relics of otherwise completely reconstituted garnet-mica schist (Fig. 7.22a). At the temperatures of peraluminous melt formation (melting of quartz, Na-plagioclase, muscovite) in many of the Eifel xenoliths, garnet has evidently adjusted its morphology by developing a euhedral habit in contact with melt. In other cases, garnet is partially reacted to magnetite (Fig. 7.22b) and at a more advanced stage of decomposition, possibly to a fine grained mixture of orthopyroxene, ?clinopyroxene and magnetite.

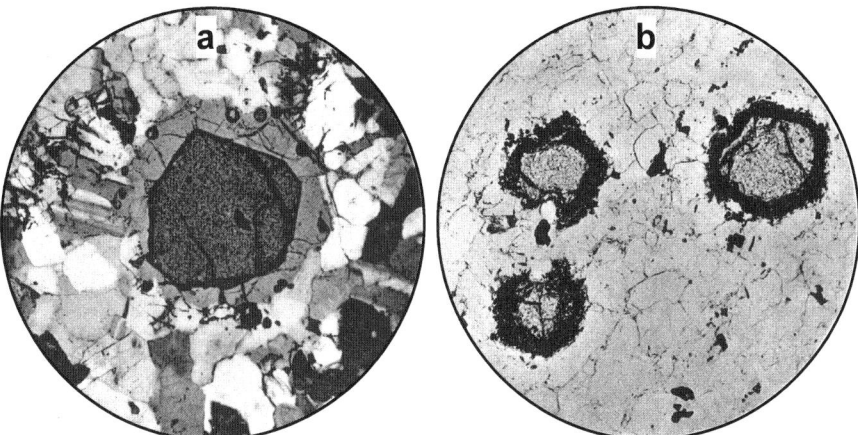

Fig. 7.22. Photomicrographs of garnet reconstitution and breakdown. **a** Euhedral garnet surrounded by glass in a sanidinite xenolith, trachyte, Dachsbusch, Eifel, Germany (× 54, crossed nicols; Plate 9, Fig. 3 of Brauns 1912a). The sanidinite consists of sanidine, plagioclase, magnetite with minor biotite, corundum and glass. In this almost completely recrystallised rock, garnet has not melted but may have adjusted its morphology to a euhedral habit. **b** Garnets in a fine grained sanidinite xenolith that have reacted to magnetite around their margins. Associated minerals are sanidine, magnetite, corundum, sillimanite and glass (× 60, plane polarised light. Dachsbusch, Eifel, Germany. Plate 9, Fig. 2 of Brauns 1912a)

The expected breakdown reaction of almandine-rich garnet with excess H_2O is

almandine + H_2O = Fe-cordierite + fayalite + hercynite$_{ss}$ + H_2O

at between ~800–825 °C / 250 bar and 900 °C / 2 kb under IQF, IM and IWü oxygen buffer conditions (Hsu 1968). Under higher oxidation conditions, i.e. QFM buffer, almandine stability is controlled by the reaction

almandine + H_2O = quartz + hercynite$_{ss}$ + magnetite$_{ss}$ + H_2O

at temperatures 100–180 °C lower than the cordierite + fayalite + hercynite-producing reaction over the same pressure range (Fig. 7.23). With increasing temperature, the two spinel phases in the latter reaction eventually merge to form a solid solution at temperatures above about 850 °C (Fig. 7.23). Fayalite and cordierite become unstable (they are replaced by Fe-oxide) at any temperature when fO_2 is equal to or above that of the QFM buffer (Hsu 1968). Oxidising conditions probably account for the replacement of almandine-rich garnet by magnetite in the Eifel sanidinite shown in Fig. 7.22B.

Schairer and Yagi (1952) have determined the high temperature metastable breakdown of almandine-rich garnet ($Alm_{91.3–69.6} Py_{2.4–21.2} Sp_{1.7–3.5} Gr_{4.6–5.7}$) under anhydrous conditions in a nitrogen atmosphere at atmospheric pressure. Evidence of a reaction to hercynite, cordierite and fayalite occurred only after heating at 900 °C for two weeks with melt first appearing at 1090 °C that was accompanied by the disappearance of cordierite and fayalite. Heating of a more Mg-rich garnet ($Alm_{40} Pyr_{43} Sp_{1.4} And_2 Gr_{1.4}$) yielded Mg-rich cordierite, spinel and clinopyroxene.

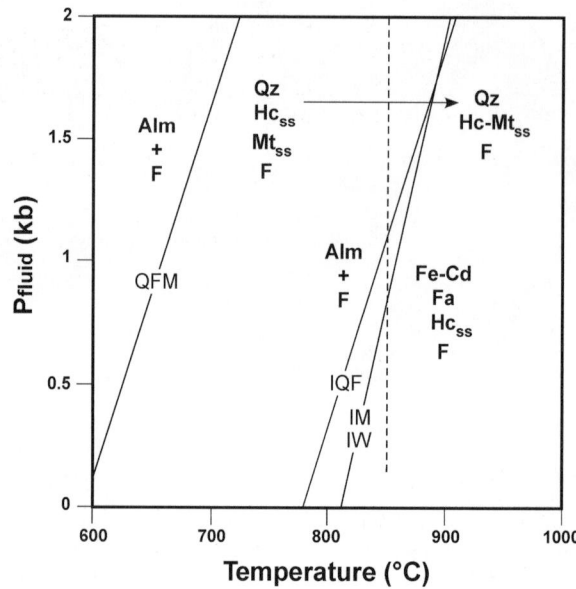

Fig. 7.23.
T-P diagram showing stability of almandine garnet + excess H_2O in terms of various oxygen buffers; Iron-Wüstite (*IW*), Iron-Magnetite (*IM*), Iron-Fayalite-Magnetite (*IFM*) and Quartz-Fayalite-Magnetite (*QFM*). The *dashed line* (QFM buffer conditions) indicates up-temperature, pressure independent change (*arrow*) to hercynite-magnetite solid solution (after Figs. 5, 6, 7 of Hsu 1968)

7.10
Staurolite

The high temperature breakdown of staurolite to a hercynite-rich assemblage together with sillimanite or cordierite and quartz has been noted by Aitkin (1978) and Cesare (1994) and is also illustrated from the breakdown of staurolite porphyroblasts in schist xenoliths from the Wehr volcano, Eifel, Germany (e.g. Worner et al. 1982; Grapes unpublished data). Evidence of the initial reaction is indicated by the growth of hercynite along cracks (Fig. 7.24) and replacement is complete when about 95% of former staurolite consists of green spinel (Hc_{53-85} Sp_{8-11}) elongated parallel to the original cleavage and also occurring as irregular, crosscutting veins (Fig. 7.25a and b). The spinel is intergrown with small amounts of orthoamphibole (aluminous ferrogedrite with 23 wt.% Al_2O_3; 30 wt.% FeO; 4 wt.% MgO) and interstitial areas to the spinel are occupied by a pale yellow glass that contains crystals quartz and aluminium silicate (43.9–47 mol% SiO_2, i.e. a composition intermediate between the most siliceous natural mullite with 43 mol% SiO_2 and the most aluminous natural sillimanite with

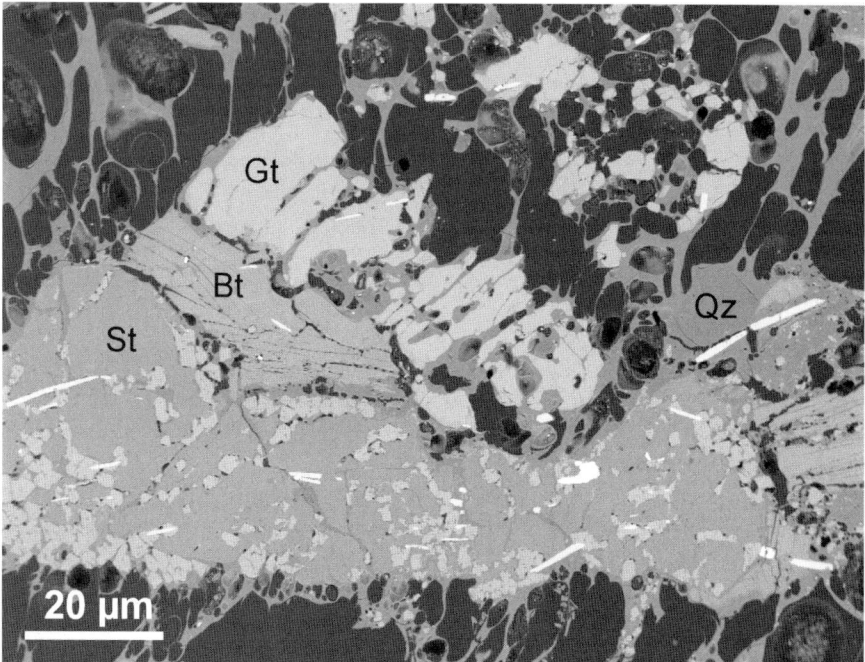

Fig. 7.24. BSE image photo showing a staurolite porphyroblast (*St*) partly reacted to hercynitic spinel (*light grey grains*) along cracks and around margins in a partially melted staurolite schist xenolith, Wehr volcano, Eifel, Germany. The staurolite is in contact with unreacted biotite (*Bt*) and garnet (*Gt*) originally intergrown with quartz (*Qz*) that is now largely melted. Elongate white crystals are ilmenite. Note the web-like pattern of glass that has resulted from vesiculation due to volatile loss when the xenolith was erupted

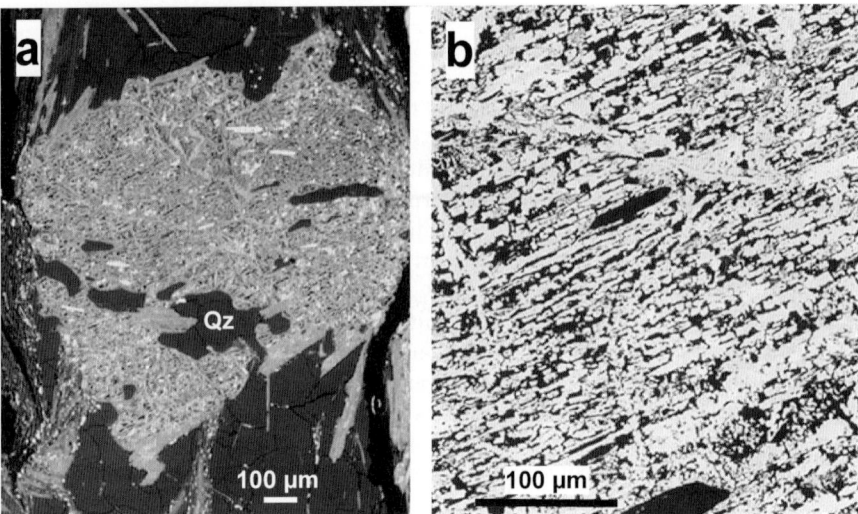

Fig. 7.25. BSE image photos. **a** Advanced stage of staurolite porphyroblast breakdown to a hercynite-rich assemblage. Elongate white crystal inclusions are ilmenite. Grains of biotite occur around the margin of the porphyroblast and are partly reacted to magnetite. **b** Detail of hercynite replacement. High temperature silicates and glass occupy the dark areas interstitial to hercynite and are shown in Fig. 7.26

Fig. 7.26.
Drawing from BSE image photo showing hercynite (*Hc*) and interstitial sillimanite (*Sil*), gedrite (*Gd*), quartz (*Qz*) reaction products after staurolite in glass (*Gl*)

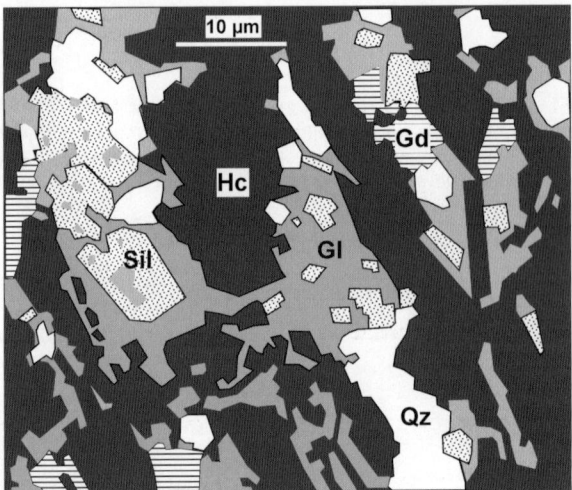

48.6 mol% SiO_2 and with 2.41–5.23 wt.% Fe_2O_3) (Fig. 7.26). The glass is siliceous (76.9–77.4 wt.% SiO_2), has low alkalis (1.4–3.1 wt.%), normative corundum between 8.2 and 9.8%, and molar $Al_2O_3/(CaO + Na_2O + K_2O)$ (A/(CNK) = 3.64–5.88, with FeO of 3.4–4.5 wt.%. The alkali content of the glass is assumed to be derived from melting of associated muscovite (providing a K-feldspar$_{ss}$ component), possibly together with some K_2O derived from biotite dehydroxylation (biotite is oxidised but unmelted) as there is no plagioclase in the xenolith.

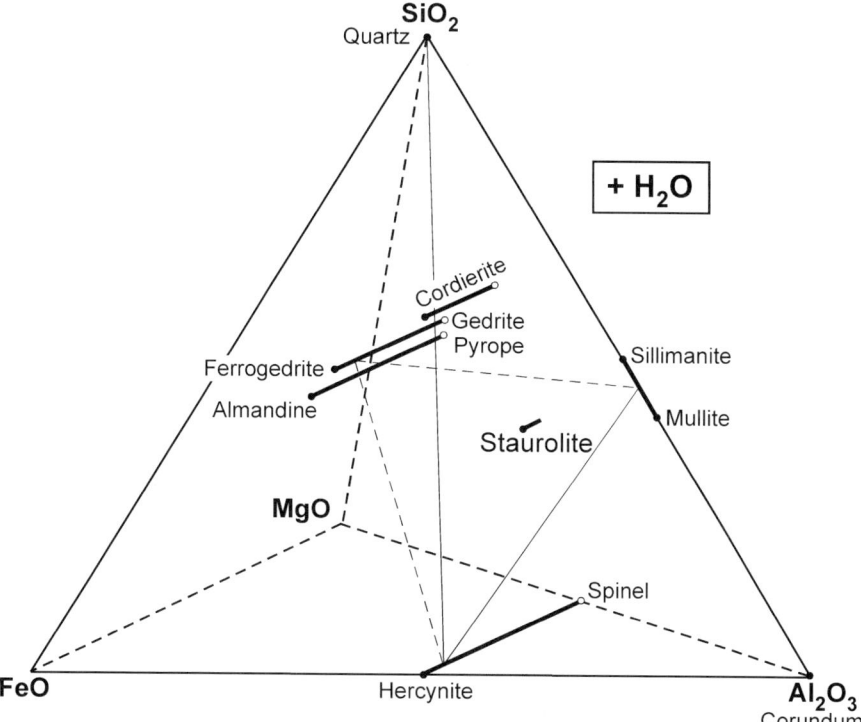

Fig. 7.27. Staurolite breakdown products, hercynite-gedrite-AS and hercynite-AS-Qz and other possible breakdown phases (cordierite and garnet), plotted in terms of mol% (Fe,Mg)O – Al_2O_3 – SiO_2. Note that the Al-silicate reaction product after staurolite in the Wehr xenolith has a composition intermediate between sillimanite and mullite. See text

Chemographic relations in terms of mol% FeO – MgO – SiO_2 – Al_2O_3 projected through H_2O (Fig. 7.27) shows that the melt-absent breakdown of Fe-staurolite in the Wehr xenolith can be represented by two three phase fields of hercynite-gedrite-Al-silicate and hercynite-Al-silicate-quartz representing the reactions

$$9\,H_4Fe_4Al_{18}Si_8O_{48} + 3.75\,O_2 = 21\,FeAl_2O_4 + 3\,Fe_5Al_4Si_6O_2(OH) + 54\,Al_2SiO_5 + 16.5\,H_2O$$
Fe-staurolite hercynite gedrite Al-silicate

$$H_4Fe_4Al_{18}Si_8O_{48} = 4\,FeAl_2O_4 + 5\,Al_2SiO_5 + 3\,SiO_2 + 2\,H_2O$$
Fe-staurolite hercynite Al-silicate quartz

Under equilibrium conditions, the high temperature stability of staurolite in the presence of quartz is controlled by the reactions

staurolite + quartz = cordierite + sillimanite

staurolite + quartz = almandine + sillimanite + quartz + H_2O

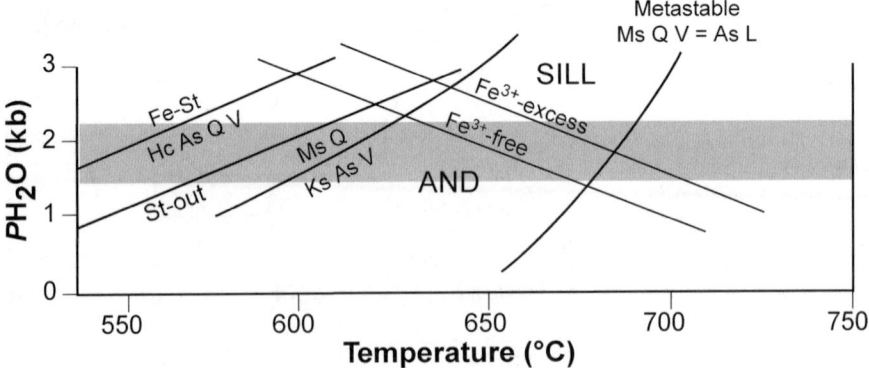

Fig. 7.28. *T-P* diagram showing calculated stabilities of staurolite compositions with respect to stable and metastable Ms + Qz reactions and Fe^{3+}-free, Fe^{3+}-excess curves for the andalusite-sillimanite reaction from Holdaway (1971). *Horizontal grey strip* = pressure estimate of pyrometamorphic reaction. See text. Fe-St = Hc AS Qz V reaction assuming a C-O-H, H_2O = 2, graphite-saturated fluid after Cesare (1994). The St-out curve is derived from a computed assemblage stability diagram of the Wehr staurolite composition $H_4(Fe_{3.06}Mn_{0.11}Mg_{0.91})_{4.08}Al_{18.64}Si_{7.89}O_{48}$ using the Domino/Theriac software of de Capitani and Brown (1987)

(e.g. Richardson 1968) and in the absence of quartz, by the reactions

staurolite = hercynite + Al_2SiO_5 + quartz + H_2O

staurolite = hercynite + corundum + quartz + H_2O

staurolite = hercynite + almandine + sillimanite + H_2O

(e.g. Loomis 1972; Atkin 1978; Cesare 1994). In addition to almandine or cordierite, an Al-member of the anthophyllite-gedrite series is also a possible breakdown product of staurolite (Richardson 1968; Grieve and Fawcett 1974). In the case of the Wehr staurolite, aluminous Fe-gedrite may have formed instead of almandine + H_2O because of the presence of Mg (~2.2 wt.% MgO) in the reactant staurolite which is preferentially partitioned into orthoamphibole rather than garnet (Greive and Fawcett 1974).

T-P conditions of the staurolite breakdown reaction in the Wehr xenolith can be evaluated from Fig. 7.28 where the calculated upper stability of a partly unreacted staurolite composition (Fig. 7.24) would have been overstepped by ~110–120 °C to intersect the disequilibrium breakdown curve of Ms Qz V = AS L within the pressure range of 1.5–2.2 kb for pyrometamorphism (Grapes 1986) at a minimum temperature of ~700 °C.

7.11
Cordierite

Because cordierite is a diagnostic mineral of pyrometamorphic rocks, particularly in buchites where it has crystallised from a melt (Chapter 3), it is not generally observed to have undergone a breakdown reaction involving melting. A rare example of the pyrometamorphic breakdown of cordierite associated with muscovite in a mica rich-

Fig. 7.29.
Drawing from a BSE image showing biotite (*Bt*), spinel (*black*), sillimanite (elongate crystals) and glass (*Gl*) after cordierite (*Cd*) and muscovite (*Ms*) in the marginal part of a mica schist xenolith, Wehr volcano, Eifel, Germany. Small spinels accompanied by suspected mullite in glass occurs along cleavage planes of the muscovite. *Dark grey areas* are holes in glass resulting from H_2O loss during eruption of the xenolith

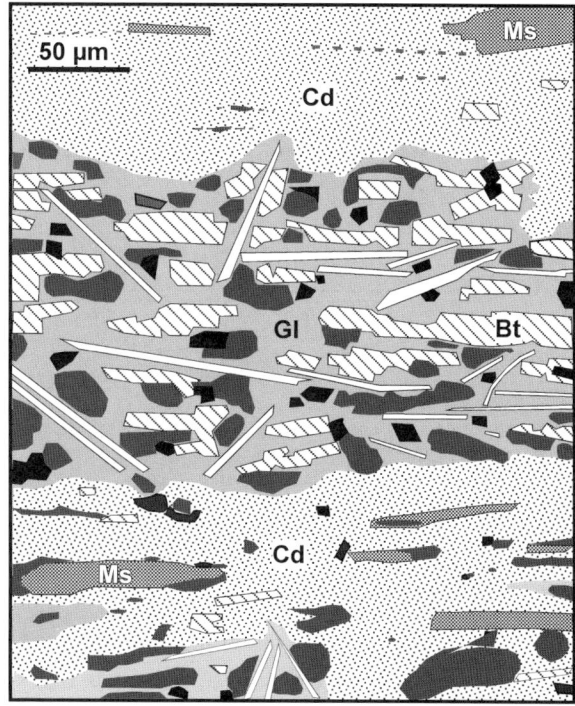

xenolith from the Eifel is described by Grapes (2003). Near the edge of the xenolith, cordierite (XMg ~ 0.72) intergrown with phengitic muscovite has been corroded by a peraluminous glass that contains newly-formed crystals of sillimanite (mol% SiO_2 = 50.8; 1–2% Fe_2O_3), biotite (XMg ~ 0.66) and spinel (XMg = 0.35; $Hc_{56.3}$ $Sp_{37.1}$ $Mt_{3.4}$) (Fig. 7.29) indicating the melting reaction

cordierite + muscovite = spinel + biotite + sillimanite + peraluminous melt

Muscovite grains within the cordierite also show evidence of initial stage reaction to mullite, spinel and glass along grain boundaries and cleavage planes. Quartz–oligoclase (An_{18}) contacts in the xenolith are typically lined with melt. Comparison of Mg/Fe ratios of spinel (0.35) and biotite (1.95) after cordierite with those derived from experimental disequilibrium melting of biotite extrapolated from Brearley (1987a) suggests that the outer part of the xenolith where cordierite + muscovite have reacted could have attained a temperature of ~770 °C.

Except for the formation of hercynitic spinel, the cordierite breakdown reaction is analogous to the reaction

cordierite + muscovite = biotite + Al-silicate + quartz + H_2O

that has a very shallow negative slope in *T-P* space (Fig. 7.30). The inferred ~770 °C temperature reached in the xenolith implies that the above reaction must have been

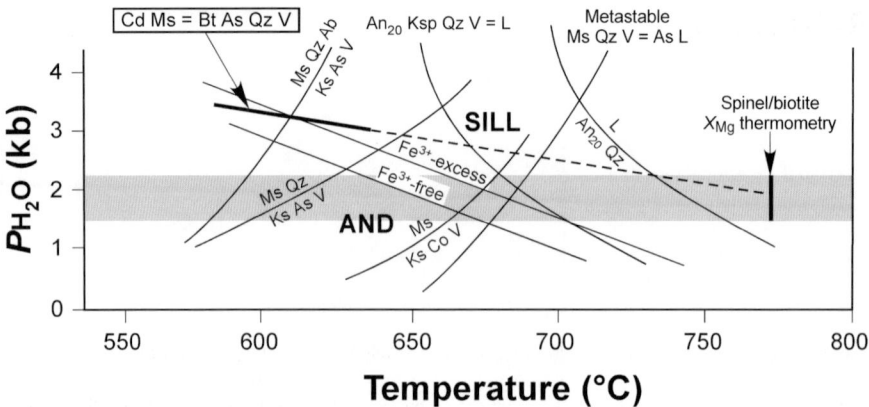

Fig. 7.30. T-P diagram showing mineral reaction curves relevant to the high temperature breakdown of cordierite + muscovite (after Fig. 4 of Grapes 2003). *Shaded strip* = inferred pressure range of pyrometamorphism. *Dashed line* = metastable extension of the Cs Ms = Bt AS Qz V reaction determined by Pattison et al. (2002) with $XMg_{biotite}$ = 0.7. Fe^{3+}-free and Fe^{3+}-excess curves for andalusite-sillimanite reaction from Holdaway (1971). Reaction curves An_{20} Ks Qz V = L and An_{20} Qz V = L after Johannes and Holz (1996). Metastable Ms Qz V = AS L curve after Rubie and Brearley (1987). See text

significantly overstepped by ~180 °C within the pressure range of 1.5–2.2 kb (Fig. 7.30) determined for the depth range of the xenolith sampling zone (5–8 km) by the Wehr magma (Fig. 3.24). The generation of melt rather than quartz as predicted by the equilibrium reaction indicates that the observed cordierite + muscovite breakdown reaction in the xenolith is metastable, with a melt, the kinetically most favourable phase, being formed instead of those predicted from thermodynamic considerations, i.e. the reaction follows Ostwald's step rule.

The temperature overstepping suggested for the breakdown of cordierite + muscovite probably occurred with fragmentation and incorporation of the cordierite-bearing schist into the Wehr trachyte magma causing a further, and presumably rapid increase in temperature. Cordierite intergrown with muscovite in the central part of the same xenolith implies that its breakdown in the marginal part of the xenolith was probably triggered by diffusion of H_2O from the magma into the xenolith along schistosity planes. Water in excess of that available from muscovite and cordierite dehydration would favour rapid metastable melting reactions rather than the simultaneous nucleation of the predicted stable assemblage biotite + sillimanite + quartz. On the scale of tens of microns, it is evident from Fig. 7.25 that the extent of reaction in the cordierite-muscovite intergrowths is highly variable, exhibiting a kind of textural disequilibrium or heterogeneity in that some parts show complete reaction to biotite, Al-silicate, spinel and melt while other parts show only the initial stages of reaction to melt or melt + spinel. Thus, areas of extensive reaction may correspond to avenues of H_2O infiltration in contrast to those showing little reaction that represent areas undergoing vapour-absent melting at the same temperature, i.e. the rates and mechanisms of cordierite + muscovite breakdown are greatly affected by the presence of excess H_2O.

References

Abraham K, Gebert W, Medenbach O, Schreyer W, Hentschel G (1983) Eifelite, $KNa_3Mg_4Si_{12}O_{30}$, a new mineral of the osumilite group with octahedral sodium. Contrib Mineral Petrol 82:252–258

Ackermann PB (1983) Vitrification of Cave Sandstone by Karroo Dolerite in the Sterkspruit Valley, Barkly East. Trans Geol Soc S Africa 8:19–35

Ackermann PB, Walker F (1959) Vitrification of arkose by Karroo Dolerite near Heilbron, Orange Free State. Q J Geol Soc Lond 116:239–254

Addison CC, Addison WE, Neal GH, Sharp JH (1962) Amphiboles, part 1: the oxidation of crocidolite. J Chem Soc 1962:1468–1471

Agee CB (1998) Phase transformation and seismic structure in the upper mantle and transition zone. In: Hemley RJ (ed) Ultrahigh-pressure mineralogy: physics and chemistry of the Earth's interior. Reviews in Mineralogy 36, Mineralogical Society of America, pp 165–204

Agrell SO (1965) Polythermal metamorphism of limestones at Kilchoan, Ardnamurchan. Mineral Mag Tilley vol, pp 1–15

Agrell SO, Langley JM (1958) The dolerite plug at Tievbulliagh, near Cushendall, Co. Antrim. Proc R Irish Acad 59:93–127

Allen JA (1874) Metamorphism produced by the burning of lignite beds in Dakota and Montana territories. Boston Soc Nat Hist Proc 16:246–262

Almond DC (1964) Metamorphism of Tertiary lavas in Straithlaird, Skye. Trans R Soc Edinb 45:13–435

Al-Rawi Y, Carmichael ISE (1967) A note on the natural fusion of granite. Am Mineral 52:1806–1814

Anderson O (1915) The system anorthite-forsterite-silica Am J Sci 39:407–454

Anovitz LM, Kalin RM, Ruiz J (1991) High temperature contact metamorphism: Marble Canyon, West Texas. Geol Soc Am Abs Progr 23

Arai S (1975) Contact metamorphosed dunite-harzburgite complex in the Chugoku District, western Japan. Contrib Mineral Petrol 52:1–16

Arnold R, Anderson R (1907) Metamorphism by combustion of the hydrocarbons in the oil-bearing shale of California. J Geol 15:750–758

Atkin BP (1978) Hercynite as a breakdown product of staurolite from within the aureole of the Andara pluton, Co. Donegal, Eire. Mineral Mag 42:237–240

Avnimelech M (1964) Remarks of the occurrence of unusual high-temperature minerals in the so-called "Mottled Zone" complex of Israel. Isr J Earth Sci 13:102–110

Baker G (1953) Naturally fused coal ash from Leigh Creek, South Australia. Trans R Soc Australia 76:1–20

Baker CK, Black PM (1980) Assimilation and metamorphism at a basalt-limestone contact, Tokatoka, New Zealand. Mineral Mag 43:797–807

Barlow SG, Manning DAC, Hill PI (2000) The influence of time and temperature on the reactions and transformations of clinochlore as a ceramic clay mineral. Int Ceram 2:5–10

Barton MD, Staude J-M, Snow EA Johnson DA (1991) Aureole systematics. In: Kerrick DM (ed) Contact metamorphism. Reviews in Mineralogy 26, Mineralogical Society of America, pp 723–847

Basi MA, Jassim SZ (1974) Baked and fused Miocene sediments from the Injana area, Hemrin South, Iraq. J Geo Soc Iraq 7:1–14

Bastin ES (1905) Note on baked clays and natural slags in eastern Wyoming. J Geol 13:408–412
Beard JS (1990) Partial melting of metabasites in the contact aureoles of gabbroic plutons in the Smartville Complex, Sierra Nevada, California. In: Anderson JL (ed) The nature and origin of Cordilleran magmatism. Geol Soc Am Mem 174:303–313
Beard TFW, Lofgren GE (1991) Dehydration melting and water-saturated melting of basaltic and andesitic greenstones and amphibolites at 1, 3 and 6.9 kb. J Petrol 32:365–401
Belakvski D (1990) The minerals of the burning coal seams at Ravat, Tadshikistan. Lapis 15, 21–26
Belikov BP (1933) Composition of some burned rocks from the Kuzbass. Tr Petrog Inst Akad Nauk SSSR 4:91–100 (in Russian)
Bentor YK, Kastner M (1976) Combustion metamorphism in southern California. Science 193:486–488
Bentor YK, Kastner M, Perlman I, Yelin Y (1981) Combustion metamorphism of bituminous sediments and the formation of melts of granitic and sedimentary composition. Geochim Cosmochim Acta 45: 2229–2255
Bergen MJ van, Barton M (1984) Complex interaction of aluminous sedimentary xenoliths and siliceous magma: an example from Mt. Amiata (Central Italy). Contrib Mineral Petrol 86:374–385
Black PM (1969) Rankinite and kilchoanite from Tokatoka, New Zealand. Mineral Mag 37:513–519
Bowen NL (1940) Progressive metamorphism of siliceous limestone and dolomite. J Geol 48:225–274
Bowen, NL, Aurousseau M (1923) Fusion of sedimentary rocks in drill-holes. Bull Geol Soc Am 34:431–448
Bowen NL, Schairer JF (1935) The system $MgO-FeO-SiO_2$. Am J Sci 29:151–217
Bowen NL, Schairer JF, van Willems HV (1930) The ternary system $Na_2SiO_3-Fe_2O_3-SiO_2$. Am J Sci 20: 405–455
Bowen NL, Schairer JF, Poshjak E (1933) The system $CaO-FeO-SiO_2$. Am J Sci 26:193–284
Brady LF, Gregg JW (1939) Note on the temperature attained in a burning coal seam. Am J Sci 237:116–119
Brauns R (1912a) Die kristallinen Schiefer des Laacher Seegebeits und ihre Umbildung zu Sanidinit. E. Schweizerbarth'sche Verlagsbuchhandlung, Stuttgart
Brauns R (1912b) Die chemische Zusammensetzung granatführender kristalliner Schiefer, Cordieritgesteine und Sanidinite aus dem Laacher Seegebiet. N Jb Mineral, Geol Pal 34:85–175
Brauns R (1922) Die Mineralien der Niederrheinischen Vulkangebiete. Schweizerbart, Stuttgart
Brearley AJ (1986) An electron microprobe study of muscovite breakdown in pelitic xenoliths during pyrometamorphism. Mineral Mag 357:385–397
Brearley AJ (1987a) A natural example of the disequilibrium breakdown of biotite at high temperature: TEM observations and comparison with experimental kinetic data. Mineral Mag 359:93–106
Brearley AJ (1987b) An experimental and kinetic study of the breakdown of aluminous biotite at 800 °C: reaction microstructures and mineral chemistry. Bull Mineral 110:513–532
Brearley AJ, Rubie DC (1990) Effects of H_2O on the disequilibrium breakdown of muscovite + quartz. J Petrol 31:925–956
Bridge TE (1966) Bredigite, larnite and γ-dicalcium silicates from Marble Canyon. Am Mineral 51: 1766–1774
Bridges JC, Grady MM (1998) Melted sediment from Mars in Nakhla. Lunar Planet Sci 29:1399–1400
Brindley GW, Maroney DM (1960) High temperature reactions of clay mineral mixtures and their ceramic properties, II. J Am Ceram Soc 43:511–516
Brindley GW, Nakahira M (1959a) The kaolinite-mullite reaction series, I: a survey of outstanding problems. J Am Ceram Soc 43:311–314
Brindley GW, Nakahira M (1959b) The kaolinite-mullite reaction series, II: metakaolin. J Am Ceram Soc 43:314–318
Brindley GW, Nakahira M (1959c) The kaolinite-mullite reaction series, III: the high-temperature phases. J Am Ceram Soc 43:319–324
Brown WL, Parsons I (1981) Towards a more practical two-feldspar geothermometer. Contrib Mineral Petrol 76:369–396
Bücking H (1900) Cordierit von Nord-Celebes und aus den sog. verglasten Sandsteinen Mitteldeutschlands. Ber Senkenb Naturf Ges Frankfurt, pp 3–20
Buist DS (1961) The composite sill of Rudh'a'Chromain, Carsaig, Mull. Geol Mag 98:67–76
Bulatov VK (1974) Experimental studies of mineral equilibria in the high-temperature part of the system $CaO-MgO-SiO_2-CO_2-H_2O$. Geochemia 8:1268–1271 (in Russian)

Burg A, Starinsky A, Bartov Y, Kolodny Y (1992) Geology of the Hatrurim Formation ("Mottled Zone") in the Hatrurim basin. Isr J Earth Sci 40:107–124

Burnham CW (1959) Contact metamorphism of magnesian limestones at Crestmore, California. Geol Soc Am Bull 70:879–920

Burnham CW (1979) Magmas and hydrothermal fluids. In: Barnes HL (ed) Geochemistry of hydrothermal ore deposits, 2nd edn. Wiley Interscience, New York, pp 71–136

Burnham CW, Nekvasil H (1986) Equilibrium properties of granite pegmatite magmas. Am Mineral 71:239–263

Bustin RM, Mathews WH (1982) In situ gasification of coal, a natural example: history, petrology, and mechanics of combustion. Can J Earth Sci 19:514–523

Butler BCM (1961) Metamorphism and metasomatism of rocks in the Moine Series by a dolerite plug at Glenmore, Ardnamurchan. Mineral Mag 32:866–897

Butler BCM (1977) Al-rich pyroxene and melilite in a blast-furnace slag and a comparison with the Allende meteorite. Mineral Mag 41:493–499

Cameron WE (1976a) Coexisting sillimanite and mullite. Geol Mag 6:497–514

Cameron WE (1976b) A mineral phase intermediate in composition between sillimanite and mullite. Am Mineral 61:1025–1026

Cameron WE (1977) Mullite: a substituted alumina. Am Mineral 62:747–755

Cann JR (1965) The metamorphism of amygdales at S' Airde Beinn, northern Mull. Mineral Mag 34:92–106

Capitanio F, Larocca F, Improta S (2004) High temperature rapid pyrometamorphism induced by a charcoal pit burning: the case of Ricetto, central Italy. Int J Earth Sci (Geol Rundsch) 93:107–118

Cawhorn RG, Walraven F (1998) Emplacement and crystallization time for the Bushveld Complex. J Petrol 39:1669–1687

Cesare B (1994) Hercynite as the product of staurolite decompression in the contact aureole of Vedrette di Ries, eastern Alps, Italy. Contrib Mineral Petrol 116:239–246

Chatterjee NN, Ray S (1946) On the burnt coal outcrop from the central Kujama coalfield, Jharia. Geol Min Metall Soc India 18:133–135

Chesnokov BV, Tsherbakova EP (1991) The mineralogy of burned coal heaps in the Chelyabinsk coal basin. Publ H Nauka, Moscow (in Russian)

Chernosky JV Jr (1974) The upper stability of clinochlore at low pressure and the free energy of formation of Mg-cordierite. Am Mineral 59:496–507

Chinner GA, Cornell DH (1974) Evidence of kimberlite-grospydite reaction. Contrib Mineral Petrol 45:153–160

Chinner GA, Dixon PD (1973) Irish osumilite. Mineral Mag 39:189–192

Chinner GA, Schairer JF (1962) The join $Ca_3Al_2Si_3O_{12}$ – $Mg_3Al_2Si_3O_{12}$ and its bearing on the system CaO-MgO-Al_2O_3-SiO_2 at atmospheric pressure. Am J Sci 260:611–634

Cho M, Fawcett JJ (1986) A kinetic study of clinochlore and its high temperature equivalent forsterite-cordierite-spinel at 2 kbar water pressure. Am Mineral 71:68–77

Church BN, Matheson A, Hora ZD (1979) Combustion metamorphism in the Hat Creek area, British Columbia. Can J Earth Sci 16:1882–1887

Cisowski SM, Fuller M (1987) The generation of magnetic anomalies by combustion metamorphism of sedimentary rock, and its significance to hydrocarbon exploration. Geol Soc Am Bull 99:21–29

Clark BH, Peacor DR (1987) Pyrometamorphism and partial melting of shales during combustion metamorphism: mineralogical, textural, and chemical effects. Contrib Mineral Petrol 112:558–568

Clemens JD, Wall VJ (1981) Origin and crystallization of some peraluminous (S-type) granite magmas. Can Mineral 19:111–131

Clocchiatti R (1990) Les fulgurites et roches vitrifiées de l'Etna. Eur J Mineral 2:479–494

Coates DA, Naeser CW (1984) Map showing fission-track ages of clinker in the Rochelle Hills, southern Campbell County, Wyoming. US Geol Surv Misc Inv Map I-1462

Cole, D (1974) A recent example of spontaneous combustion of oil-shale. Geol Mag 111:355–356

Cole WF, Segnit ER (1963) High-temperature phases developed in some kaolinite-mica-quartz clays. Trans Brit Ceram Soc 62:375–395

Comer JJ (1960) Electron microscope studies of mullite derivement in fired kaolinites. J Am Ceram Soc 43:375–384

Comer JJ (1961) New electron-optical data on the kaolinite-mullite transformation. J Am Ceram Soc 44:561–563

Connolly JA, Holness MB, Rubie DC, Rushmer T (1997) Reaction-induced micro-cracking: an experimental investigation of a mechanism for enhancing anatectic melt extraction. Geology 25:591–594

Cosca M, Peacor D (1987) Chemistry and structure of esseneite, (CaFe^{3+}AlSiO$_6$), a new pyroxene produced by pyrometamorphism. Am Mineral 72:148–156

Cosca M, Rouse RR, Essene EJ (1988) Dorrite [Ca$_2$(Mg,Fe$^{3+}_4$)(Al$_2$Si$_2$)O$_{20}$], a new member of the aenigmatite group from a pyrometamorphic melt-rock. Am Mineral 73:1440–1448

Cosca M, Essene EJ, Geissman JW, Simmons WB, Coates DA (1989) Pyrometamorphic rocks associated with naturally burned coal beds, Powder River Basin, Wyoming. Am Mineral 74:85–100

Crickmay CH (1967) A note on the term boccane. Am J Sci 265:626–627

Cultrone G, Rodriguez-Navarro C, Sebastian E, Cazalla O, De La Torre MJ (2001) Carbonate and silicate reactions during ceramic firing. Eur J Mineral 13:621–634

Damon RF (1884) Geology of Weymouth, Portland and the coast of Dorsetshire. Edward Stanford, London

Davies GR, Tommasini S (2000) Isotopic disequilibrium during rapid crustal anatexis: implications for petrogenetic studies of magmatic processes. Chem Geol 162:169–191

Davis GL, Tuttle OF (1952) Two new crystalline phases of the anorthite composition, CaO · Al$_2$O$_3$ · 2 SiO$_2$. Am J Sci 250:107

Dawson J (1951) The Brockley dolerite plug and the Church Bay volcanic vent, Rathlin island, Co. Antrim. Irish Nat J 10:156–162

de Capitani C, Brown TH (1987) The computation of chemical equilibrium in complex systems containing non-ideal solutions. Geochim Cosmochim Acta 51:2639–2652 (updated version Theriak-Domino v. 140205. Inst. Mineral. Petrol. Univ. Basel Switzerland)

Delaney PT, Pollard DD (1982) Solidification of a basaltic magma during flow in a dyke. Am J Sci 282:854–885

Devine JD, Sigurdsson H (1980) Garnet-fassaite calc-silicate nodule from La Soufrière, St. Vincent. Am Mineral 65:302–305

Diaz-Martínez E, Ormö J (2003) An alternative hypothesis for the origin of ferroan ringwoodite in the pumice of El Gasco (Cáceres, Spain). Lunar Planet Sci 34:1318

Dondi M, Ercolani G, Farbri B, Marsigli M (1998) An approach to the chemistry of pyroxenes formed during the firing of Ca-rich silicate ceramics. Clay Mineral 33:443–452

Ermankov NP (1935) Pasrud-Yagnob coal deposit and burning mines of Kan-Tag mountain. In: Geology of Tadjikistan coal deposits. PH of Acad Sci USSR, Moscow, pp 47–66 (in Russian)

Eskola P (1920) The mineral facies of rocks. Norsk Tidsskr 6:143–194

Eskola P (1939) Die metamorphosen Gesteine. In: Barth TFW, Correns CW, Eskola P (eds) Die Entstehung der Gesteine. Springer, Berlin, pp 263–407

Essene, EJ (1980) The stability of bredigite and other Ca-Mg silicates. J Am Ceram Soc 63:464–466

Essene EJ, Fisher DC (1986) Lightning strike fusion: extreme reduction and metal-silicate liquid immiscibility. Science 234:189–193

Eyles VA (1952) The composition and origin of the Antrim laterites and bauxites. Mem Geol Surv North Ireland

Faure F, Trolliard G, Montel J-M, Nicollet C (2001) Nano-petrographic investigation of a mafic xenolith (maar de Beaunit, Massif Central, France). Eur J Mineral 13:27–40

Fawcett JJ, Yoder HS (1966) Phase relations of chlorites in the system MgO-Al$_2$O$_3$-SiO$_2$-H$_2$O. Am Mineral 51:353–380

Fermor LL (1914) Geology and coal resources of Korea State, C.P. Mem Geol Surv India 61:158–160

Fermor LL (1918) Preliminary note on the burning of coal seams at the outcrop. Trans Min Geol Metall Inst India 12:50–63

Fermor LL (1924) Discussion on Tilley's paper. Q J Geol Soc Lond 80:70–71

Ferry JM, Mutti LJ, Zuccala J (1987) Contact metamorphism/hydrothermal alteration of Tertiary basalts from the Isle of Skye, northwest Scotland. Contrib Mineral Petrol 95:166–181

Flett JS (1911) In: Cunningham Craig EH and others. Geology of Colonsay and Oronsay, with part of the Ross of Mull. Mem Geol Surv Scotland

Foit FF, Hooper RL, Rosenberg PE (1987) Unusual pyroxene, melilite, and iron oxide mineral assemblage in a coal-fire buchite from Buffalo, Wyoming. Am Mineral 72:137–147

Frechen J (1947) Vorgänge der Sanidinit-Bildung im Laacher See-Gebiet. Fortschr Mineral 26:147–166

Frenzel G, Stähle V (1982) Fulgurite glass on peridotite from the Frankenstein near Darmstadt. Chem Erde 41:111–119

Frenzel G, Stähle V (1984) On aluminosilicate glass with inclusions of lechatelierite from a fulgurite tube on the Hahnenstock Mt. (Glarner Freiburg, Switzerland). Chem Erde 43:17–26

Frenzel G, Irouschek-Zumthor A, Stähle V (1989) Shock wave compression, melting and vaporization by the formation of fulgurites on exposed summits. Chem Erde 49:265–286

Frost BR (1975) Contact metamorphism of serpentinite, chlorite blackwall and rodingite at Paddy-Go-Easy Pass, Central Cascades, Washington. J Petrol 16:272–313

Fujii T (1976) Solubility of Al_2O_3 in enstatite coexisting with forsterite and spinel. Carnegie Inst. Washington Ann Rpt Div Geophysics Lab 1975–1976:566–571

Fyfe AS, Turner FJ, Verhoogen J (1959) Metamorphic reactions and metamorphic facies. Geol Soc Am Mem 7

Fyfe AS, Price NJ, Thompson AB (1978) Fluids in the Earth's crust – their significance in metamorphic, tectonic and chemical transport. Elsevier, Amsterdam

Gaines GL Jr, Vedder W (1964) Dehydroxylation of muscovite. Nature 201:495

Ghose H, Weidner JR (1971) Oriented transformation of grunerite to clinoferrosilite at 775 °C and 500 bars argon pressure. Contrib Mineral Petrol 30:64–71

Gifford AC (1999) Clay soil fulgurites in the Eastern Goldfields of Western Australia. J Roy Soc Western Australia 82:165–168

Glen DC (1873) Notes from the Isle of Bute, I: On a tract of columnar sandstone, and a perched boulder near Kilchattan. Trans Geol Soc Glasg 5:154–158

Graham IJ, Grapes RH, Kifle K (1988) Buchitic metagreywacke xenoliths from Mount Ngauruhoe, Taupo Volcanic Zone, New Zealand. J Volc Geotherm Res 35:205–216

Grant JA (1986) The isocon diagram – a simple solution to Gresen's equation for metasomatic alteration. Econ Geol 81:1976–1982

Grapes RH (1986) Melting and thermal reconstitution of pelitic xenoliths, Wehr volcano, East Eifel, Germany. J Petrol 27:343–396

Grapes RH (1991) Aluminous alkali feldspar-bearing xenoliths and the origin of sanidinite, East Eifel, Germany. N Jb Mineral Ab 3:129–144

Grapes RH (2003) Pyrometamorphic breakdown of cordierite-muscovite intergrowths. Mineral Mag 67:653–663

Greenwood HJ (1967) Mineral equilibria in the system MgO-SiO_2-H_2O-CO_2. In: Abelson PH (ed) Researches in geochemistry 2. John Wyllie, New York, pp 542–567

Gresley WS (1883) A glossary of terms used in coal mining. London

Grieve RAF, Fawcett JJ (1974) The stability of chloritoid below 10 kb P_{H_2O}. J Petrol 16:113–139

Gross S (1977) The mineralogy of the Hatrurim Formation, Israel. Geol Surv Isr Bull 70

Gross S, Mazor E, Sass E, Zak I (1967) The Mottled Zone of Nahal Ayalon (central Israel). Isr J Earth Sci 16:84–96

Guggenheim S, Chang Y-H, Koster van Groos AF (1987) Muscovite dehydroxylation: high temperature studies. Am Mineral 72:537–550

Guppy EM, Hawkes L (1925) A composite dyke from eastern Iceland. Q J Geol Soc Lond 81:325–343

Gur D, Steinitz G, Kolodny Y, Starinsky A, McWilliams M (1995) $^{40}Ar/^{39}Ar$ dating of combustion metamorphism ("Mottled Zone", Israel). Chem Geol 122:171–184

Gustafson WI (1974) The stability of andradite, hedenbergite and related minerals in the system Ca-Fe-Si-O-H. J Petrol 15:455–496

Haggerty SE (1967) Opaque oxides in terrestrial igneous rocks. In: Rumble D III (ed) Oxide minerals. Reviews in Mineralogy 3, Mineralogical Society of America, pp 101–300

Haggerty SE, Lindsley DH (1970) Stability of the pseudobrookite (Fe_2TiO_5)-ferropseudobrookite ($FeTi_2O_5$) series. Carnegie Inst Washington Year Book 68:247–249

Harker A (1904) The Tertiary igneous rocks of Skye. Mem Geol Surv Scotland
Harker RJ (1959) The synthesis and stability of tilleyite $Ca_5Si_2O_7(CO_3)_2$. Am J Sci 257:656–667
Harley SL (1989) The origin of granulites: a metamorphic perspective. Geol Mag 126:215–247
Hayden HH (1918) General report of the director for 1918. Geol Surv India Records 50:8
Heffern EL, Coates A, Naeser CW (1993) Distribution and age of clinker in Northern Powder River Basin, Montana. Am Assoc Petrol Geol 67:1342 (Abs)
Hensen BJ, Gray DR (1979) Clinohypersthene and hypersthene from a coal fire buchite near Ravensworth, N.S.W., Australia. Am Mineral 64:131–135
Hentschel G (1964) Mayanite, 12 CaO · 7 Al_2O_3, und Brownmillerite, 2 CaO · $(Al,Fe)_2O_3$, zwei neue Minerale in den Kalksteineinschüssen der Lava des Ettringer Bellerberges. N Jb Mineral Mh, pp 22–29
Hentschel G (1977) Neufunde seltener Minerale im Laacher Vulcangebiet. Aufschluss 28:129–133
Hentschel G, Abraham K, Schreyer W (1980) First terrestrial occurrence of roedderite in volcanic ejecta of the Eifel, Germany. Contrib Mineral Petrol 73:127–130
Holdaway MJ (1971) Stability of andalusite and the aluminium silicate phase diagram. Am J Sci 271:97–131
Holgate N (1978) A composite tholeiite dyke at Imachar, Isle of Arran: its petrogenesis and associated pyrometamorphism. Mineral Mag 42:141–142
Holm JL, Kleppa OJ (1966) The thermodynamic properties of the aluminium silicates. Am Mineral 51: 1608–1622
Holness MB (1999) Contact metamorphism and anatexis of Torridonian arkose by minor intrusions of the Rum Igneous Complex, Inner Hebrides, Scotland. Geol Mag 136:527–542
Holness MB, Isherwood CE (2003) The aureole of the Rum Tertiary igneous complex, Scotland. J Geol Soc Lond 160:15–27
Holness MB, Watt GR (2002) The aureole of the Traigh Bhàn na Sgùrra Sill, Isle of Mull: reaction-driven microcracking during pyrometamorphism. J Petrol 43:511–534
Holness MB, Dane K, Sides R, Richardson C, Caddick M (2005) Melting and melt segregation in the aureole of the Glenmore plug, Ardnamurchan. J Metamorph Geol 23:29–43
Hoover JD (1977) Melting relations of a new chilled margin sample from the Skaergaard intrusion. Carnegie Inst Washington Geophys Lab Rpt, pp 739–743
Hsu LC (1968) Selected phase relationships in the system Al-Mg-Fe-Si-O-H: a model for garnet equilibria. J Petrol 71:40–83
Huckenholz HG, Lindhuber W, Springer J (1974) The join $CaSiO_3$-Al_2O_3-Fe_2O_3 of the CaO-Al_2O_3-Fe_2O_3-SiO_2 quaternary system and its bearing on the formation of granditic garnets and fassaitic pyroxenes. N Jb Mineral Abh 121:160–207
Huckenholz HG, Hölzl E, Lindhuber W (1975) Grossularite, its solidus and liquidus relations in the CaO-Al_2O_3-SiO_2-H_2O system up to 10 kbar. N Jb Mineral Abh 124:1–46
Huppert HE, Sparks RSJ (1985) Cooling and contamination of mafic and ultramafic magmas during ascent through continental crust. Earth Planet Sci Lett 74:371–386
Hussak (1883) Über den Cordierit in vulkanischen Auswürflingen. Sitz Akad Wiss Wein B 87
Ito K, Kennedy GC (1967) Melting and phase relations in a natural peridotite to 40 kilobars. Amer J Sci 265:519–538
Jaeger JC (1964) Thermal effects of intrusions. Rev Geophys 2:443–465
Jaeger JC (1968) Cooling and solidification of igneous rocks. In: Hess HH, Poldervaart A (eds) Basalts – the Poldervaart treatise on rocks of basaltic composition 2. Interscience Publishers, New York, pp 503–536
James, RS, Turnock AC, Fawcett JJ (1976) The stability and phase relations of iron chlorite below 8.5 kb PH_2O. Contrib Mineral Petrol 56:1–25
Jamtveit B, Dahlgren S, Austrheim H (1997) High-grade contact metamorphism of calcareous rocks from the Oslo Rift, Southern Norway. Am Mineral 82:1241–1254
Jasmund K, Hentschel J (1964) Seltene Mineralparagenesen in den Kalksteineinschlüssen der Lava des Ettringer Bellerberges bei Mayen (Eifel). Beitr Mineral Petrogr 10:296–314
Jenkins DM, Chernosky JV Jr (1986) Phase equilibria and crystallochemical properties of Mg-chlorite. Am Mineral 71:924–936
Joesten R (1974) Local equilibrium and metasomatic growth of zoned calc-silicate nodules from a contact aureole, Christmas Mountains, Big Bend region, Texas. Am J Sci 274:876–901

Joesten R (1976) High-temperature contact metamorphism of carbonate rocks in a shallow crustal environment, Christmas Mountains, Big Bend region, Texas. Am Mineral 61:776–781

Johannes W (1984) Beginning of melting in the granite system Qz-Or-Ab-An-H_2O. Contrib Mineral Petrol 86:264–273

Johannes W (1989) Melting of plagioclase-quartz assemblages at 2 kbar water pressure. Contrib Mineral Petrol 103:270–276

Johannes W, Holz F (1996) Petrogenesis and experimental petrology of granitic rocks. Springer

Johannes W, Koepke J, Behrens H (1994) Partial melting of plagioclases and plagioclase-bearing systems. In: Parsons I (ed) Feldspars and their reactions. Kluwer Academic Publications, pp 161–194

Jones AH, Geissman JW, Coates DA (1984) Clinker deposits, Powder River Basin, Wyoming and Montana: A new source of high-fidelity paleomagnetic data from the Quaternary. Geophys Res Lett 11:1231–1234

Jugovics L (1933) Einschlüsse von Basaltjaspis in dem Basalte des Sag-Berges (Ugarin). Min Petrol Mitt 44:68–82

Kaczor SM, Hanson GN, Peterman ZE (1988) Disequilibrium melting of granite at the contact of a basic plug: a geochemical and petrological study. J Geol 96:61–78

Kalb G (1935) Beiträge zur Kenntnis der Auswürflinge, im besonderen der Sanidinite des Laacher Seegebietes. Min Petrogr Mitt 46:20–55

Kalb G (1936) Beiträge zur Kenntnis der Auswürflinge des Laacher Seegebietes, II: Zwei Arten von Umbildungen kristalliner Schiefer zu Sanidiniten. Mineral Petrogr Mitt 47:185–210

Kalugin IA, Tret'yakov GA, Bobrov VA (1991). Iron-ore basalts in fused rocks of Eastern Kazakhstan. Novosibirsk (in Russian)

Kang S, Yoo H (2000) Phase stability of the system Mg-Fe-O. J Solid State Chem 149:33–40

Kays MA, McBirney AR, Goles GG (1981) Xenoliths of gneisses and conformable, clot-like granophyres in the marginal Border Group, Skaergaard Intrusion, East Greenland. Contrib Mineral Petrol 45:265–244

Kays MA, Goles GG Grover TW (1989) Precambrian sequence bordering the Skaergaard Intrusion. J Petrol 30:321–361

Kelsey DE, White RW, Powell R (2005). Calculated phase equilibria in K_2O-FeO-MgO-Al_2O_3-SiO_2-H_2O for silica-undersaturated sapphirine-bearing mineral assemblages. J Metamorph Geol 23:217–239

Kennedy GC, Wasserburg FJ, Heard HC, Newton RC (1962) The upper three-phase region in the system SiO_2-H_2O. Am J Sci 260:501–521

Kerrick DM, Spear JA (1988) The role of minor element solid solution on the andalusite-sillimanite equilibrium in metapelites and peraluminous granitoids. Am J Sci 288:152–192

Kifle K (1992) High temperature-low pressure, water-saturated disequilibrium melting experiments of quartzofeldspathic rock compositions. Unpub. PhD thesis, Research School of Earth Sciences, Victoria University of Wellington, New Zealand

Killie IC, Thompson RN, Morrison MA, Thompson RF (1986). Field evidence for turbulence during flow of basalt magma through conduits from southwest Mull. Geol Mag 123:693–697

Kitchen D (1984) Pyrometamorphism and the contamination of basaltic magma at Tieveragh, Co. Antrim. J Geol Soc 141:733–745

Kitchen D (1985) The partial melting of basalt and its enclosed mineral-filled cavities at Scawt Hill, Co. Antrim. Mineral Mag 49:655–662

Knopf A (1938) Partial fusion of granodiorite by intrusive basalt, Owens Valley, California. Am J Sci 36: 373–376

Kolodny Y, Gross S (1974) Thermal metamorphism by combustion of organic matter: isotope and petrological evidence. J Geol 82:489–506

Kolodny Y, Bar M, Sass E (1971) Fission track age of the "Mottled Zone Event" in Israel. Earth Planet Sci Lett 11:269–272

Koritnig S (1955) Die Blaue Kuppe bei Eschwege mit ihren Kontakterscheinungen. Heidelbg Beitr Mineral Petrogr 4:504–521

Kühnel RA, Scarlett B (1987) Criteria for recognition of thermal conditions during the underground gasification: experience from WIDCO field test in Centralia, USA. In: Proc 13th UCG Symp Laramie WY USA, USDOE/METC-88/6095:60–71

Kühnel RA, Schmit CR, Eylands KE, McCarthy GJ (1993) Comparison of the pyrometamorphism of clayey rocks during underground coal gasification and firing of structural ceramics. App Clay Sci 8:129–146

Lacroix A (1893) Les enclaves des roches volcaniques. Protat, Mâcon
Larsen ES, Switzer G (1939) An obsidian-like rock formed by the melting of a granodiorite. Am J Sci 237:562–568
Leake BE, Skirrow G (1960) The pelitic hornfelses of the Cashellough Wheelaun Intrusion, County Galway, Eire. J Geol 68:23–40
Lemberg J (1883) Zur Kenntniss der Bildung und Umwandlung von Silicaten. Zeit Deutsch Geol Ges 35: 557–618
Le Maitre RW (1974) Partial fused granite blocks from Mt. Elephant, Victoria, Australia. J Petrol 15:403–412
Levin EM, McMurdie HF, Hall FP (1956) Phase diagrams for ceramists. Am Ceram Soc, Columbus, Ohio
Lin HC, Foster WR (1966) Stability relations of bredigite ($5\,CaO \cdot MgO \cdot 3\,SiO_2$). J Am Ceram Soc 58:73 (MA 76-1616)
Lindsley DH, Brown GM, Muir ID (1969) Conditions of the ferrowollastonite-ferrohedenbergite inversion in the Skaergaard Intrusion, East Greenland. Spec Pap Mineral Soc Am 2:193–201
Liou JG (1974) Stability relations of andradite+quartz in the system Ca-Fe-Si-O-H. Am Mineral 59: 1016–1025
Longhi J, Bertka CM (1996) Graphical analysis of pigeonite-augite liquidus equilibria. Am Mineral 81: 685–695
Loomis TP (1972) Contact metamorphism of pelitic rock by the Ronda ultramafic intrusion, southern Spain. Geol Soc Am Bull 83:2449–2479
Lovering TS (1938) Temperatures in a sinking xenolith. Trans Am Geophys Union 19th Ann Meeting, pp 274–277
Manning CE, Bird DK (1991) Porosity evolution and fluid flow in the basalts of the Skaergaard magmahydrothermal system, East Greenland. Am J Sci 291:201–257
Manning CE, Ingebritsen SE, Bird DK (1993) Missing mineral zones in contact metamorphosed basalts. Am J Sci 293:894–938
Markl G (2005) Mullite-corundum-spinel-cordierite-plagioclase xenoliths in the Skaergaard Marginal Border group: multi-stage interaction between metasediments and basaltic magma. Contrib Mineral Petrol 149:196–215
Mason B (1957) Larnite, scawtite, and hydrogrossular from Tokotoka, New Zealand. Am Mineral 42: 379–392
Matthews A, Gross S (1980) Petrologic evolution of the "Mottled Zone" (Hatrurim) metamorphic complex of Israel. Isr J Earth Sci 29:93–106
Matthews A, Kolodny Y (1978) Oxygen isotope fractionation and decarbonation metamorphism: the Mottled Zone Event. Earth Planet Sci Lett 39:179–192
Matthews WH, Bustin RM (1984) Why do the Smoking Hills smoke? Can J Earth Sci 21:737–742
McCarthy GJ, Stevenson RJ, Oliver RL (1989) Mineralogy of the residues from an underground coal gasification test. In: Fly ash and coal conversion by-products: characterization, utilization and disposal V. Mat Res Soc Symp Proc 136:113–130
Macdonald GA, Katsura T (1965) Eruption of Lassen Peak, Cascade Range, California, in 1915: example of mixed magmas. Geol Soc Am Bull 76:475–482
McLintock WFP (1932) On the metamorphism produced by the combustion of hydrocarbons in the Tertiary sediments of south-west Persia. Mineral Mag 23:207–227
McOne AW, Fawcett JJ, James RS (1975) The stability of intermediate chlorites of the clinochlore-daphnite series at 2 kbar PH_2O. Am Mineral 60:1047–1062
Mehnert KR (1979) Composition and abundances of common metamorphic rock types. In: Wedepol KH (ed) Handbook of geochemistry 1. Springer, pp 272–296
Melluso L, Conticelli S, D'Antonio M, Mirco NP, Saccani E (2003) Petrology and mineralogy of wollastonite-and melilite-bearing paralavas from the Central Appenines, Italy. Am Mineral 88:1287–1299
Michaud V (1995) Crustal xenoliths in recent hawaiites from Mount Etna, Italy: evidence for alkali exchanges during magma-wall rock interaction. Chem Geol 122:21–42
Milch A (1922) Die Umwandlung der Gesteine. Grundzüge der Geologie I
Miyashiro A (1973) Metamorphism and Metamorphic Belts. Allen & Unwin, London
Miyashiro A, Iiyama T (1954) A preliminary note on a new mineral, indialite, polymorphic with cordierite. Proc Imp Acad Japan 30:746–751

Miyashiro A, Yamasaki M, Miyashiro T (1955) The polymorphism of cordierite and indialite. Am J Sci 253:185–208

Mohl H (1873) Die südwestlichsten Ausläufer des Vogelsberges. Ber Offenbacher Verein Naturkunde 14: 51–101

Mohl H (1874) Mikromineralogische Mitteilungen. N Jb Mineral Geol Pal 687–710:785–804

Moranville-Regourd M (1998) Cements made from blastfurnace slag. In: Hewlett PC (ed) Lea's chemistry of cement and concrete, 4th edn, pp 637–678

Morey GW, Bowen NL (1925) The ternary system sodium metasilicate-calcium metasilicate-silica. J Soc Glass Tech 9:226–64T

Morosevicz J (1898) Experimentelle Untersuchungen über die Bildung der Minerale im Magma. Zsch Min Petrol Mitt 18:1–90

Muan A (1956) Phase equilibria at liquidus temperatures in the system iron oxide-Al_2O_3-SiO_2 in air atmosphere. J Am Ceram Soc 40:121–133

Muan A, Osborn EF (1956) Phase equilibria at liquidus temperatures in the system MgO-FeO-Fe_2O_3-SiO_2. J Am Ceram Soc 39:121–140

Muan A, Hauck J, Löfall T (1972) Equilibrium studies with a bearing on lunar rocks Proc. Lunar Sci Conf 3:185–196

Müller G, Schuster AK, Zippert Y (1988) Spinifex textures and texture zoning in fayalite-rich slags of medieval iron-works near Schieder Village, NW-Germany. N Jb Mineral Mh 3:111–129

Nawaz R (1977) Pyrometamorphic rocks at the contact of a dolerite plug near Bunowen, Co. Galway, Ireland. Irish Nat J 19:101–105

Nelson BW, Roy R (1958) Synthesis of the chlorites and their structural and chemical constitutions. Am Mineral 43:707–725

Nicholls, IA (1971) Calcareous inclusions in lavas and agglomerates of Santorini volcano. Contrib Mineral Petrol 30:261–276

Nose CW (1808). In: Nöggerath J (ed) Mineralogische Studien über die Gebirge am Niederrhein. JC Hermann, Frankfurt/M.

Novikov VP (1993) Derivative organic substances of coal-fire on Fan-Yagnob deposit. Proc Tadjikistan Rep Acad Sci Earth Sci Ser 4:91–58 (in Russian)

Novikov VP, Suprychev VV (1986) Parameters of modern mineral-forming processes associated with underground coal combustion at Fan-Yagnob deposit. Mineral Tadjikistan 7:97–104 (in Russian)

Nzalii T, Duchesnei JC, Jacquemin C, Vander Auweri J (1999) Pyrométamorphisme induit par la gazéification souterraine de niveaux charbonneux du Westphalien dans le bassin de Mons (Belgique). Geol Belgica 2/3-4:221–134

O'Hara MJ (1968) The bearing of phase equilibria studies on the origin and evolution of basic and ultrabasic rocks. Earth Sci Rev 4:69–133

Olesch M, Seifert F (1981) The restricted stability of osumilite under hydrous conditions in the system K_2O-MgO-Al_2O_3-SiO_2-H_2O. Contrib Mineral Petrol 76:362–367

Osborn EF (1942) The system $CaSiO_3$-diopside-anorthite. Am J Sci 240:751–758

Osborn EF, Muan A (1960) The system MgO-CaO-SiO_2. Columbus, Ohio, Am Ceram Soc, Pl 2

Osborn EF, Tait DB (1952) The system diopside-forsterite-anorthite. Am J Sci Bowen vol, pp 413–433

Ostrovsky IA (1966) P-T-diagram of the system SiO_2-H_2O. Geol J 5:127–134

Owens BF (2000) High-temperature contact metamorphism of calc-silicate xenoliths in the Kiglapait Intrusion, Labrador. Am Mineral 85:1595–1605

Pan V, Longhi J (1990) The system Mg_2SiO_4-$CaSiO_4$-$CaAl_2O_4$-$NaAlSiO_4$-SiO_2: one atmosphere liquidus equilibria of analogs of alkaline mafic lavas. Contrib Mineral Petrol 105:569–584

Pan Y, Stauffer MR (2000) Cerium anomaly and Th/U fractionation in the 1.85 Ga Flin Flon Paleosol: clues from REE- and U-rich accessory minerals and implications for paleoatmospheric reconstruction. Am Mineral 85:898–911

Parodi GC, Ventura GD, Lorand J-P (1989) Mineralogy and petrology of an unusual osumilite + vanadium-rich pseudobrookite assemblage from the Vico Volcanic Complex (latinum, Italy). Am Mineral 74:1278–1284

Patterson EM (1955) The Tertiary lava succession on the northern part of the Antrim plateau. Proc R Irish Acad 57:79

Pattison DRM (1992) Stability of andalusite and sillimanite and the Al_2SiO_5 triple point: constraints from the Ballachulish aureole, Scotland. J Geol 100:423–446

Pattison DRM (2001) Instability of Al_2SiO_5 'triple-point' assemblages in muscovite + biotite + quartz-bearing metapelites, with implications. Am Mineral 86:1414–1422

Pattison DRM, Spear FS, Debuhr CL, Cheney JT, Guidotti CV (2002) Thermodynamic modeling of the reaction muscovite + cordierite = Al_2SiO_5 + biotite + quartz + H_2O: constraints from natural assemblages and implications for the metapelitic petrogenetic grid. J Metamor Geol 20:99–118

Pavlov DI (1970) Halite anatectite and some of its less highly metamorphosed analogs. Doklady 195:171–173

Pedersen AK (1978) Non-stoichiometric magnesian spinels in shale xenoliths from a native iron-bearing andesite at Asuk, Disko, Central West Greenland. Contrib Mineral Petrol 67:331–340

Pederson AK (1979) A shale buchite xenolith with Al-armalcolite and native iron in a lava from Asuk, Disco, Central West Greenland. Contrib Mineral Petrol 69:83–94

Pederson AK, Nygaard E, Rønsbo JG, Bender Koch C, Buchwald VF (1992) Drilling induced pyrometamorphism of clastic sediments in the Lavø-1 well, Denmark. Scientific Drilling 3:127–137

Peng WX, van Genderen JL, Kang GF, Guan HY, Tan YJ (1997) Estimating the depth of underground coal fires using data integration techniques. Terra Nova 9:180–183

Pertsev NN (1977) High-Temperature Metamorphism and Metasomatism of Carbonate Rocks. Akad Nauk Moscow (in Russian)

Peters Tj, Iberg R (1978) Mineralogical changes during firing of calcium-silicate brick clays. Am Ceram Soc Bull 57:503–505, 509

Peters Tj, Jenni JP (1973) Mineralogical study of the firing characteristics of brick clays. Beitr Geol Schweiz Geotechn Ser 50

Pettijohn FJ (1949) Sedimentary rocks. Harper & Brothers, New York

Petty JJ (1936) The origin and occurrence of fulgurites in the Atlantic Coastal Plain. Am J Sci 31:188–201

Phemister J (1942) Note on fused spent shale from a retort at Pumpherston, Midlothian. Trans Geol Soc Glasgow 20:238–247

Philpotts AR, Pattison EF, Fox JS (1967) Kalsilite, diopside and melilite in a sedimentary xenolith from Brome Mountain. Nature 214:1322–1323

Piwinskii (1968) Experimental studies of igneous rock series central Sierra Nevada batholith, California. J Geol 76:548–570

Poddar MC (1952) Notes on the columnar structure of a sandstone of Bhuj (Middle Cretaceous) age, Kutch. Q J Geol Soc India 24:71–73

Preston RJ, Dempster TJ, Bell BR, Rogers G (1999) The petrology of mullite-bearing peraluminous xenoliths: implications for contamination processes in basaltic magmas. J Petrol 40:549–573

Prohaska A (1885) Über den Basalt von Kollnitz im Labentthale und dessen glasige Cordierit-führende Einschlüsse. Sitz Akad Wiss Wein B 62

Putnis A, McConnell, JDC (1980) Principles of mineral behaviour. Blackwell Scientific Publications, Oxford

Qi Q, Taylor LA, Zhou X (1994) Petrology and geochemistry of an unusual tridymite-hercynite xenolith in tholeiite from southeastern China. Mineral Petrol 50:195–207

Raeside JD (1968) A note on artificial fulgurites from a soil in south-east Otago. NZ J Geol Geophys 11:72–76

Rankin GA, Merwin HE (1918) The ternary system $MgO-Al_2O_3-SiO_2$. Am J Sci 45:301–325

Rattigan JH (1967) Phenomenon about Burning Mountain, Wingen, New South Wales. Aust J Sci 30:183–184

Reverdatto VV (1965) Paragenetic analysis of carbonate rocks of the spurrite-merwinite facies. Geochem Internat 5:1038–1053

Reverdatto VV (1970) Pyrometamorphism of limestones and the temperature of basaltic magmas. Lithos 3:135–143

Richardson, Sir John (1851) Arctic Searching Expedition: a journal of a boat journey through Rupert's Land and the Arctic Sea, in search of the discovery ships under the command of Sir John Franklin, 2 vols. Longman, Green & Longmans, London

Richardson SW (1968) Staurolite stability in part of the system Fe-Al-Si-O-H. J Petrol 9:467–488

Richardson SW, Gilbert MC, Bell PM (1969) Experimental determination of kyanite-andalusite and andalusite-sillimanite equilibria: the aluminium silicate triple point. Am J Sci 267:259–272

Ricker RW, Osborn EF (1954) Additional phase data for the system CaO-MgO-SiO$_2$. Am Ceram Soc J 37: 133–139

Rinne F (1895) Ueber rhombischen Augit als Contactproduct, chondrenartige Bildungen aus künstlichen Schmelzen und über Concretionen in Basalten. N Jb Mineral Geol Pal 2:167–299

Robie RA, Hemingway BS (1995) Thermodynamic properties of minerals and related substances at 298.15 K and 1 bar (10^5 Pascals) pressure and at higher temperatures. US Geol Surv Bull 2131

Roedder E (1983) Fluid inclusions. Reviews in Mineralogy 12, Mineralogical Society of America

Rogers GS (1917) Baked shale and slag formed by the burning of coal beds. US Geol Surv Prof Pap 108-A: 1–10

Roy DM, Roy R (1955) Synthesis and stability of minerals in the system MgO-Al$_2$O$_3$-SiO$_2$-H$_2$O. Am Mineral 40:147–178

Rubie DC (1998) Disequilibrium during metmamorphism: the role of nucleation kinetics. In: Treloar PJ, O'Brian PJ (eds)What drives metamorphism and metamorphic reactions? Geol Soc Lond Sp Pub 138:199–214

Rubie DC, Brearley AJ (1987) Metastable melting during the breakdown of muscovite + quartz at 1 kbar. Bull Mineral 110:533–549

Rubie DM, Brearley AJ (1990) A model for rates of disequilibrium melting during metamorphism. In: Ashworth JR, Brown M (eds) High-temperature metamorphism and crustal anatexis. Unwin Hyman, London, pp 57–86

Rushmer T (2001) Volume change during partial melting reactions: implications for melt extraction, melt geochemistry and crustal rheology. Tectonophys 34:389–405

Sabine PA (1975) Metamorphic processes at high temperature and low pressure: the petrogenesis of the metasomatised and assimilated rocks at Carneal, Co. Antrim. Roy S Lond Phil Trans 280:225–269

Sabine PA, Styles MT, Young BR (1985) The nature and paragenesis of natural bredigite and associated minerals from Carneal and Scawt Hill, Co. Antrim. Mineral Mag 49:663–670

Sanchez-Navas A (1999) Sequential kinetics of a muscovite-out reaction: a natural example. Am Mineral 84:1270–1286

Sanchez-Navas A, Galindo-Zaldivar J (1993) Alteration and deformation microstructures of biotite from plagioclase-rich dykes (Ronda Massif, S. Spain). Eur J Mineral 5:245–256

Saraf AK, Prakash A, Sengupta S, Gupta RP (1995) Landsat-TM data for estimating ground temperature and depth of subsurface coal fire in the Jharia coalfield, India. Int J Remote Sensing 16:2111–2124

Sarkar SL, Jeffrey JW (1978) Electron microprobe analysis of the Scawt Hill bredigite-larnite rock. Am Ceram Soc J 61:177–178

Sassi R, Mazzoli C, Spiess R, Cester T (2004) Towards a better understanding of the fibrolite problem: the effect of reaction overstepping and surface energy anisotropy. J Petrol 45:1467–1479

Sauvage J-F, Sauvage M (1992) Tectonique, néotectonique et phénomènes ignés a l'extremité est dur fossé de Nara (Mali): Daounas et lac Faguibine. J Afr Earth Sci 15:11–33

Schairer JF (1942) The system CaO-FeO-Al$_2$O$_3$-SiO$_2$, 1: Results of quenching experiments on five joins. J Am Ceram Soc 25:241–274

Schairer JF (1954) The system K$_2$O-MgO-Al$_2$O$_3$-SiO$_2$, I: Results of quenching experiments on four joins in the tetrahedron cordierite-mullite-potash feldspar. J Am Ceram Soc 37:501–533

Schairer JF, Bowen NL (1947a) The system anorthite-leucite-silica. Bull Comm Géol Finlande 140:67–87

Schairer JF, Bowen NL (1947b) Melting relations in the systems Na$_2$O-Al$_2$O$_3$-SiO$_2$ and K$_2$O-Al$_2$O$_3$-SiO$_2$. Am J Sci 245:193–204

Schairer JF, Yagi K (1952) The system FeO-Al$_2$O$_3$-SiO$_2$. Am J Sci Bowen vol, pp 471–512

Schairer JF, Yoder HS (1967) The nature of residual liquids from crystallization, with data on the system nepheline-diopside-silica. Am J Sci 258A:273–283

Schiffman P, Lofgren G (1981) Dynamic crystallization studies on the Grande Ronde pillow basalts, Central Washington. J Geol 90:49–78

Schlaudt CM, Roy DM (1966) The join Ca$_2$SiO$_4$-CaMgSiO$_4$. J Am Ceram Soc 49:430–432

Schmulovich KI (1969) Stability of merwinite in the system CaO-MgO-SiO$_2$-CO$_2$. Dokl Earth Sci Sect 184:125–127

Schreyer W, Schairer JF (1961) Compositions and structural states of anhydrous Mg-cordierites: a reinvestigation of the central part of the system MgO-Al$_2$O$_3$-SiO$_2$. J Petrol 2:324–406

Schreyer W, Hentschel G, Abraham K (1983) Osumilith in der Eifel und die Verwendung dieses Minerals als petrogenetischer Indikator. Tschermaks Mineral Petrog Mitt 31:215–234

Schreyer W, Maresch WV, Daniels P, Wolfsdorff P (1990) Potassic cordierites: characteristic minerals for high-temperature, very low-pressure environments. Contrib Mineral Petrol 105:162–172

Schulling RD (1961) Formation of pegmatitic carbonatite in a syenite-marble contact. Nature 192:1280

Scott PW, Critchley SR, Wilkinson FCF (1986) The chemistry and mineralogy of some granulated and pelletized blast furnace slags. Mineral Mag 50:141–147

Searle EJ (1962) Xenoliths and metamorphosed rocks associated with the Auckland basalts. NZ J Geol Geophys 5:384–403

Segnit ER (1950) New data on the slag-minerals nagelschmidtite and steadite. Mineral Mag 24:173–190

Segnit ER, Anderson CA (1971) An SEM study of fired kaolinite. Am Ceram Soc Bull 50:480–491

Seifert F (1974) Stability of sapphirine: a study of the aluminous part of the system $MgO-Al_2O_3-SiO_2-H_2O$. J Geol 82:173–204

Sen Gupta S (1957) Petrology of the para-lavas of the eastern part of Jharia coalfield. Q J Geol Min Metall Soc India 29:79–101

Sharp ZD, Essene EJ, Anovitz LM, Metz GW, Westrum EF Jr, Hemingway BS, Valley JW (1986) The heat capacity of monticellite and phase equilibria in the system $CaO-MgO-SiO_2-CO_2$. Geochim Cosmochim Acta 50:1475–1484

Sharygin VV, Sokol EV, Nigmatulina EN, Lepezin GG, Kalugin VM, Frenkel AE (1999) Mineralogy and petrography of technogenic parabasalts from the Chelyabinsk brown-coal basin. Russ Geol Geophys 40:879–899

Shaw HR (1963) The four-phase curve sanidine-quartz-liquid-gas between 500 and 4000 bars. Am Mineral 48:883–896

Shaw HR (1974) Diffusion of H_2O in granitic liquids; part I: experimental data, part II: mass transfer in magma chambers. In: Hoffman AW, Giletti BJ, Yoder HS Yund RA (eds) Geochemical transport and kinetics. Carnegie Inst Washington Publ 634:139–170

Shelby JE (1997) Introduction to glass science and technology. Roy Soc Chem, Cambridge UK

Sigsby RJ (1966). The present general lack of "scoria" in two burning lignite areas in North Dakota. Ann Proc North Dakota Acad Sc: 7–14

Sigurdsson H (1968) Petrology of acid xenoliths from Surtsey. Geol Mag 105:440–453

Sigurdsson H (1971) Feldspar relations in a composite magma. Lithos 4:231–238

Sigurdsson H, Sparks RSJ (1981) Petrology of rhyolitic and mixed magma ejecta from the 1875 eruption of Askja, Iceland. J Petrol 22:41–84

Smith DGW (1965) The chemistry and mineralogy of some emery-like rocks from Sithean Sluaigh, Strachur, Argyllshire. Am Mineral 50:1982–2022

Smith DGW (1969) Pyrometamorphism of phyllites by a dolerite plug. J Petrol 10:20–55

Smith A (2003) BCR: Optimizing firing through TTT analysis. Ceramicindustry 01/01/2002

Smith DM, Griffin JJ, Goldberg ED (1973) Elemental carbon in marine sediments: A baseline for burning. Nature 241:268–270

Smulikowski W, Desmons J, Harte B, Sassi FP, Schmid R (1997) Towards a unified nomenclature of metamorphism, 3: types, grade and facies. TOWAR97.11

Sokol EV, Volkova NI, Lepezin GG (1998) Mineralogy of pyrometamorphic rocks associated with naturally burned coal-bearing spoil-heaps of the Chelyabinsk coal basin, Russia. Eur J Mineral 10:1003–1014

Sokol EV, Maksimova NV, Volkova NI, Nigmatulina EN, Frenkel AE (2000) Hollow silicate microspheres from fly ashes of the Chelyabinsk brown coals (South Urals, Russia). Fuel Proc Tech 67:35–52

Sokol EV, Nigmatulina EN, Volkova NI (2002a) Fluorine mineralisation from burning coal-heaps in the Russian Urals. Mineral Petrol 75:23–40

Sokol EV, Sharygin V, Kalugin V, Volkova N, Nigmatulina E (2002b) Fayalite and kirschsteinite solid solutions in melts from burned spoil-heaps, south Urals, Russia. Eur J Mineral 14:795–807

Sokol EV, Kalugin VM, Nigmatulina EN, Volkova NI, Frenkel AE, Maksimova NV (2002c) Ferrospheres from fly ashes of Chelyabinsk coals: chemical composition, morphology and formation conditions. Fuel 81:867–876

Speakman K, Taylor HFW, Bennet JM, Gard J (1967) Hydrothermal reactions of dicalcium silicate. Chem Soc (London) Jour 1967A:1052–1060

Spear FS (1981) An experimental study of hornblende stability and compositional variability in amphibolite. Am J Sci 281:697-734
Speight JG (1983) The Chemistry and Technology of Coal. Marcel Dekker Inc, New York
Spray JG (1999) Shocking rocks by cavitation and bubble implosion. Geology 27:695-698
Spry AH, Solomon M (1964) Columnar buchites at Apsley, Tasmania. Q J Geol S Lond 120:519-545
Steiner A (1958) Petrogenetic implications of the 1954 Ngauruhoe lava and its xenoliths. NZ J Geol Geophys 1:325-363
Svensen H, Dysthe DK, Bendlien EH, Sacko S, Coulibaly H, Planke S (2003) Subsurface combustion in Mali: refutation of the active volcanism hypothesis in West Africa. Geology 31:581-584
Switzer G, Melson WG (1969) Partially melted kyanite eclogite from the Roberts Victor Mine, South Africa. Smithsonian Contrib Earth Sci 1:1-9
Thomas, HH (1922) On certain xenolithic Tertiary minor intrusions in the Island of Mull (Argyllshire). Q J Geol Soc Lond 78:229-259
Thwaites RG (ed) (1969) Original journals of the Lewis and Clark expedition, 1804-1806. Arno Press, New York
Tilley CE (1929) On larnite and its associated minerals from the contact zone of Scawt Hill, Co. Antrim. Mineral Mag 22:77-86
Tilley CE (1942) Tricalcium disilicate (rankinite), a new mineral from Scawt Hill, Co. Antrim. Mineral Mag 28:190-196
Tilley CE (1947) The gabbro-limestone contact of Camus Mor, Muck, Inverness-shire. Bull Comm Géol Finlande 20:97-105
Tilley CE (1951) A note on the progressive metamorphism of siliceous limestones and dolomites. Geol Mag 88:175-178
Tilley CE, Alderman AR (1934) progressive metasomatism in the flint nodules of the Scawt Hill contact-zone. Mineral Mag 23:513-518
Tilley CE, Harwood HF (1931) The dolerite-chalk contact of Scawt Hill, Co. Antrim. The production of basic alkali-rich rocks by assimilation of limestone by basaltic magma. Mineral Mag 132:439-468
Tilley CE, Vincent HCG (1948) The occurrence of an orthorhombic high-temperature form of Ca_2SiO_4 (bredigite) in the Scawt Hill contact zone, and as a constituent of slags. Mineral Mag 28:255-271
Tomkeieff SI (1940) The dolerite plugs of Tieveragh and Tievebulliagh near Cushendall, Co. Antrim, with a note on buchite. Geol Mag 7:54-64
Tommasini S, Davies GR (1997) Isotope disequilibrium during anatexis: a case study of contact melting, Sierra Nevada, California. Earth Planet Sci Lett 148:273-285
Traber D, Mäder Urs K, Eggenberger Urs (2002) Petrology and geochemistry of a municipal solid waste incinerator residue treated at high temperature. Schweiz Mineral Petrogr Mitt 82:1-14
Tracy RJ, Frost BR (1991) Phase equilibria and thermobarometry of calcareous, ultramafic and mafic rocks, and iron formations. In: Kerrick DM (ed) Contact metamorphism. Reviews in Mineralogy 26, Mineralogical Society of America, pp 207-280
Tracy RJ, McLellan EL (1985) A natural example of the kinetic controls of compositional and textural equilibration. In: Thompson AB, Rubie DC (eds) Metamorphic reactions: kinetics, textures and deformation. Advances in Physical Geochemistry 4, Springer, New York, pp 118-137
Treiman AH, Essene EJ (1983) Phase equilibria in the system $CaO-SiO_2-CO_2$. Am J Sci 283-A:97-120
Tropper P, Recheis A, Konzett J (2004) Pyrometamorphic formation of phosphorous-rich olivines in partially molten metapelitic gneisses from a prehistoric sacrificial burning site (Ötz Valley, Tyrol, Austria). Eur J Mineral 16:631-640
Tsuchiyama A, Takahashi E (1983) Melting kinetics of plagioclase feldspar. Contrib Mineral Petrol 84: 345-354
Tulloch AJ, Campbell JK (1993) Clinoenstatite-bearing buchites possibly from combustion of hydrocarbon gases in a major thrust zone: Glenroy Valley, New Zealand. J Geol 101:404-412
Turner FJ (1948) Mineralogical and structural evolution of the metamorphic rocks. Geol Soc Am Mem 30
Turner FJ (1967) Thermodynamic appraisal of steps in progressive metamorphism of siliceous dolomitic limestones. N Jb Mineral Mh:1-21
Turner FJ (1968) Metamorphic petrology. McGraw-Hill Book Company
Turner FJ, Verhoogen J (1960) Igneous and metamorphic rocks. McGraw-Hill Book Company

Turnock AC (1960) The stability of iron chlorites. Carnegie Inst Washington Yb 59:98–103

Turnock AC, Eugster HP (1962) Fe-Al oxides: phase relations below 1000 °C. J Petrol 3:533–565

Tuttle OF, Bowen NL (1958) Origin of granite in the light of experimental studies in the system $NaAlSi_2O_8$-$KAlSi_2O_8$-SiO_2-H_2O. Geol Soc Am Mem 74

Tuttle OF, Harker RI (1957) Synthesis of spurrite and the reaction wollastonite + calcite = spurrite + carbon dioxide. Am J Sci 255:226–234

Tyrrell GW (1926) The principles of petrology. Methuen & Co. Ltd, London

Uman MA (1969) Lightning. McGraw-Hill, New York

Vedder W, Wilkins RWT (1969) Dehydroxylation and rehydroxylation, oxidation and reduction of micas. Am Mineral 54:482–509

Venkatesh V (1952) Development and growth of cordierite in paralavas. Am Mineral 37:831–848

Viti C, Mellini M, Di Vincenzo G (2003) Nanotextures of laser-heated biotites. Ninth International Symposium on Experimental Mineralogy, Petrology and Geochemistry (Abs)

Walker D, Kirkpatrick RJ, Longhi J, Hays JF (1976) Crystallization history of lunar picritic basalt sample 2002 – phase equilibria and cooling rate studies. Geol Soc Am Bull 87:646–656

Wallmach T, Hatton CJ, Droop GTR (1989) Extreme facies of contact metamorphism developed in calc-silicate xenoliths in the eastern Bushveld Complex. Can Mineral 27:509–523

Walter LS (1963) Experimental studies on Bowen's decarbonation series, I. P-T univariant equilibria of the "monticellite" and "akermanite" reactions. Am J Sci 261:488–500

Wartho J-A, Kelley SP, Blake S (2001) Magma flow regimes in sills deduced from Ar isotope systematics of host rocks. J Geophys Res 106:4017–4035

Weeks WF (1956) A thermochemical study of equilibrium relations during metamorphism of siliceous carbonate rocks. J Geol 64:245–270

Weill DF (1966) Stability relations in the Al_2O_3-SiO_2 system calculated from solubilities in the Al_2O_3-SiO_2-Na_3AlF_6 system. Geochim Cosmochim Acta 30:223–237

Wenzel T, Baumgartner LP, Brügmann GE, Konnikov EG, Kislov EV, Orsoev DA (2001) Contamination of mafic magma by partial melting of dolomitic xenoliths. Terra Nova 13:188–196

Wenzel T, Baumgartner LP, Brügmann GE, Konnikov EG, Kislov EV (2002) Partial melting and assimilation of dolomitic xenoliths by mafic magma: the Iolo-Dovyren Intrusion (north Baikal region, Russia). J Petrol 43:2049–2074

Werner AG (1789) Bergm J 1, p. 375

White JC, Gray JJ, Town JW (1968) Emery and emery-like rocks of the west-central Cascade Range, Oregon. Ore Bin 30:213–223

Whitworth HF (1958) The occurrence of some fused sedimentary rocks at Ravensworth, N.S.W. J Roy Soc New South Wales 92:204–210

Willemse J, Bensch JJ (1964) Inclusions of original carbonate rocks in gabbro and norite of the eastern part of the Bushveld Complex. Trans Geol Soc S Africa 67:1–87

Willemse J, Viljoen EA (1970) The fate of argillaceous material in a gabbroic magma of the Bushveld complex. Geol Soc S Africa Special Pub 1:336–366

Wimmenauer W, Wilmanns O (2004) Neue Funde von Blitzsprengung und Fulguritbildung im Schwarzwald. Ber Naturf Ges Freiburg i Br 94:1–22

Winkler HGF (1979) Petrogenesis of metamorphic rocks, 5th edn. Springer, Berlin Heidelberg

Winkler HGF, von Platten H (1961) Experimentalle Gesteinsmetamorpose, III. Bildung anatektischer Schmelzen aus metamorphisierten Grauwacken. Geochem Cosmochim Acta 18:294–316

Wolf, K-HAA, Kühnel RA, de Pater CJ (1987) Laboratory experiments on thermal behaviour of roof and floor rocks of Carboniferous coal seams. In: Moulijin JA (ed) International conference on coal science, Maastricht. Elsevier, Amsterdam, pp 911–914

Wood CP (1994) Mineralogy at the magma-hydrothermal system interface in andesite volcanoes, New Zealand. Geology 22:75–78

Wood CP, Browne PRL (1996) Chlorine-rich pyrometamorphic magma at White Island volcano, New Zealand. J Volc Geotherm Res 72:21–35

Worden RH, Champness PE, Droop GTR (1987) Transmission electron microscopy of the pyrometamorphic breakdown of phengite and chlorite. Mineral Mag. 359:107–122

Wörner G, Schmincke H-U, Schreyer W (1982) Crustal xenoliths from the Quaternary Wehr volcano (East Eifel). N Jb Mineral Mh H1:39–47
Wright FW Jr (1999) Florida's fantastic fulgurite find. Rocks Minerals 74:157–159
Wyllie PJ (1959) Microscopic cordierite in fused Torridonian arkose. Am Mineral 44:1039–1046
Wyllie PJ (1961) Fusion of a Torridonian sandstone by a picrite sill on Soay (Hebrides). J Petrol 2:1–37
Wyllie PJ (1965) Melting relationships in the system CaO-MgO-CO_2-H_2O with petrological applications. J Petrol 6:101–123
Wyllie PJ, Haas JL Jr (1966) The system CaO-SiO_2-CO_2-H_2O, II. The petrogenetic model. Geochim Cosmochim Acta 30:525–543
Wyllie PJ, Tuttle OF (1960) the system CaO-CO_2-H_2O and the origin of carbonatites. J Petrol 1:1–46
Wyllie PJ, Tuttle OF (1961) Hydrothermal melting of shales. Geol Mag 98:56–66
Wys EC de, Foster WR (1958) The system diopsdie-anorthite-åkermanite. Mineral Mag 31:736–743
Xu H, Veblen DR, Luo G, Xue, J (1996) Transmission electron microscopy study of the thermal decomposition of tremolite into clinopyroxene. Am Mineral 81:1126–1132
Yavorsky VI, Radugina LV (1932) Coal-fire combustion and attendant events in the Kuznetsky basin. Mining J 10:55–59 (in Russian)
Yoder HS (1952) The MgO-Al_2O_3-SiO_2-H_2O system and the related metamorphic facies. Am J Sci Bowen vol:569–627
Yoder HS, Eugster HP (1955) Synthetic and natural muscovites. Geochim Cosmochim Acta 8:225–280
Yoder HS, Tilley CE (1962) Origin of basaltic magmas: an experimental study of natural and synthetic rock systems. J Petrol 3:342–532
Yorath CJ, Balkwill HR, Klassen RW (1969) Geology of the eastern part of the Northern Interior and Arctic coastal plains, Northwest Territories. Geol Survey Canada Paper 68-27
Yorath, CJ, Balkwill, HR, Klassen, RW (1975) Franklin Bay and Malloch Hill map-areas, District of Mackenzie. Geol Survey Canada Paper 74-36
Zacek V, Skála R, Chulupácova M, Dvorák Z (2005) Ca-Fe^{3+}-rich, Si-undersaturated buchite from Zelensky, North-Bohemian Brown Coal Basin, Czech Republic. Eur J Mineral 4:623–634
Zambonini (1935) Mineralogia Vesuviana (2nd ed), E Quercigh. Naples, p 66
Zhang X, Kroonenberg SB, de Boer CB (2004) Dating of coal fires in Xinjiang, north-west China. Terra Nova 16:68–74
Zbarskiy MI (1963) Mineralogical and petrographical features of burned rocks from central Asia. Zapiski Kirgiz Otdel Vses Mineral Obshch 4:53–67 (in Russian)
Zharikov VA, Shmulovich KI (1969) High temperature mineral equilibria in the system CaO-SiO_2-CO_2. Geochem Int 6:853–869
Zharikov VA, Shmulovich KI, Bulatov VK (1977) Experimental studies in the system CaO-MgO-Al_2O_3-CO_2-H_2O and conditions of high-temperature metamorphism. Tectonophys 43:145–162
Zirkel F (1872) Mikromineralogische Mitteilungen. N Jb Mineral Geol Pal:1–25
Zirkel F (1891) Cordieritbildung in verglasten Sandsteinen. N Jb Mineral Geol Pal 1:109–113
Zhou J, Hsu LC (1992) The stability of merwinite in the system CaO-MgO-SiO_2-H_2O-CO_2 with CO_2-poor fluids. Contr Miner Petrol 112:385–392
Zhou H, McKeegan KD, Xisheng X, Zindler A (2004) Fe-Al-rich tridymite-hercynite xenoliths with positive cerium anomalies: preserved lateritic paleosols and implications for Miocene climate. Chem Geol 207:101–116

Index

A

aegirine 3, 57, 113, 157, 161, 163, 173, 187
–, -augite 3, 57, 157, 161, 163, 173, 187
åkermanite 117, 119, 121, 126, 128, 134, 136, 139, 140, 148–153, 157, 174, 195, 196, 215, 216
albite 14, 47, 56–57, 65, 82, 84, 87, 116, 165, 186, 187, 195, 218–220, 234, 242
Aldrich Creek (Canada) 103, 104
alginite 26
almandine 1, 49, 245, 246, 249, 250
alum 111, 112, 208
alunite 102
alunogen 110
Amiata, Mt. (Italy) 84, 85
amphibole 113, 167–169, 174, 238, 239
amygdule 54, 102, 172–175
analcite (analcime) 102, 172
anatase 41, 178
andalusite 1, 4, 8, 48, 81, 84, 107, 231, 242–244, 250, 252
andradite 99, 119, 126, 129, 140, 144, 145, 147, 157, 158, 160, 161, 163, 209
anhydrite 160, 161, 166, 186, 187, 203, 208
anorthite 41, 46, 47, 60, 63, 65, 68, 70, 99, 100, 103, 109, 117, 119, 128, 146, 157, 161, 163–165, 178, 195, 196, 200, 201, 203–205, 208, 211, 212, 214–216, 239
anorthoclase 1, 3, 5, 86, 91, 102
anthophyllite 184, 185, 250
apatite 18, 41, 47, 49, 65, 72, 88, 89, 91, 99, 101, 102, 109, 148, 163, 165, 169, 173, 188, 201, 203, 205, 215, 217, 230
apophyllite 160
Apsley (Tasmania) 33–35, 52
aragonite 102, 160
Ardnamurchan (Scotland) 39, 53, 55
argillite 72, 225–227
arkose/arkosic 12, 14, 16, 32, 49, 50, 53, 54, 110, 209, 238
armalcolite 64, 65, 211

Arran (Scotland) 57, 58
assimilation 1
Auckland (New Zealand) 63, 64
augite 32, 81, 88, 92, 102, 104, 109, 166, 167, 170, 173, 175, 205, 216, 240, 241
aureole 2, 8, 12, 14, 39, 40, 53–55, 59, 66, 67, 75, 79, 131–133, 137, 138, 144, 169–171, 233, 238

B

baghdadite 144
Baksteen (South Africa) 49, 50
barite 160
barium hexaferrite 165
basalt 2, 4, 8, 31, 32, 52–54, 57, 60, 61, 63, 64, 84, 87, 94, 105, 123, 126–128, 134, 139, 167, 170–181, 183, 205, 223, 224, 235, 240, 241
–, temperature 2, 4, 54, 57, 59, 60, 84, 94, 105, 126, 128, 134, 139, 170–174, 178, 205, 240, 241
basaltic andesite 62, 187
bauxite 4, 178, 180–182, 214
bayerite 160
Beaunit maar (France) 240
bentorite 160
biotite 1, 3, 5, 12, 20, 37, 41, 47, 48, 52–54, 57, 60, 63, 65, 72–74, 76–78, 81–84, 86–89, 98, 113, 160, 165, 170, 171, 201, 208, 215, 216, 221, 228–230, 234–237, 239, 242, 245, 247, 248, 251, 252
bituminous sediment 4, 101, 156
Blaue Kuppe (Germany) 14, 15, 17
bloating 38, 191, 195, 208
boccanes 26
bog iron 214
boehmite 160, 177
Bokaro Coalfield (India) 4, 92
bredigite 126, 136, 139–142, 144
bricks 1, 97, 101, 164, 191
Brome Mountain (Canada) 151
brownmillerite 126, 148, 149, 157, 159
brucite 122, 142, 148, 153, 155, 160, 166, 233

buchite 2, 3, 14, 32–36, 39, 47, 49, 54, 56–64, 77, 80, 82, 95, 100, 108, 109, 223, 243, 244
Bunowen (Northern Ireland) 169
Burning Mountain (Australia) 106
burning process 22, 159
burnt rock 2, 21, 24, 25, 109, 156
Bushveld Intrusion (South Africa) 67, 69, 148, 150, 243, 244
bytownite 61, 105, 106, 109, 161, 179, 212

C

calcio-olivine 122, 136
calcite 23, 42, 47, 65, 102, 103, 115–117, 119, 121, 124–126, 131–133, 138–142, 144, 146, 148, 149, 151–155, 157, 158, 160–163, 165, 166, 172–175, 195, 196, 199, 200, 209, 215
calcium disilicate 157
calcium ferrite 165
calcium-dialuminate 157
calc-silicate rock 5, 122, 139, 195, 222
 –, composition 126, 128, 129, 132, 134–136, 141, 144–148, 151–153, 158, 165, 194–197, 213, 215
California 87, 88, 90, 91, 101, 122, 168, 169, 209
carbon 1, 21, 26, 27, 65, 101, 112, 113, 160, 211, 218
carbonaceous sediment 2, 4, 9, 24, 65
carbonaceous shale 38, 64
Carneal (Northern Ireland) 130, 139–141
Centralia (USA) 205–207
ceramics 191
chalcopyrite 153, 165, 209, 214
Chelyabinsk (Russia) 4, 200–204
chemical potential 28, 124, 133
chimney 23, 98
China 24, 25, 94, 181, 183
chlorite 16, 33, 35, 41, 42, 47, 52, 58, 65, 74, 98, 102, 116, 121, 160, 169, 172, 178, 184, 185, 190, 195, 196, 208, 210, 211, 214, 218–220, 232–234
chlorophaeite 172, 173
chondrodite 203
Christmas Mountains (USA) 131, 132, 142
chromatite 160
chromite 211
clay 2, 24, 30, 31, 33, 38, 41, 49, 52, 53, 97–99, 102, 117, 161, 172, 175, 176, 191, 194–196, 200, 203, 208, 209, 218
claystone 32, 41, 97, 203, 204
clinker 2, 4, 21, 23, 26, 94, 96–99, 103, 109, 111, 165, 201, 202
clinoenstatite 107, 109
clinoferrosilite 46, 109
clinohypersthene 4, 105, 106
clinopyroxene 3–5, 41, 46, 49, 54, 57, 58, 60, 63–65, 68, 70, 84, 99, 100, 111, 113, 151, 157, 161–165, 168–173, 176, 179, 187, 196, 199, 203, 208, 211, 216, 238–242, 245, 246
clinozoisite 216
coal seam 1, 2, 4, 22, 37, 38, 92, 96, 97, 105, 109, 111, 205, 214
cohenite 64, 65
columnar jointing 11, 24, 31–33, 52
combustion (= burning) metamorphism 2, 22, 27, 42, 47, 96, 122, 126, 156
composition plots of minerals, glass and rocks
 –, ACF 42, 43, 117, 119, 120, 162, 168, 202, 212
 –, Al_2O_3-[SiO_2,$(Na,K)_2O$] -[$(Fe,Mg,Mn,Ca)O,TiO_2$] 190
 –, CaO-$(Mg,Fe)O$-$(Al,Fe)_2O_3$-SiO_2-CO_2 117
 –, CaO-$(MgFe)O$-$(Al,Fe)_2O_3$-SiO_2 143
 –, CaO-Al_2O_3-Fe_2O_3-SiO_2 128, 129, 147
 –, CaO-Al_2O_3-SiO_2 45, 46, 128, 213
 –, CaO-MgO-Al_2O_3-SiO_2 129
 –, CaO-MgO-SiO_2 115, 118, 127, 135
 –, CaO-MgO-SiO_2-CO_2 115, 118, 127
 –, CMAS 176
 –, $FeAl_2O_4$-$MgAl_2O_4$-Fe_3O_4 238
 –, FeO-MgO-SiO_2-Al_2O_3 249
 –, Na_2O-K_2O-$(Ca,Mg,Fe)O$ 91
 –, SiO_2-Al_2O_3-$(Fe,Mg)O$ 83
 –, SiO_2-Al_2O_3-MgO 42, 184, 212
cordierite 1, 3–5, 18, 36, 37, 41, 42, 45–49, 52–54, 57–61, 63–66, 68–70, 72–74, 77, 78, 84, 86, 89, 92, 94–96, 99, 100, 102, 103, 105, 106, 109, 111, 113, 170, 178–180, 183–185, 188, 201, 202, 204, 208, 210–212, 232–234, 244, 246, 247, 249–252
Cortland complex 5
Crestmore (USA) 122
cristobalite 8, 18, 41, 47, 99, 100, 102, 103, 106, 107, 109, 111, 113, 161, 178–182, 185–187, 191, 193, 195, 201, 208
cuspidine 122, 135, 140, 144, 205
Czech Republic 165

D

dacite 145, 146
decarbonation 38, 116, 119, 122, 125, 126, 131, 133–134, 137, 142, 148, 149, 151, 159, 160, 173, 174
dehydration 32, 33, 38, 67, 77, 160, 169, 208, 220, 221, 228, 252
delamination 98, 99, 228
devitrification 40, 51–53, 89, 215
devolatilisation 27, 38, 145
diatomite 92, 101, 110, 112
dilation 11, 37, 38
diopside 5, 65, 99, 103, 109, 116–119, 126, 128, 144, 147–149, 151–154, 157, 161, 163, 167, 190, 195, 196, 199, 213, 242

Index 271

Disco (Greenland) 64, 79
dmisteinbergite 203
dolerite 14, 15, 33-35, 39, 49-51, 53, 54, 58-60, 65, 67, 129, 130, 138, 139, 142, 166, 169, 172, 173, 175, 178-180, 233
dolomite 38, 65, 101, 103, 115, 116, 119, 126, 134, 139, 142, 148, 150-155, 160, 166, 195, 199, 200, 204, 212
dorrite 99

E

eclogite 242, 245
edenite 167, 170
Eifel (Germany) 1, 75-83, 113, 125, 148, 149, 227, 229, 230, 235, 236, 239, 243-247, 251
ellestadite 205
emery 2, 3, 5, 65-67, 178
enstatite 99, 109, 183, 184, 232, 233
epidote 41, 51, 65, 163, 169, 214, 218
epsomite 101
essenite 64, 100, 165
ettringite 160
evaporites 166

F

fassaite (fassaitic pyroxene) 109, 117, 119, 126, 128, 144, 146, 147, 153, 157, 162, 163, 165, 195, 196, 213
fayalite 94-96, 99, 100, 109, 146, 167, 183, 202, 203, 205, 208, 209, 214, 232, 246
ferroalloys 208
ferrosphere 189
fingerprint texture 86, 89
Flekkeren (Norway) 144
fluid/fluid phase 9, 14, 37, 39, 40, 60, 88, 119, 121, 122, 124, 126, 133, 134, 137, 139, 141, 145, 153, 154, 160, 170, 172, 219, 231, 250
fluorallestadite 203
fluorite 102, 201-203
forsterite 5, 119, 126, 128, 142, 148, 149, 151-154, 166, 183-185, 203, 232-234
foshagite 160
Frankenstein (Germany) 190
fulgurite 2, 4, 28-31, 112, 113, 217
-, sand fulgurite 30

G

gabbro 14, 16, 70, 72-74, 131, 132, 134-136, 142, 148, 151, 168, 170, 171, 173, 189, 190, 240, 241
garnet 1, 4, 5, 41, 49, 52, 53, 60, 65, 72, 74, 81, 82, 102, 126, 128, 144, 146, 147, 162, 245-247, 249, 250

Gasco (Spain) 217
gedrite 232, 248-250
gehlenite 99, 100, 102, 117, 119, 134, 139-143, 146, 148, 149, 157, 158, 160, 163, 165, 195, 196, 199, 200, 204, 215, 216
gibbsite 160, 177-179
giuseppetite 145
glass 1, 3-5, 14, 18, 23, 28, 30, 31, 33, 35, 36, 39-41, 49-55, 57-61, 64, 66, 72, 78-80, 83, 84, 86-89, 91, 92, 94, 95, 99-103, 105-107, 109, 111-113, 161-163, 165, 167, 169, 175, 181, 182, 187, 189-191, 194, 195, 201-205, 208-218, 223-225, 227, 229, 230, 235, 236, 239, 240, 242, 244, 245, 247, 248, 251
-, composition 28, 33, 57, 58, 64, 65, 78, 83, 87-89, 94, 95, 100, 165, 167, 175, 182, 189, 190, 194, 195, 204, 205, 210, 211, 213, 215, 225, 239, 240, 244, 247
-, preservation 1, 4, 17, 39, 58, 215
-, transformation temperature 39
Glenmore (Scotland) 39, 40, 53-55
Glenroy Valley (New Zealand) 107, 108
gneiss 41, 49, 70, 72, 84, 113, 169, 215, 216
goethite 3, 41, 49, 102, 172, 177, 178, 208, 214
grandite 146, 147
granite (granitoid) 8, 12, 17, 19, 20, 41, 47, 88-91, 184, 185, 235
-, peraluminous 87, 89
granodiorite 12, 41, 47, 86-88, 168
granophyre 65, 67, 74
granulite 5, 8
graphic intergrowth 169, 244
graphite 52, 64, 65, 84, 100, 112, 144, 145, 203, 216, 250
greigite 159
greywacke 47, 48, 72, 73, 218, 225-227, 240
grit 54-57, 110, 161
grossite 157
grossular/grossularite 5, 119, 126, 129, 134, 135, 140, 144, 146, 147, 157, 161, 163
gypsum 23, 24, 26, 102, 160, 161, 163, 165, 186
gyrolite 172-174

H

halite 166, 186, 187
halloysite 52, 177-179, 181
halotrichite-pickeringite 110
hardystonite 214
Hat Creek (Canada) 102
Hatrurim Basin (Israel) 156-160
hatrurite 122, 157
heat transfer 11, 14, 60
-, conduction 12, 17, 24, 33, 67, 105, 133, 171
-, convection 12, 33, 34

Index

hedenbergite 109, 119, 157, 159, 161, 187, 188, 195, 205
hematite 3, 26, 41, 42, 52, 55, 58, 59, 66, 99, 100, 102, 103, 107, 109, 113, 159, 165, 177–180, 185, 189, 195, 201–203, 208, 214, 232
hercynite 41, 57, 58, 60, 72, 83, 99, 100, 109, 159, 179, 180–183, 201, 208, 211, 217, 230, 232, 237, 246–250
hillbrandite 144, 160
hornblende 47, 84, 88, 89, 168–170, 172, 189, 190, 214, 238, 239
hornfels 5, 77, 83, 167, 168, 170, 171, 173
humite 205
hydrogrossular 145
hydromagnesite 160
hydrotalcite 160

I

Iceland 84, 86
illite 23, 41, 52, 98, 99, 102, 103, 160, 191, 195, 196, 199, 200, 208, 210, 211
ilmenite 3, 5, 41, 47, 49, 58, 60, 61, 64–66, 70, 72–74, 76–78, 80, 81, 83, 84, 89, 99, 107, 109, 116, 163, 168–170, 175–179, 181, 183, 204, 230, 239–241, 247, 248
indialite 94, 113, 185, 187, 211
in-situ gasification 205
Ioko-Dovyren (Russia) 153, 155, 166
Iran 27, 28, 160, 162, 214
Iraq 27, 162, 163
Irkutsk (Russia) 166
iron (native) 64, 165, 203, 205, 208, 214

J

jarosite 26, 102

K

kalsilite 116, 126, 144, 145, 148, 151, 152
karroo 49
Kenderlyk Depression (Kazakhstan) 108, 109
K-feldspar 5, 12, 37, 41, 47, 49, 51–54, 56, 58–60, 68, 70, 73, 74, 84, 87–89, 98, 99, 102, 109, 116, 126, 144, 152, 163, 169, 195, 203, 208, 211, 228–231, 235, 237, 242, 248
Kilchoan (Scotland) 134–136
kilchoanite 134, 135, 137
kimberlite 242, 245
Kirghiza 111
kirschsteinite 203, 205
kogenite 203
Kuznetskiy coal basin (Russia) 109
kyanite 8, 242, 245

L

La Soufrière (Lesser Antilles) 146, 147
larnite 116, 118, 119, 122, 124, 129–131, 134, 136–142, 144, 148, 149, 157–159, 163–164, 173, 174, 203, 204
laterite 4, 107, 170
leucite 4, 116, 148, 151, 161, 162, 203–205
lightning strike metamorphism/fusion 2, 189, 190
lignite 22–24, 26, 97, 98, 110, 162
lime 115, 116, 122, 125, 148, 195, 196, 199, 203, 204, 215
limestone 101, 110, 119, 122, 125, 126, 129, 131, 134–136, 139, 142, 144, 148, 157, 158, 161, 212
lithomarge 2, 167, 178–182, 214

M

mafic rock 167–170, 178
magnesioferrite 99, 109, 159, 165, 189, 203, 204
magnetite 3, 16–18, 24, 32, 41, 42, 46, 51, 52, 55, 57–61, 66, 73, 74, 77–80, 86–89, 92, 96, 99, 100, 102–105, 107, 109, 111, 117, 138–140, 144–146, 153, 157, 159, 162, 165, 167–171, 175, 177–181, 185, 189, 195, 201, 203, 205, 208, 209, 211, 214, 216, 232, 234–237, 239, 245, 246, 248, 251
Mali 110, 111
Marble Canyon (USA) 135–137
marl/marly 2, 5, 116, 119, 126, 139, 156, 157, 160–163, 191, 195, 198, 201, 214, 215
mayenite 126, 148, 149, 157, 204
melanterite 110
melilite 99, 100, 118, 119, 125, 126, 128, 135, 136, 138–142, 144, 145, 148, 149, 152–153, 157, 161–165, 172–174, 204, 212–216
melt 1, 4, 5, 14, 16–19, 23, 24, 31, 35–40, 42, 45, 47, 49, 51, 54, 55, 58, 62, 63, 65, 70–75, 78, 80, 81, 83, 84, 92, 98–100, 102, 109, 112, 113, 126, 154, 166, 168, 169, 175, 178, 187, 189, 191, 193–195, 200, 203, 205, 208, 209, 216, 219, 221, 223, 225–231, 234, 235, 237, 239–242, 244–246, 249–252
merwinite 118, 119, 121, 126, 134, 136, 138–142, 144, 148, 149, 151, 157, 212, 213
mesostasis 172, 175, 176
metal 41, 112
metastable reactions 231
metatalc 42, 45
microcracking 36, 37
millosevichite 110
monticellite 5, 118, 119, 126, 134, 142, 144, 148, 149, 151–154, 157, 173, 174, 209
montmorillonite 172, 178, 195, 196
Mottled Zone (Israel) 33, 156–160

Muck (Scotland) 142, 143
mud pellet 36
mudstone 27, 101, 102, 201–204, 210
Mull (Scotland) 16, 17, 36, 60–62, 68, 73, 74, 79, 172, 245
mullite 2, 3, 5, 35, 37, 41, 42, 45–48, 53, 58–66, 68–70, 72–74, 76–78, 81, 83, 84, 86, 92, 94–96, 98–100, 102, 105–107, 109, 111–113, 170, 177–183, 185–187, 191, 193–195, 201, 208–212, 221, 228–232, 243–245, 247, 249, 251
muscovite 12, 35–37, 41, 47, 48, 52–55, 58, 63, 65, 66, 70, 73, 74, 76, 78, 98, 121, 191, 193, 195, 201, 210, 214, 221, 227–231, 235, 242, 245, 248, 250–252

N

nagelschmidtite 157
natroalunite 186, 187
natrolite 172
nepheline 3, 81, 99, 116, 144, 242
Ngauruhoe (New Zealand) 72
norbergite 203
nucleation rate 223
nuclei 39, 65, 222, 223

O

oil shale 109, 110, 211, 212
oldhamite 204, 205, 212, 213
oligoclase 41, 52, 60, 77, 82, 215, 251
olivine 4, 5, 57, 58, 63, 64, 87, 111, 136, 142, 153–155, 166–176, 179, 180, 184, 185, 187, 190, 203, 205, 211, 216, 232–235, 239
omphacite 242
optalic metamorphism 1
Oregon 181, 182
orthoclase 14, 49, 55, 87, 89, 214
orthopyroxene 3, 5, 41, 42, 45–49, 54, 57–60, 65, 72, 84, 86–88, 91, 92, 105, 107, 117, 163, 167–179, 184, 188, 211, 212, 232–240, 245
ostwald step rule 219, 221, 231
osumilite 59, 113, 188
Ötz Valley (Austria) 215
oxidation 1, 21, 22, 24, 26, 66, 67, 79, 100, 105, 106, 111, 113, 159, 161, 165, 172, 189, 218, 234, 235, 237, 246
oxidation ratio 66, 67
oxygen buffer 246
  –, HM 100, 154, 165
  –, NNO 146
  –, QFM 100, 146, 165, 167, 205, 246
  –, WM 146
oxygen fugacity 145, 211

P

parabasalt 4, 94, 104, 164, 203–205, 208
paragonite 195
paralava 2, 4, 21, 23, 39, 42, 92, 94–96, 98–101, 107, 109, 111, 164, 187, 188, 201–203, 214
pelite (pelitic) 2, 4, 5, 11, 12, 37, 41, 49, 53, 54, 61, 65, 67, 70, 72, 73, 75, 77, 84, 111, 187, 205, 227–230, 236
pentlandite 153
periclase 119, 126, 142, 148, 151, 153–155, 166, 183, 199, 201, 204
peridotite 14, 16, 190
permeability 33, 60, 107, 145, 170–172
perovskite 116, 138, 141, 144, 153, 157, 162, 165, 173, 181
petrified wood 201–204
phlogopite 113, 116, 126, 144, 148, 151, 152, 188, 203
phonolite 1, 5, 75, 81, 230
phosphide 112
phosphorite 101, 102
phosphoritic sediment 41
phyllite 41, 65, 66
picrite 56, 57
picrodolerite 178
pigeonite 4, 5, 167, 175, 241
plagioclase 4, 12, 18, 41, 47, 49, 52, 54, 55, 57–59, 61–66, 68, 70–72, 74, 76–78, 84, 86–89, 91, 92, 94, 102–104, 109, 111, 126, 162–173, 175–179, 181, 185, 189, 195, 203, 209–211, 214–216, 223–227, 235–237, 239, 240, 242, 245, 248
pleonaste 68, 77–79, 154, 165, 175, 203, 228, 235, 237, 238
porcellanite 1, 2, 64, 106, 179–182, 187, 223
portlandite 160, 203, 204
Powder River Basin (USA) 98, 99, 101, 165
protoenstatite 57
pseudobrookite 3, 41, 65–67, 99, 113, 178–181, 188, 201, 211
pseudowollastonite 119, 122, 128, 148, 157, 160, 161, 163, 195
pyrite 21, 23, 24, 26–28, 102, 105, 111, 162, 165, 179, 186, 190, 214
pyrometamorphic magma 186, 187
pyrophyllite 98, 187, 232
pyroxene 3–5, 46, 49, 58–60, 63, 65, 74, 84, 88, 94, 102, 103, 105, 113, 146, 152, 153, 161, 163, 167, 168, 170–173, 175, 176, 183, 185, 187, 195, 196, 213, 237–242
pyroxene hornfels facies 4, 183
pyrrhotite 72, 100, 109, 139, 144, 148, 162, 165, 203, 205, 209

Q

quartz 3, 5, 8, 12, 14, 17, 18, 23, 28, 31–33, 36–38, 40–42, 47, 49, 51–55, 57–60, 64, 65, 68, 72–74, 77, 78, 84, 86–89, 91, 94, 98–100, 102, 103, 105–107, 112, 113, 115, 116, 118, 123, 131–133, 135, 146, 152–154, 160, 161, 163, 165–170, 177, 178, 181, 184, 191, 193–196, 199–201, 208–211, 214–221, 223–225, 227–230, 232, 234–236, 242, 244–252
 -, paramorph after tridymite 52, 65, 73
quartzite 12, 17, 35, 57, 217
quartzofeldspathic rock 41, 42, 169
 -, composition 41, 42, 169

R

rankinite 116, 119, 122, 124, 130–134, 136, 137, 140–142, 144, 157, 170, 173, 174
Rathlin Island (Northern Ireland) 180, 181
Ravensworth (Australia) 105
reduction 21, 24, 28, 35, 55, 65, 112, 113, 119, 144, 145, 154, 183, 200, 211, 221
REE pattern 183
rhyodacite 84
rhyolite 86, 87, 205
Rincorn volcano (California) 101
Ringstead Bay (England) 111
ringwoodite 217
roedderite 113
Ruapehu, Mt. (New Zealand) 188
Russia 108, 109, 153, 155, 200–204
rustumite 135
rutile 18, 41, 49, 52, 64, 65, 68, 70, 72, 89, 107, 116, 163, 179, 204, 211, 214, 230, 237

S

sandstone 2, 4, 14, 15, 17, 18, 31–36, 41, 47–49, 51, 52, 54, 55, 58, 60, 72, 92, 103, 105–107, 109–111, 156, 161, 163, 178, 201, 208, 209
sanidine 1, 3–5, 41, 49, 53, 54, 59, 63–66, 73, 76–78, 81, 82, 84, 87, 89, 91, 102, 103, 113, 170, 185, 186, 188, 193, 195, 196, 200, 203, 211, 229, 230, 235, 242, 243, 245
 -, microcline-sanidine transition 16
sanidinite 2–5, 8, 37, 41–43, 46, 47, 81–85, 113, 115, 116, 121, 122, 126, 136, 141–144, 150, 151, 153, 157, 167, 168, 171, 174, 177, 178, 183–185, 188, 245, 246
 -, facies 4, 5, 8, 37, 41–43, 46–48, 81, 115, 116, 121, 122, 126, 136, 141–144, 150, 151, 153, 157, 167, 168, 171, 174, 177, 178, 183–185, 188
Santorini (Greece) 145, 146
sapphirine 5, 41, 68, 69, 243, 245
schist 1, 17, 41, 49, 60, 61, 63, 65, 70, 75–77, 80, 82–84, 169, 218, 229, 230, 242, 245, 247, 251, 252
schreibersite 65
scoria 4, 21, 92, 98
sellaite 201
sericite 42, 51, 52, 55, 102
shale 2, 4, 26, 27, 38, 41, 47, 64, 65, 92, 98, 99, 101, 105, 106, 111, 126, 144, 156, 158, 163, 165, 208, 209, 212
shrinkage 33, 35, 36, 38, 99, 196, 197, 199
Siberian Traps (Russia) 139, 140
siderite 23, 38, 42, 65, 96, 105, 109, 116, 160, 165, 166, 201, 203, 204, 208, 209
Sierra Nevada (USA) 87, 88, 90, 91, 168
silicides 112, 208
sillimanite 1, 3, 5, 8, 41, 48, 53, 60, 61, 65, 67–70, 72, 76–78, 81, 83, 84, 89, 92, 106, 107, 113, 117, 178, 180–182, 208, 209, 228, 229, 231, 243–245, 247–252
siltstone 18, 38, 47, 48, 92, 101, 103, 109–111, 201, 210
Sithean Sluaigh (Scotland) 65–67, 75, 79, 178, 233, 245
Skaergaard (Greenland) 70–72, 170, 171
Skye (Scotland) 170, 171
slag 4, 21, 23, 100–102, 105, 106, 109, 157, 161–165, 202, 208, 211–217
Smartville complex (USA) 168, 169
smectite 98, 99, 172, 195, 208, 210, 211
Smoking Hills (Canada) 24, 26
Soay (Scotland) 54–58
sodalite 144
solid waste 214, 215
spinel 1, 3, 5, 18, 35, 41, 45–47, 49, 51, 53, 54, 60–74, 76–79, 81, 84, 87, 92, 99, 102, 106, 107, 109, 113, 117, 126, 138–142, 148, 149, 152–155, 165, 166, 171, 175, 178, 181–184, 189–191, 203, 205, 208, 221, 228, 229, 232–234, 236–239, 242, 246, 247, 251, 252
spoil heaps 200–203
spurrite 33, 118, 119, 121, 122, 124, 129–136, 138–142, 144, 157, 159, 170, 173, 174, 203
srebrodolskite 165, 203
staurolite 247–250
Sterkspruit Valley (South Africa) 51
sulphides 21, 41, 99, 116, 165, 205, 208, 215
sulphur 64, 101, 103, 106, 110, 111, 186, 188, 208
svyatoslavite 203
sylvite 166
system
 -, Al_2O_3-CaO-SiO_2 215
 -, Al_2O_3-SiO_2 41, 42, 45, 46, 57, 68, 107, 178, 183, 185, 199, 232, 242, 244, 249
 -, Al_2SiO_5-Al_2O_3 244
 -, CaO-Al_2O_3-Fe_2O_3-SiO_2 128, 129, 147

-, CaO-Al$_2$O$_3$-SiO$_2$-H$_2$O 128
-, CaO-Al$_2$O$_3$-SiO$_2$-H$_2$O-CO$_2$ 128
-, CaO-MgO-SiO$_2$-CO$_2$ 115, 118, 127, 153
-, CaO-SiO$_2$-CO$_2$ 122–124, 130
-, FeO-Al$_2$O$_3$-SiO$_2$ 45–47, 53, 58, 72, 94–96, 111, 180, 183, 214
-, K$_2$O-Al$_2$O$_3$-SiO$_2$ 191, 193, 229
-, K$_2$O-Al$_2$O$_3$-SiO$_2$-H$_2$O 229
-, KAlO$_2$-CaO-MgO-SiO$_2$-C-O-H 145
-, KAlO$_2$-CaO-MgO-SiO$_2$-H$_2$O-CO$_2$ 144, 145
-, larnite-forsterite-silica 163, 164
-, melilite-anorthite-leucite 213
-, MgO-CaO-SiO$_2$-H$_2$O-CO$_2$ 155
-, Na$_2$O-Al$_2$O$_3$-SiO$_2$ 187
-, Na$_2$O-CaO-SiO$_2$ 187

T

talc 122, 233
Taupo Volcanic Zone (New Zealand) 73
tephrite 148, 149
thermometry 59, 80, 84, 86, 168, 170, 171, 175, 185, 187
-, magnetite-ilmenite 18, 59, 60, 78–80
-, two-pyroxene 59, 187
thompsonite 172
Thulin (Belgium) 208, 209
Tievbulliagh (Northern Ireland) 178
Tieveragh (Northern Ireland) 58–60
tilleyite 116, 121, 122, 124, 131–134, 138–140, 142, 144, 173
titanite 18, 49, 65, 89, 113, 173, 188, 203, 218
tobermorite 160
topaz 113, 201
tourmaline 41, 51, 52, 65, 211, 214
trachyandesite 19, 20, 88–91
trachyte 1, 5, 75, 79, 80, 242, 245, 252
Traigh Bhàn na Sgùrra (Scotland) 36, 72
tridymite 2, 3, 8, 18, 32, 41, 42, 45–47, 49, 52–54, 57–61, 63, 70, 74, 84, 86–88, 91, 94–96, 99, 100, 102, 103, 105–107, 109, 111, 113, 123, 139, 161, 169, 170, 178–188, 191, 201, 208, 209, 223, 224
tridymite-cristobalite inversion 123
troilite 64, 65, 203, 208, 211
tschermakite 167, 170
tschermigite 110

U

ultramafic rock 5, 185, 190
Uzbekistan 111

V

vapour phase 37, 119, 124, 125, 172
vapourisation 113, 217
vaterite 160
vesuvianite 122, 135, 142, 144, 159
viscosity 12, 16, 17, 39, 58, 208, 217
vitrification 38, 49, 50, 52, 53, 195, 197, 217
vitrinite 21, 26, 105, 210, 211
vivianite 214
volkonskoite 160

W

wagnerite 205
weathered mafic rock 177
Wehr volcano (Germany) 76, 77, 79–82, 247, 251
Western United States 96
White Island (New Zealand) 185–187
whitlockite 216
wilkeite 203
Winans Lake (USA) 112, 113
wollastonite 46, 72, 99, 102, 117–119, 122, 124, 128, 130, 131, 133–135, 139, 140, 142, 144–149, 151, 152, 157, 160–164, 173, 174, 185–187, 195, 196, 199–201, 203, 209, 214
wüstite 211, 214, 246

X

xenolith 17–20, 56, 57, 61, 63, 64, 68–72, 76, 80, 81, 84, 85, 90, 91, 122, 144–148, 150–152, 183, 187, 222–224, 227, 229, 230, 235, 236, 239–245, 247–252
xenotime 99
xonolite 160

Z

zeolite 102, 145, 160, 172
zircon 18, 41, 47, 49, 51, 52, 57, 64, 65, 72, 87–89, 98, 99, 112, 163, 211, 214, 215, 218

Printing: Krips bv, Meppel
Binding: Stürtz, Würzburg